Biochemistry

Biochemistry

Essential Concepts

Charles C. Hardin and James A. Knopp

Department of Molecular and Structural Biochemistry,
North Carolina State University

New York   Oxford
OXFORD UNIVERSITY PRESS

Oxford University Press, Inc., publishes works that further
Oxford University's objective of excellence
in research, scholarship, and education.

Oxford New York
Auckland   Cape Town   Dar es Salaam   Hong Kong   Karachi
Kuala Lumpur   Madrid   Melbourne   Mexico City   Nairobi
New Delhi   Shanghai   Taipei   Toronto

With offices in
Argentina Austria Brazil   Chile   Czech Republic   France   Greece
Guatemala Hungary Italy Japan Poland Portugal Singapore
South Korea Switzerland Thailand Turkey Ukraine Vietnam

For titles covered by Section 112 of the U.S. Higher Education
Opportunity Act, please visit www.oup.com/us/he for the latest
information about pricing and alternate formats.

Published by Oxford University Press, Inc.
198 Madison Avenue, New York, New York 10016
www.oup.com

Library of Congress Cataloging-in-Publication Data
Hardin, Charles.
Biochemistry : essential concepts / Charles Hardin and James Knopp.
p. cm.
ISBN 978-0-19-976562-1 (cl : acid-free paper)   1. Biochemistry—Outlines, syllabi, etc.
I. Knopp, James.   II. Title.
QP518.3.H37 2011
572—dc23

2011044803

Brief Contents

Contents

Preface

Using *Biochemistry: Essential Concepts*

Learning biochemistry is difficult for many life science students because, in essence, they are asked to learn a very complicated language, filled with many new concepts, all within a sixteen-week time frame. This "field manual" is a concise guide to biochemistry concepts and is intended as an efficient, pared-down aid to help students assimilate the key ideas. It presents a self-contained sixteen-week course, at a level that will help students proceed successfully to professional and medical school course work.

Biochemistry: Essential Concepts (BEC) has evolved over many years of teaching introductory biochemistry. In one concise volume it contains *(a)* a textual summary of the essential information distilled from a standard encyclopedic biochemistry textbook, and *(b)* relevant review questions and sample tests with answers. *BEC* thus serves as a complete and self-contained handbook, notebook, and study guide. Because *BEC* presents material in the same sequential order as most biochemistry textbooks, it may easily be used alongside another text. The content in *BEC* is intended to provide a backbone. It is not intended to replace a full textbook but, rather, through concurrent use, is designed to assist students in the learning process by presenting to them a clear, pared-down presentation of the basics together with problem-solving and review tools.

We have taught graduate- and undergraduate-level biochemistry and biophysics courses at North Carolina State University for over twenty-five years. The main challenges we have experienced are: *(1)* students arriving to take a biochemistry course who have not retained basic concepts from freshman and organic chemistry; and *(2)* the vastness of the field of biochemistry itself. Many students have relied on memorization in their previous science courses without grasping fundamental concepts such as pH, pK_a, nucleophilic attack, equilibrium, and thermodynamics. Reliance on rote memorization in a biochemistry course is a very ineffective approach and does not result in a deep understanding of more subtle aspects of biochemical processes, such as assessing the poise of a reaction equilibrium, predicting enzyme reactivity and substrate-dependent regulation. Our goals in writing *BEC* were therefore twofold:

(1) To carefully extract and present the essential core concepts imbedded in a typical biochemistry textbook,

(2) to reiterate in a variety of contexts those most fundamental chemical concepts that are essential to *understand* the processes of biochemistry and related biological science, not simply memorize them.

The text contains several key features:

Textbook Flexibility. The approach used in *BEC* focuses on teaching the fundamental structure of the field within a single semester time frame. It is not based on a single textbook. As a result, it is compatible for use as a study guide with any of a wide variety of much more definitive tomes, several of which are cited in context within the text.

Integration of Concepts Into The Big Picture. *BEC* is a clear, concise guide to biochemical concepts, which is readily accessible and provides a wealth of well-chosen examples. A key strength is that one is rapidly orientated regarding a given subject, with emphasis on the big picture, and then shown how fundamental concepts become integrated to produce more complex linked processes. These lessons are reinforced by an extensive set of practice exercises and tests designed to reinforce key concepts and relationships, highlighting techniques from medicine and other biotechnological fields.

Many Real-World Applications. These notes are a map so that students can continually look at the big picture and see how the subject of the moment fits. The same fundamental principles govern many aspects of biological processes, so once students have built a set of models into their memory, they can use this knowledge to dissect some new, yet related, biological setting, predict the pertinent chemistry and sort out the processes that matter. Students can then use this kind of circumspect viewpoint to realistically understand the complexity of new situations. In the long view, this knowledge can be used to understand a process, develop new procedures, troubleshoot methodological problems, design a new pharmaceutical, and otherwise be applied to use in medicine, agriculture, and biotechnology.

A Concise, Clear Format. How does a professor decide what to keep and what to leave out when faced with a twelve-hundred-page textbook and sixteen weeks to teach the course? Current textbooks tend to be encyclopedic, requiring careful choice of the material to emphasize if one is to effectively transmit both a working knowledge of the fundamental tools of the discipline as well as their breadth and importance in medicine, materials science, genetics, cell biology, and so forth.

In the streamlined presentation of *BEC*, the focus is on concepts that govern and regulate biological processes. Those concepts include equilibrium, pK_a, K_d, K_M, pH, nucleophilic attack, the relation between bond polarity and reactivity, the enthalpic and entropic contributions to ΔG, and so forth. This emphasizes the more difficult-to-learn physical and mechanistic concepts and leaves the more digestible, qualitative and familiar biological foci for class discussion.

Reiteration of Core Concepts. The best way to show students that *the same fundamental principles govern many aspects of biological processes* is to reiterate those principles, where applicable, throughout the course in various contexts. If students grasp these core concepts, they can understand at a fundamental level the biochemical processes that underlie much of biological science.

Reiteration is built into *BEC* at several levels. For example, the concept of pK_a is reiterated a total of fourteen times: nine times in various text sections, twice in review material, and three times in the sample tests. The concept is first discussed on in the context of acid-base ionization and the relation to protonation and deprotonation. It reappears again in the context of the bicarbonate blood buffering mechanism and again a discussion of the functional groups of amino acids. It occurs twice in a table of pK_a values and in a section that defines the isoelectric point and explains how it is calculated. It appears again during the explanation of pK_a shifts at the C-terminus of amino acids on incorporation into a polypeptide, in a discussion of the allosteric control of oxygen binding to hemoglobin, in an explanation of enzyme activity at different pH values, and in a discussion of the charge of backbone phosphates in nucleic acids. The concept occurs in several contexts in the review sessions. Sample Exam 1 makes use of the pK_a concept in a question that requires students to draw a specified oligopeptide structure and the corresponding pH titration curve. Sample Exam 1 also contains a multiple-choice question that focuses on the comparison of pK_a with pH and the partial pressure of oxygen (pO_2). The final occurrence of the concept is in the sample Final Exam in a question regarding factors that control the catalytic capability of an enzyme. Similar reiteration strategies are employed for other fundamental principles.

The material in *BEC* produces an interwoven set of reiterated ideas that shows how analogous chemical principles govern a large number of biological reactions. The intent of this reiteration is to help students retain the basics, understand how they apply to a variety of biological processes, and develop the ability to dissect new analogous situations on their own.

Reinforcement Through Review. Several review tools have been incorporated into *BEC,* complete with answers, to foster integration of the textbook material with the questions. A single volume contains a concise summary of the lecture material, the students' own class notes, and directly linked review questions and sample tests, with answers. This integrated format encourages students to work problems and think about results in a well-organized and efficient way.

The first six sets of review questions do not have an answer key *per se*, but the answers are easily located in the text of *BEC.* The expectation is that students can read the relevant portions of the text and *BEC* and use the information to answer the assigned questions. However, the material covered in review sessions 7–13 is more challenging and requires a more concept-driven approach. We've also found that students begin to feel frustration and fatigue at this stage of the course. We've therefore provided answers for each of the questions in these sections. In addition, four sample tests and a final exam are supplied with complete answer keys.

These review tools provide a completely integrated curriculum that stresses the core concepts of biochemistry and greatly facilitates student comprehension. By writing and drawing out the answers to the practice questions in *BEC,* students exercise and refine their use of the tools and language of biochemistry.

Acknowledgments

It is a pleasure to extend our deep appreciation to all the students and colleagues at North Carolina State who have helped, and provided suggestions and encouragement. We especially thank Pamela Cook, Mike Lisanke, Brad Moffitt, Brad Kearney, Meredith Ellis, Lindsey Wright, Daniel Zaetz, and Nathanial Sorenson for their suggestions regarding both scientific and editorial issues, and Dr. Dennis Brown for his support. We thank Jason Noe, senior editor, Katie Naughton and Caitlin Kleinschmidt, editorial assistants, Marianne Paul, production editor, Kim Howie, senior graphic designer, and Lisa Grzan, managing editor, at Oxford University Press for their clarity, encouragement, and patience. Thanks also to Jason Kramer, marketing manager, Frank Mortimer, director of marketing, Patrick Lynch, editorial director, and John Challice, vice president and publisher. We also wish to express our appreciation for the efforts of the dedicated individuals who provided detailed content and accuracy reviews of the text:

Ruth E. Birch, *St. Louis University*

Gary J. Blomquist, *University of Nevada, Reno*

Edward J. Carroll, *California State University, Northridge*

Anjuli Datta, *The Pennsylvania State University*

Matthew Gage, *Northern Arizona University*

Peter Gegenheimer, *University of Kansas*

Tamara Hendrickson, *Wayne State University*

Christine A. Hrycyna, *Purdue University*

Holly A. Huffman, *Arizona State University*

Harry van Keulen, *Cleveland State University*

C. Martin Lawrence, *Montana State University*

Shelley L. Lusetti, *New Mexico State University*

Victoria Mariani, *James Madison University*

John J. Mitnick, *Quinnipiac University*

Jerry L. Phillips, *University of Colorado*

Rachel Roberts, *Texas State University – San Marco*

Andrew Shiemke, *West Virginia University*

Kevin R. Siebenlist, *Marquette University*

Maxim Sokolov, *West Virginia University*

Madhavan Soundararajan, *University of Nebraska*

Chalet Tan, *Mercer University*

Sandra L. Turchi-Dooley, *Millersville University*

Ales Vancura, *St. John's University*

Laura S. Zapanta, *University of Pittsburgh*

Charles C. Hardin

James A. Knopp

July 2011

Chapter 1

Biochemistry: Subject Overview

Definition: Study of the *chemical reactions* and linked *metabolic processes* that sustain a *reproducing line* of *viable organisms*.

1.1 Central Themes

Energy. Organisms break down food (substrates) to produce the common molecular currencies that drive energy-requiring transformations in cells. Important examples include adenosine triphosphate (ATP), reduced nicotinamide adenine dinucleotide (NADH), reduced coenzyme Q and phosphoenolpyruvate.

Thermodynamics. Ground state differences in the energy of the reactants versus that of products. The sign and magnitude of the Gibbs free energy tells one about the degree of spontaneity and poise (extent of reaction) of a metabolic step. The focus is on the effect of mass action on the reaction equilibrium.

Kinetics. Rates and resulting excited-state energies required for a reaction to proceed, which depends on details of the mechanism and any drive imposed by the reverse reaction. The focus is on how the reaction depends on time and how to use this dependence to uncover details of the mechanism.

Enzymes. Typically, but not always, a protein that catalyzes a reaction by lowering the *transition-state energy* of the key energy-requiring step. The catalyst is regenerated in its original form after catalysis is completed. Several metals, RNAs, carbohydrate complexes and lipid aggregates have also been shown to support catalysis.

Genetics. Characterizes the transfer of information from one generation to the next. Stable storage occurs with DNA. RNA participates in transient use to mobilize new proteins when required and then to get rid of the message. The genetic material in a "pluripotent" cell contains all of the information necessary to recreate the new organism. An overview of the processes involved is captured by an overarching concept called the *central dogma of molecular biology*.

1.2 Central Dogma of Molecular Biology

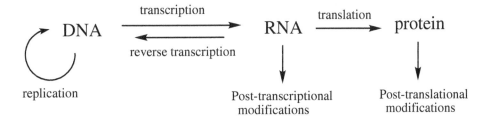

This scheme summarizes the respective transfers of genetic information that are required to support the life cycles and process of reproducing life forms. Replication involves reproducing the two daughter chromosomes from the maternal copy of double helical DNA. Transcription involves making RNA, based on the template information residing in the replicated DNA sequence. These processes are orchestrated by the DNA and RNA polymerase complexes, respectively. Translation of messenger RNAs to form proteins is catalyzed by the ribosome, a huge complex composed of 2/3 RNA and 1/3 protein.

Chapter 2

Cell Biology Review

2.1 The Animal Cell. (Adapted from McKee and McKee, *Biochemistry The Molecular Basis of Life* [5th ed.], *2012,* p. 43; Fig. 2.11.)

While eukaryotic cells have a wide range of morphologies, with unique and distinguishing properties, most cells have comparable structural features.

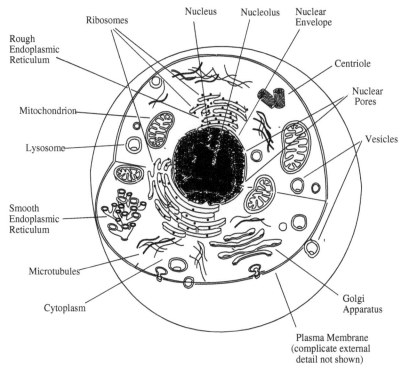

One specific example, fibroblasts, synthesizes the *extracellular matrix* (ECM) and collagen. The ECM is composed of structural proteins and complex carbohydrates that are secreted from fibroblasts to form a viscous surface that aids in cell-cell adhesion. It also serves functions such as providing support and anchorage for the cell, protection and regulation of intercellular communication. Membrane receptors control various chemical and mechanical signaling processes.

Rat liver cells have been used to characterize many biochemical systems. The following table shows the percentage of membrane in several cellular compartments and substructures.

Table 1. Membranes of the Rat Liver Cell

Membrane Type	% of total membrane in cell	Membrane Type	% of total membrane in cell
Plasma membrane	5	Lysosomes, peroxisomes and other compartments	6
Rough endoplasmic reticulum	30	Mitochondria	
Smooth endoplasmic reticulum	15	Outer membrane	7
Nuclear membrane	1	Inner membrane	30
Golgi apparatus	6		

(Adapted from Moran *et al.*, *Biochemistry* [2nd ed.], 1994, p. 20-10; Table 2.2.)

2.2 The Plant Cell

Two crucial biochemical processes that distinguish plant cells from animal cells are photosynthesis and carbon fixation, which both occur in plants.

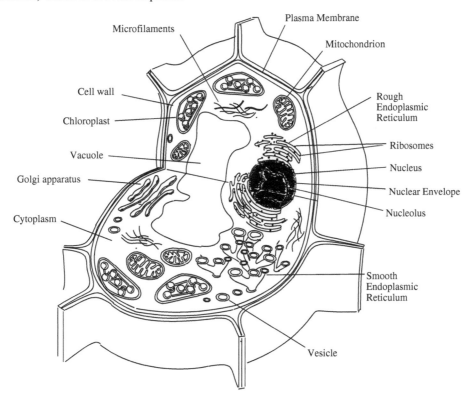

(Adapted from McKee and McKee, *Biochemistry The Molecular Basis of Life* [5th ed.], *2012*, p. 43; Fig. 2.12.)

2.3 Selected Organelles

Endoplasmic Reticulum

Rough and smooth variants exist. The rough ER is rough due to the presence of *ribosomes*, the ribonucleoprotein machines that synthesize proteins (see the *Protein Maturation* section). The smooth ER does not. A variety of post-translational modifications occur there, for example, disulfide bond formation, proteolysis, addition of carbohydrate and lipid molecules, acetylation, and so on.

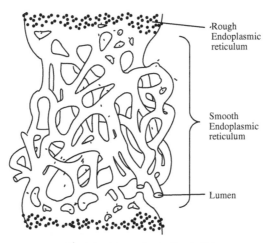

(Adapted from Moran *et al., Biochemistry* [2nd ed.], *1994*, p. 2.16; Fig. 2.13.)

Golgi Apparatus

This organelle is the location of lipid and steroid biosynthesis and specific post-translational modifications. *Transfer vesicles* transport modified proteins to the extracellular interface.

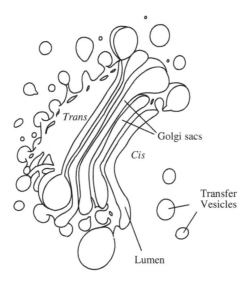

(Adapted from Moran *et al.*, *Principles of Biochemistry* [5th ed.], *2012*, p. 21.)

Mitochondria

This organelle is used to produce cellular adenosine triphosphate (ATP). This process, called oxidative phosphorylation is driven by a proton gradient, which is generated as a result of the electron transport process. Reducing equivalent carriers, such as nicotinamide adenine dinucleotide (NADH) and reduced coenzyme Q (CoQH$_2$) provide the electrons (reducing power, reducing equivalents) that drive electron transport.

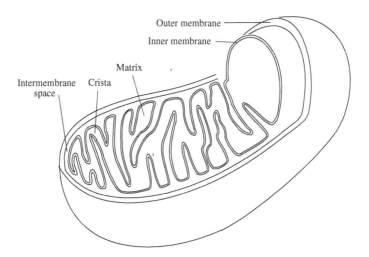

(Adapted from McKee and McKee, *Biochemistry The Molecular Basis of Life* [5th ed.], *2012*, p. 54; Fig. 2.23.)

Evolutionary Origins. Both mitochondria and chloroplast were at one point independent blue-green bacterial cells with their own chromosome. They were captured by other cells and subjugated. Much of the deoxyribonucleic acids (DNA) was eliminated but they still function genetically like prokaryotic cells, with their own polymerases, ribosomes and transfer RNAs. *Mitochondrial DNA is inherited from one's mother*, thus providing a way to reconstruct *maternal lineages*.

Chloroplast

This organelle is involved with light-driven ATP production, which is coupled to electron transport. Carbon fixation occurs in the Calvin Cycle pathway.

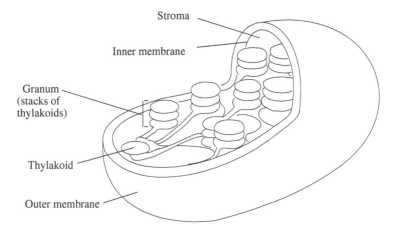

Functional aspects of the key features, the grana, thylakoid membrane, and stroma, are described in the Photosynthesis section. (Adapted from Moran *et al.*, *Biochemistry The Molecular Basis of Life* [5th ed.], *2012*, p. 50; Fig. 2.25.)

The Cytoskeleton

The cytoskeleton is a protein scaffold network that consists of actin filaments, intermediate filaments, and microtubules. The nature of incorporated biomolecules is very different in prokaryotes and eukaryotes. The components support movement and traction. They are very dynamic, building and disassembling at the two ends, in a controlled manner.

Membranes

These ubiquitous structures separate a variety of biochemical functions from each other by forming separate *compartments*. They can both limit and direct the trafficking of biomolecules, and are crucial to the function of signal transduction, active transport, cell division, nerve function, and many other processes.

2.4 The Cell Cycle: Mitotic Cell Division

The purpose of the cell cycle is to replicate the DNA and segregate it into two genetically identical daughter cells. The S phase (for synthesis) requires ten to twelve hours and requires about half of the typical mammalian cell cycle. M phase involves chromosome condensation, nuclear breakdown, attachment of the mitotic spindle to microtubules, in order to align the chromosomes, then segregation to form the two daughter cells. Cytokinesis (division) completes the process.

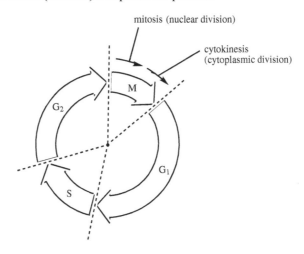

Specific stages in the mitosis process are shown in detail below:

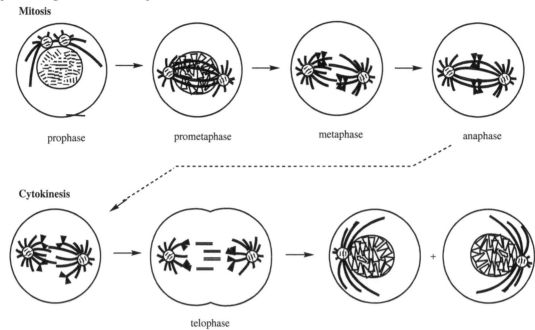

(Adapted from Alberts *et al. Molecular Biology of the Cell* [4th ed.], 2002, p. 985; Fig. 17-3.)

The regulation of mitosis in eukaryotic cells is described in the *Phosphorylation and Dephosphorylation* section. This *cyclin-dependent kinase* mechanism involves a complex set of threonine- and tyrosine-specific reactions. This system regulates progression through a series of *checkpoints* in the cell cycle.

2.5 Viruses

The general structure of a virus with a membrane envelope is shown below. A detailed example is provided by the Human Immunodeficiency Virus (HIV), which is described in the next section.

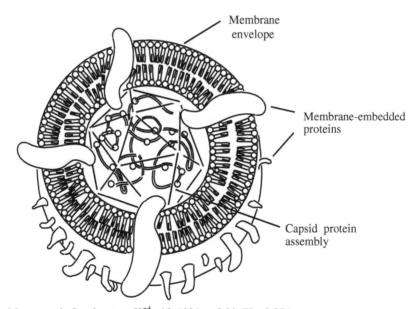

(Adapted from Moran *et al., Biochemistry* [2nd ed.], 1994, p. 2.30; Fig. 2.27.)

(*1*) The following structure shows a *T-even bacteriophage*, a virus that infects bacterial cells, such as *Esheria coli*, the grand old work horse of biochemists and molecular biologist alike.

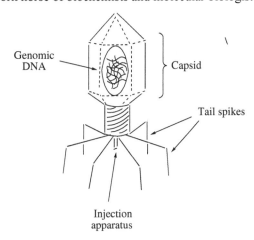

Structural elements include the viral *capsid* (the *head*), which contains the DNA, the tail spikes and the injection apparatus at the bottom center of the virus.

(*2*) Viruses are *symbiotic* with their hosts. They require their host's functions to complement their typically very limited capability to grow and reproduce on their own.

(*3*) In some cases, a payload of extra DNA can be included in the viral DNA, piggy-backing with the normal genomic DNA. Following injection into the host bacteria, the (enlarged) DNA sequence is replicated by (*e.g.*) *E. coli* DNA polymerase. The DNA is cloned in the viral genome and propagated in the bacterium. *Plasmids* can carry much more payload DNA than these less efficient viral systems, so they are generally preferred in cloning strategies.

Human Immunodeficiency Virus Structure

The HIV virus is surrounded by the viral envelope which is composed of two layers of phospholipids. Embedded within the viral envelope are proteins from the host cell known as HLA (human leukocyte antigens) that protect the particle from the immune system and copies of a complex HIV protein such as gp120 and gp41. Within the envelope is a matrix composed of the viral protein p17 that enclose the capsid and guarantee the viral particle's integrity. The capsid contains two copies of single stranded RNA coding for the virus's nine genes and the viral proteins p6 and p24. The RNA is bound to p7 and p9. The enzymes involved in the replication process are reverse transcriptase (RT), integrase, and protease.

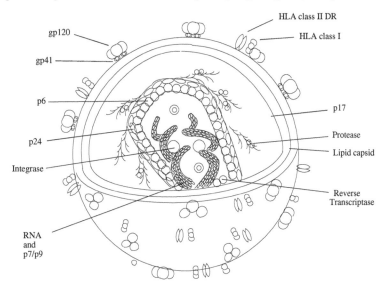

(Adapted from McKee and McKee, *Biochemistry the Molecular Basis of Life* [5th ed.], 2012, p.642; Fig. 17K.)

Reproductive Cycle of the Retrovirus HIV

The viral particle binds to the surface receptors of the host cell (1) and the viral envelope fuses with the cell membrane (2), releasing the HIV capsid into the cell (3). The contents of the capsid, RNA and many viral enzymes, also enter the cytoplasm (4). An enzyme called reverse transcriptase frees the single-stranded RNA from the viral proteins and makes a copy into complementary DNA. DNA polymerase then uses this DNA strand to make a complementary DNA sense strand from this antisense strand (5). This double-stranded viral DNA travels to the nucleus where (6) another viral enzyme, integrase, incorporates it into the host cell chromosome. This newly formed provirus gets replicated with each new DNA synthesis of the cell.

As this viral DNA undergoes transcription (7), it forms two types of RNA transcript: (8) RNA molecules that are used in the viral genome and (9) molecules making up the components that are used to remake the viral protein such as reverse transcriptase, capsid proteins, envelope proteins, and viral integrase. (10) The protein molecules combine with the RNA genome, creating a new virus which (11) makes budding on the surface of the host cell and (12) goes on to infect other cells.

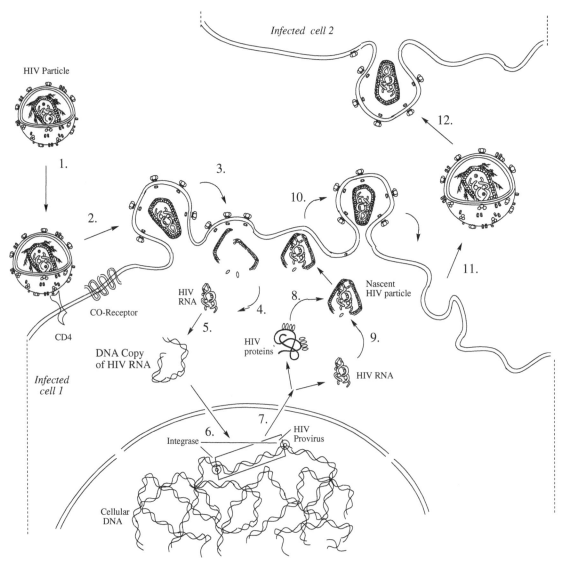

(Adapted from McKee and McKee, *Biochemistry the Molecular Basis of Life* [5th ed.], 2012, p. 642; Fig. 17L.)

Chapter 3

Chemistry Review

This review begins with the functional groups in organic chemistry and the relation between structure and properties in different atomic contexts, which is the foundation of biochemistry. It then reviews physical chemistry ideas one typically learns in basic chemistry course, applied to biochemical molecules and principles.

3.1 Organic Compounds

The following functional groups play central roles in biochemistry. You will review them in the Review Sessions and in the context of drawing structures throughout the course.

3.1.1 Functional Groups

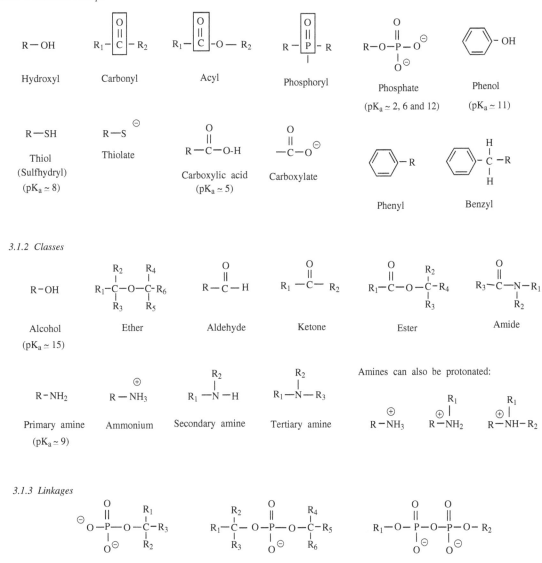

3.1.2 Classes

3.1.3 Linkages

3.2 Chirality

Specification of Molecular Configuration: The Designations R and S
(Adapted from Morrison and Boyd *Organic Chemistry* [3rd ed.], 1973, Allyn & Bacon, NY.)

Chirality has had a profound effect on the structural and functional properties of biomolecules. Molecular configurations can be determined in a more simple way than drawing out the structure. A generally useful way is the use of the prefixes R and S.

   Step 1. Following a set of *sequence rules* (detailed in the following section), we assign a sequence of priority to the four atoms or groups of atoms attached to the *chiral center*.
   In the case of C, H, Cl, Br and I, for example, the four atoms attached to the chiral center are all different and priority depends simply on atomic number, the atom of higher number having higher priority: I > Br > Cl > H.

Bromochloroiodomethane

I   *II*

   Step 2. We visualize the molecule oriented so that the group of *lowest* priority is directed *away* from us, and observe the arrangement of the remaining groups. If, in proceeding from the group of highest priority to the group of second priority and thence to the third, our eye travels in a clockwise direction, the *configuration* is specified **R** (Latin: *rectus,* right); if counterclockwise, the configuration is specified **S** (Latin: *sinister,* left).
   Thus, configuration I and II are viewed as follows:

R   *S*

and are specified R and S, respectively.
   A complete name for an optically active compound reveals—if they are known—both configuration and direction of rotation, as, for example, (S)-(+)-*sec*-butyl chloride. A *racemic mixture* can be specified by the prefix RS, as, for example, (RS)- *sec*-butyl chloride.

Sequence Rules
For ease of reference and for convenience in reviewing, we shall set down here those sequence rules we shall have need of. The student should study Rules 1 and 2 now, and Rule 3 later when the need for it arises.
   Sequence Rule 1. If the four atoms attached to the chiral center are all different, priority depends on atomic number, with the atom of higher atomic number getting higher priority. If two atoms are isotopes of the same element, the atom of higher mass number has the higher priority.
   For example, in chloroiodomethanesulfonic acid the sequence is I, Cl, S, H; in α-deuterioethyl bromide it is Br, C, D, H.

Chloroiodomethanesulfonic acid   *α-Deuterioethyl bromide*

Sequence Rule 2. If the relative priority of two groups cannot be decided by Rule 1, it shall be determined by a similar comparison of the next atoms in the groups (and so on, if necessary, working outward from the chiral center). That is to say, if two atoms attached to the chiral center are the same, we compare the atoms attached to each of these first atoms.

For example, take *sec*-butyl chloride, in which two of the atoms attached to the chiral center are themselves carbon. In CH_3, the second atoms are H, H and H.

$$H_3C-CH_2-\overset{\overset{\displaystyle H}{|}}{\underset{\underset{\displaystyle Cl}{|}}{C}}-CH_3$$

sec-Butyl Chloride

In C_2H_5 they are C, H and H. Since carbon has a higher atomic number than hydrogen, C_2H_5 has the higher priority. A complete sequence of priority for *sec*-butyl chloride is therefore Cl, C_2H_5, CH_3, H.

In 3-chloro-2-methylpentane the C, C, H of isopropyl takes priority over the C, H, H of ethyl, and the complete sequence of priority is Cl > isopropyl > ethyl > H.

$$H_3C-\overset{\overset{\displaystyle CH_3}{|}}{CH}-\overset{\overset{\displaystyle H}{|}}{\underset{\underset{\displaystyle Cl}{|}}{C}}-CH_2-CH_3 \qquad\qquad H_3C-\overset{\overset{\displaystyle CH_3}{|}}{CH}-\overset{\overset{\displaystyle H}{|}}{\underset{\underset{\displaystyle Cl}{|}}{C}}-CH_2Cl$$

3-Chloro-2-methylpentane   1,2-Dichloro-3-methylbutane

In 1, 2-dichloro-3-methylbutane the Cl, H, H of CH_2Cl takes priority over the C, C, H of isopropyl. Chlorine has a higher atomic number than carbon, and the fact that there are *two* C's and only *one* Cl does not matter. (One higher number is worth more than two - or three - of a lower number.)

Sequence Rule 3.

Where there is a double or triple bond, both atoms are considered to be duplicated or triplicated. Thus,

$$-\overset{|}{C}=A \quad\text{equals}\quad -\overset{|}{\underset{\underset{\displaystyle A}{|}}{C}}-A \quad\text{and}\quad -C\equiv A \quad\text{equals}\quad -\overset{\overset{\displaystyle A\ \ C}{|\ \ |}}{\underset{\underset{\displaystyle A\ \ C}{|\ \ |}}{C}}-\overset{C}{\underset{A}{}}$$

For example, in glyceraldehydes the OH group has the highest priority of all,

$$\overset{\overset{\displaystyle H}{|}}{\underset{}{C}}{=}O$$
$$H-\overset{|}{\underset{\underset{\displaystyle CH_2OH}{|}}{C}}-OH \qquad\qquad \overset{\overset{\displaystyle H}{|}}{C}{=}O \quad\text{equals}\quad \overset{\overset{\displaystyle H}{|}}{\underset{\underset{\displaystyle O\ \ C}{|\ \ |}}{C}}{-}O$$

Glyceraldehyde

The O, O, H of –CHO takes priority over the O, H, H of –CH$_2$OH. The complete sequence is then –OH, –CHO, –CH$_2$OH, and –H.

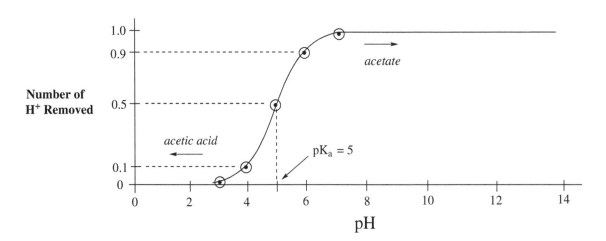

3.3. Chemical Reactions

Acid-Base Chemistry

The key idea with acid-base reactions is that H^+ is gained by one species and lost by another. The following scheme shows an example of a protonation-deprotonation reaction:

Acetate $\quad + \quad H^+ \quad \rightleftharpoons \quad$ Acetic Acid

Assume that the pK_a that governs this equilibrium is 5. Draw a plot that shows the number of H^+ atoms removed as the y axis and pH as the x axis. Recall that:

$$pH = -\log [H^+] \quad \text{and} \quad [H^+] = 10^{-pH} \tag{1}$$

The *Henderson-Hasselbalch equation* is expressed in terms of the conjugate base (A) and conjugate acid (HA) as follows:

$$pH = pK_a + \log\left(\frac{[A^-]}{[HA]}\right) \tag{2}$$

In the case of the equilibrium shown above, this equation is:

$$pH = pK_a + \log\left(\frac{[\text{acetate}^-]}{[\text{acetic acid}]}\right)$$

This equation allows one to calculate the following relations between protonated and deprotonated species:

(*1*) When the pH is equal to the pK_a, 50% of the population is protonated and 50% is deprotonated.

(*2*) When the pH is 1 unit less than the pK_a, 90% of the population is protonated and 10% is deprotonated

(*3*) When the pH is 2 units less than the pK_a, 99% of the population is protonated and 1% is deprotonated

(*4*) When the pH is 1 unit more than the pK_a, 90% of the population is deprotonated and 10% is protonated

(*5*) When the pH is 2 units more than the pK_a, 99% of the population is deprotonated and 1% is protonated.

The logarithmic plot has a sigmoidal shape and gives equal consideration to each concentration range, that is, molar (M), millimolar (mM), micromolar (μM), nanomolar (nM), and picomolar (pM) each correspond to a pH unit. In contrast, the nonlogarithmic curve is hyperbolic and emphasizes the high concentration. Lower concentrations are "crunched" into the left portion of the curve.

Oxidation Reduction Reactions

The following scheme illustrates the relation between the different classes of two-carbon organic compounds.

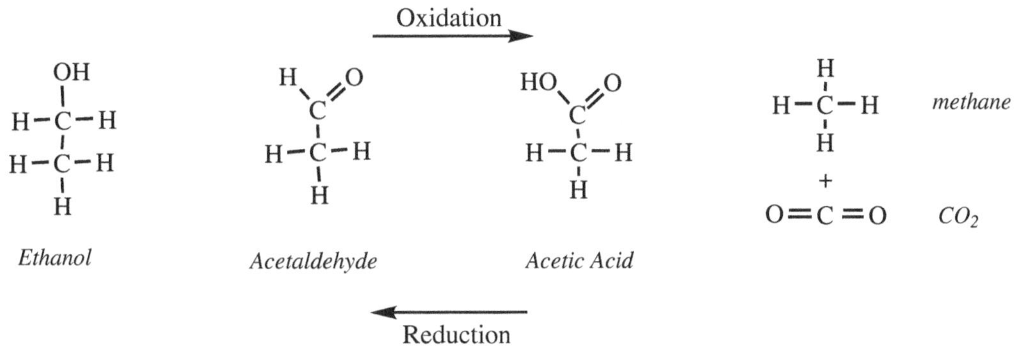

The key idea is that electrons are gained by one compound and lost by another. This is captured by the mnemonic acronym OIL RIG, which abbreviates the phrase: Oxidation Is the Loss of electrons; Reduction Is their Gain. The electrons are sometimes attached to a hydrogen nucleus, forming a species called a *hydride*.

Applying the redox series shown above to the 3-carbon compound glycerol:

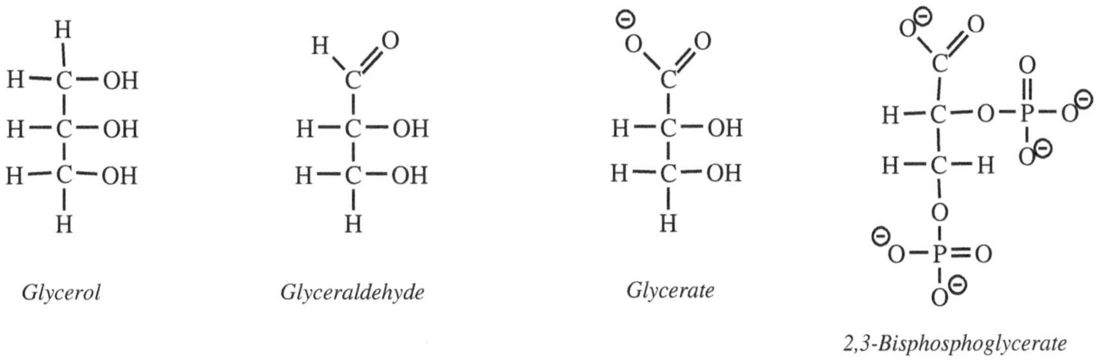

The compound shown to the right, 2, 3-bisphosphoglycerate (BPG), will be discussed as an allosteric regulator of oxygen binding to hemoglobin in the *Ligand Binding and Functional Control* section. (Note that the prefix is *bis* not *bi*.)

3.4 Physical Chemistry Concepts

Water: Structure and Properties

Water is composed of a tetrahedrally linked lattice of hydrogen-bonded molecules. The connections resemble the methane molecule, but each oxygen is connected to 2 hydrogens by covalent bonds and 2 hydrogens by H-bonds.

Lone pair electrons donate and bond to H acceptors (δ^+ sinks). The lone pair of one H_2O's oxygen hydrogen bonds with another's hydrogen. The gradient of electronegativity (electron retention) produces an overall molecular *dipole*, which is indicated by the symbol to the left of the water molecule below.

The *polarity* of water produces a property called *hydrophilic*ity, in which polar biomolecule solutes are soluble in the solvent. The high electron density on oxygen, nitrogen, and metals make them polarizable, which, in turn, leads to effective ionic and polar bonding with water and other fully or partially charged compounds.

Acid-Base "Ionization" of water occurs, with a pK_a of 14. This is the basis of the pH scale, which ranges over 14 orders of magnitude in H^+ concentration—from 1 M to 10^{-14} M. As a result, at pH 7, both H^+ and OH^- are present at 10^{-7} M. This becomes important in biochemical reactions because it means that a reasonably large concentration (in biochemical terms) of each is always around to *aid in the catalytic reaction mechanisms*.

Three types of reactions can lead to the release either hydroxylate OH^-, a proton H^+ or both.

$$2\,H_2O \; \rightleftharpoons \; H_3O^+ \; + \; OH^- \qquad\qquad H_2O \text{ disproportionation}$$

$$R\text{-}H \; \rightleftharpoons \; H^+ \; + \; R^- \qquad\qquad H^+ \text{ and } R^- \text{ anion formation}$$

$$R^- \; + \; H_2O \; \rightleftharpoons \; R\text{-}H \; + \; OH^- \qquad\qquad R^- \text{ uptake and } OH^- \text{ formation}$$

Water is a dynamic moldable liquid, so it is well adapted to filling grooves and gaps formed on the exterior surface of biomolecules. Moreover, the large enthalpy associated with hydrogen bonding between water molecules provides the energetic drive that leads to the folding of most native biomolecular structures.

Hydrophobicity

Water binds other H_2Os, which leads to the exclusion of nonpolar molecules; for example benzene, alkanes, alkenes, including the aromatic and aliphatic *side chains* of biomolecules. Carbon-bound H's are not very polarizable, producing the important incompatibility of "oil" and "water" which is responsible for the separation of different compartments—a key requirement for molecules to organize together into cells, capsids, organelles, and so on.

The *hydrophobicity* of nonpolar molecules can drive assembly of a variety of biomolecular structures, that is, folded proteins, assembled membranes and vesicles, and nucleic acid double helices.

External waters act like a cage, at temperatures below about 60°C, enclosing the components of the folded biomolecule within their collective volume. Formation of the maximal density of hydrogen bonds among waters in the external solvent network leads to a large advantage in terms of gained *enthalpic energy* (ΔH). This offsets the *entropic contribution to the Gibbs free energy* , that is, $-T\Delta S$, which typically opposes folding. Recall that the Gibbs free energy (ΔG) is defined as:

$$\Delta G = \Delta H - T\Delta S \qquad\qquad (3)$$

These ideas will be described in detail in the context of protein folding in the *Protein Structure* section (Sect. 6.8).

Van der Waals Forces

One must consider the effects of all atoms within 20 Å.

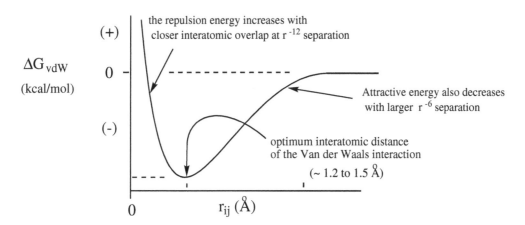

The Van der Waals Interaction Free Energy (ΔG_{vdW}) is proportional to the interatomic distance. More negative values indicate more stable interatomic distances.

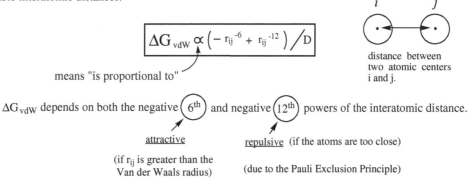

(*1*) *Attraction*: Coordinated alignment of the dipoles, like a synchronized dance.

(*2*) *Repulsion*: Atom-atom electronic overlap is opposed by the *Pauli Exclusion Principle*, that is, two electrons with the same spin cannot coexist in the same space.

(*3*) The parameter **D** is the *dielectric constant*. H_2O has a large D ($\sim$ 80) so molecular *charges are well shielded from each other* in H_2O. This is much less true in *nonpolar solvents*, which enhances the effectiveness of charge-charge interactions.

Hydrogen Bonding
They form between polar atoms such as oxygen and hydrogen or nitrogen and hydrogen.

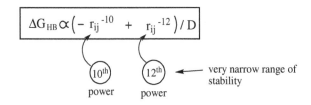

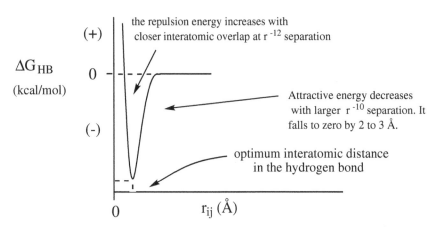

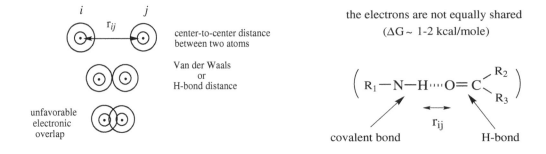

Note the much narrower range of stability of hydrogen bonds relative to Van der Waals interactions.

Some Typical Biochemical Hydrogen Bonds

Amide-carbonyl:

$$R_1\!\!\diagdown \atop R_2\!\!\diagup \!\!N\!-\!H \text{-----} O\!=\!C\!\! {\diagup R_3 \atop \diagdown R_4}$$

Hydroxyl-hydroxyl:

$$R_1\!-\!O\!-\!H \text{-----} {O\!-\!R_2 \atop H}$$

Amide-hydroxyl:

$$R_1\!\!\diagdown \atop R_2\!\!\diagup \!\!N\!-\!H \text{-----} {O\!-\!R_3 \atop H}$$

Hydroxyl-carbonyl:

$$R_1\!-\!O\!-\!H \text{-----} O\!=\!C\!\!{\diagup R_2 \atop \diagdown R_3}$$

Amide-imidazole nitrogen:

$$R_1\!\!\diagdown \atop R_2\!\!\diagup \!\!N\!-\!H \text{-----} N \quad NH$$

Low-Barrier Hydrogen Bonds

The *Catalytic Triad* mechanism of *proteolysis* by *Serine Proteases* is *driven by* formation of this type of bond.

LBHBs are characterized by the very far downfield chemical shift values in the proton nuclear magnetic resonance spectra (> 18 ppm). Examples have been studied in small molecule model sytems.

LBHB formation is essential to formation of the aspartate-histidine hydrogen bond that plays a key role in the 'charge relay' mechanism used by serine protease. LBHB formation activates the strong nucleophilic capability of the histidine to initiate catalysis. The histidine nitrogen is activated to remove the proton from the serine, which, as a hydroxylate, subsequently attacks the carbonyl oxygen of the peptide bond that will be cleaved in the target protein.

hybridized (simultaneous) covalent-hydrogen bonds

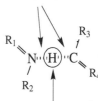

high energy: ΔG > 20 kcal/mole   more equally shared

Salt Bridges (also called *Ionic Bonds*) form between two fully or partially charged atomic species. Two examples are the *Glutamate–COO⁻* • • • *⁺H₃N–Lysine* interaction and the *Mg²⁺ - ATP⁴⁻* complex (see the Coenzyme section).

$$R_1\!-\!X^{\oplus} \text{....} \overset{\ominus}{Y}\!-\!R_2$$

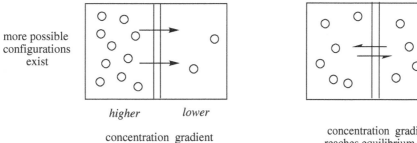

Osmotic Pressure

This is not a charge-dependent phenomenon. It is driven by the tendency of the molecules to equalize the number of configurations in the two compartments (*entropy*). When more and less concentrated molecules are released on two sides of a permeable barrier, they will try to equalize concentrations.

more possible configurations exist

higher   *lower*

concentration gradient

concentration gradient reaches equilibrium poise

Remember the *Ideal Gas Law*: $P V = n R T$ (In doing concentration-pressure-volume
                                                                         work.)

Rearranging produces: $P (V/n) = R T$

The *molar volume* (V/n) *balances* the *pressure* (P) at constant temperature. If the volume changes and concentration remains constant, the density (n/V) changes. The pressure that drives densities to equilibrate is the osmotic pressure.

Water goes into cells from blood in an attempt to dilute the cellular salts and other solutes, which are more concentrated within the cell. This creates the *"turgor pressure"* that maintains cells in the inflated state.

Amphipathicity
This occurs when a molecule is composed of subcomponents that prefer formation of two different segregated phases, one hydrophobic and one hydrophilic. For example, some alpha helices in proteins have one hydrophobic face and one hydrophilic face.

(*1*) One important example is a fatty acid:

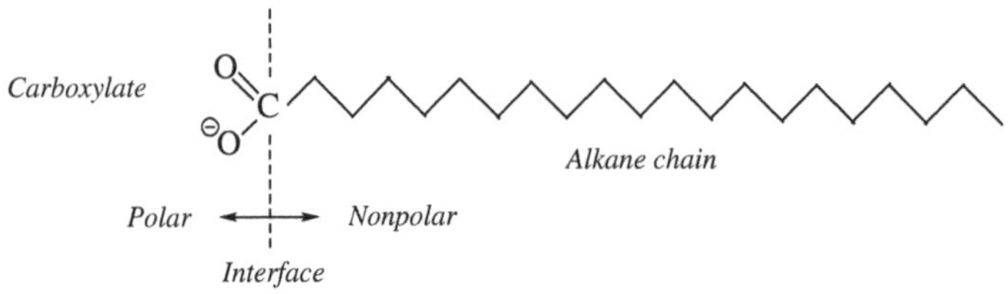

(*2*) *Sodium Dodecyl Sulfate* (*SDS*): This compound denatures proteins by disturbing the 'hydrophobic effect,' so it is a common additive in *Polyacrylamide Gel Electrophoresis* (SDS PAGE) samples.

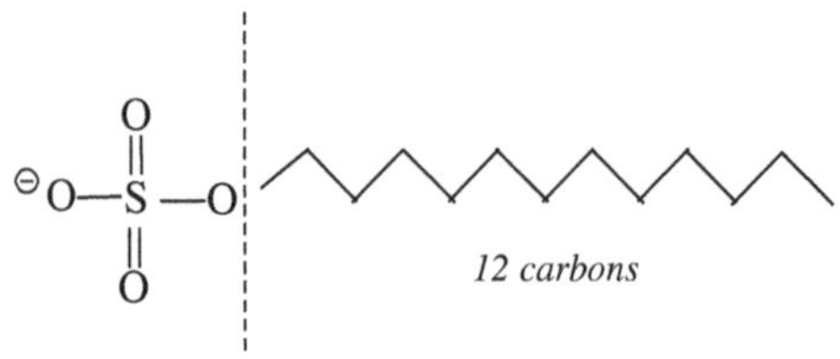

(*3*) *Diacylglycerol phosphate*

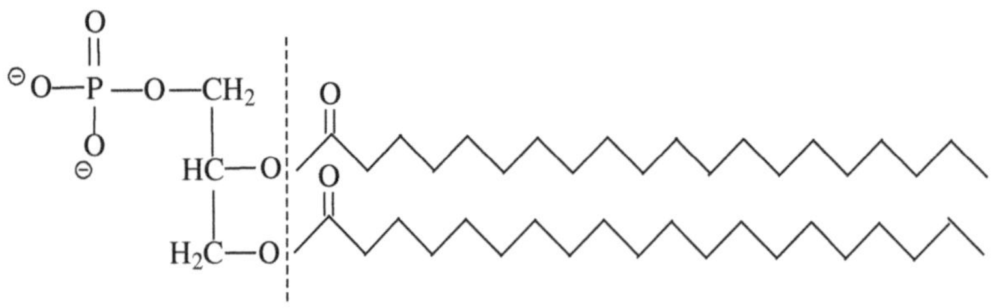

3.5 Buffering of Blood: The Bicarbonate System

The blood is buffered by dissolved bicarbonate anion. The following scheme shows the three forms of dissolved carbon dioxide:

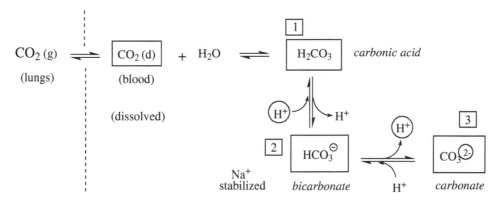

The two pK_a values for deprotonation of *carbonic acid* (1) to form *bicarbonate* (2) and then *carbonate* (3) are shown in the following pH profile. At pH 7, bicarbonate predominates.

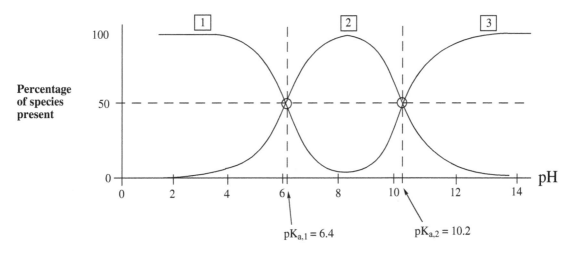

Garrett and Grisham provide a careful quantitative analysis of blood buffering. (See *Biochemistry,* 3rd ed., 2005, Brooks-Cole Publishing Company, p. 47.). Conversion of dissolved CO_2 (d) and H_2O to carbonic acid is catalyzed by the enzyme *Carbonic Anhydrase*. The equilibrium constant for hydration of CO_2 (K_h) is 0.003. This reaction is coupled to the $H_2CO_3 \leftrightarrow HCO_3^- + H^+$ equilibrium ($K_a = 2.69 \times 10^{-4}$). The overall equilibrium constant for deprotonation of carbonic acid in equilibrium with dissolved CO_2 is:

$$K_{overall} = K_a K_h = \frac{[H^+][HCO_3^-]}{K_h [CO_2 (d)]} = 8.07 \times 10^{-7}; \quad pK_{overall} = 6.1$$

The resulting *Henderson-Hasselbalch equation* is:

$$pH = pK_{overall} + \log \frac{[HCO_3^-]}{[CO_2 (d)]}$$

The total concentration of the "*Carbonic Acid Pool,*" CO_2 (d) + H_2CO_3, in blood is ~ 1.2 mM. Gaseous CO_2 at a partial pressure of 40 mm Hg in lung alveoli drives CO_2 (d) and H_2CO_3 formation, which produces a ~ 24 mM HCO_3^- concentration. Gaseous CO_2 stabilizes the pH of blood by contributing a large buffering effect on the total "carbonic acid pool."

Chapter 4

Amino Acids

4.1 Basic Structures

All of the amino acids typically found in proteins have the *L configuration* about the *alpha carbon* (C_α). Some less common structures contain D amino acids, for example, the mold-generated antibiotic valinomycin (see *Ionophores*).

pKa = 2.2

Both charges are favored at pH 7, in part due to self-attraction, *i.e.* salt bridge formation.

pKa = 9.5

In the case of *proline* the R-group is connected intramolecularly to the amine nitrogen.

"aa$_{Root}$" is the root structure of all amino acids.

The "*R group*" (containing C_β, C_δ, *etc.*)

Zwitterion. The word "zwei" means "two" in German. Both charges are favored at pH 7, due to self-attraction between the positive charge of the ammonium group and negative charge of the carboxylate. (Incidently, "zwitter" means "hermaphrodite.")

The convention for naming the isomers derives from L-glyceraldehyde.

(levorotatory) ⟶ *L-glyceraldehyde*

(sinistra) ⟶ (*S* form conformational isomer)

4.2 Amino acid "R Groups"

Twenty standard *R groups* occur in most natural proteins. The *Root* and R group atoms are often modified in real proteins for specific biochemical reasons. The standard set is like a biochemical equivalent to the alphabet. These modifications constitute a language. Deencrypting how the language is stamped into the structures is often the key to understanding the regulation of intermediary metabolism, genetic functions, cellular and organism-wide developmental strategies, modes of adaptation, and so forth.

Alkyl Side Chains
The following amino acids are aliphatic. Alanine, valine, leucine, and isoleucine are hydrophobic.

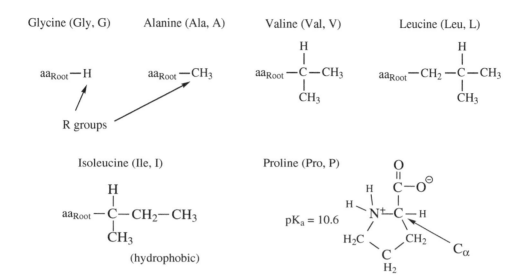

Glycine (Gly, G)   Alanine (Ala, A)   Valine (Val, V)   Leucine (Leu, L)

R groups

Isoleucine (Ile, I)   Proline (Pro, P)

(hydrophobic)

pK$_a$ = 10.6

Proline is a cyclized amino acid. It reinforces *U-turns* in protein chain structures.
Aromatic Side Chains. The following R groups occur in the *aromatic amino acids*. They are all *hydrophobic.*

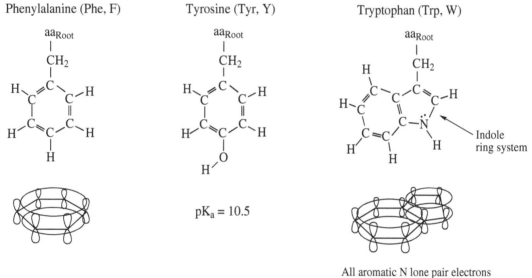

Phenylalanine (Phe, F)   Tyrosine (Tyr, Y)   Tryptophan (Trp, W)

Indole ring system

pK$_a$ = 10.5

All aromatic N lone pair electrons are in the π system.

Ultraviolet Absorbance at 280 nm

The aromatic amino acids *absorb ultraviolet light* at a wavelength of *~280 nm.*

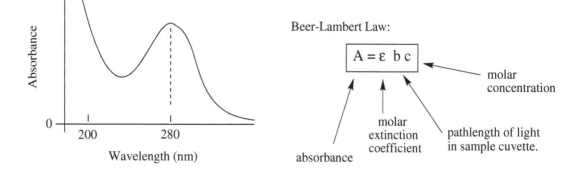

Beer-Lambert Law:

$$A = \varepsilon \ b \ c$$

molar concentration

molar extinction coefficient

pathlength of light in sample cuvette.

absorbance

The *Beer-Lambert relation* ("Beer's Law") is commonly used to determine protein and nucleic acid concentrations. If one knows the value of ε at the wavelength of interest, one can measure the absorbance and calculate the molar concentration from it.

Hydroxyl-containing Side Chains. The following amino acids contain alcohol side chains:

Serine (Ser, S)   Threonine (Thr, T)

aa_{Root} aa_{Root}
| |
CH_2 H–C–OH   [a secondary (2°) alcohol]
| |
OH CH_3

$pK_a \sim 15$

Sulfur-containing Amino Acids

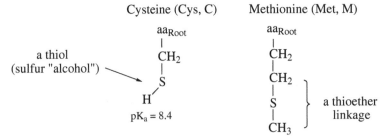

Cysteine (Cys, C)   Methionine (Met, M)

a thiol
(sulfur "alcohol")

$pK_a = 8.4$ a thioether linkage

Carboxylate-containing Side Chains

Aspartate (Asp, D)   Glutamate (Glu, E)

aa_{Root} aa_{Root}
| |
CH_2 CH_2
| |
                                                  CH_2
                                                  |
$O=C-O^{\ominus}$ $pK_a = 3.9$ $O=C-O^{\ominus}$ $pK_a = 4.1$

Amide-containing Amino Acids

Asparagine (Asn, N)   Glutamine (Gln, Q)

aa_{Root} aa_{Root}
| |
CH_2 CH_2
| |
                                                  CH_2
                                                  |
$O=C-NH_2$ $O=C-NH_2$

Basic Side Chains

Lysine (Lys, K)   Arginine (Arg, R)   Histidine (His, H)

aa_{Root} aa_{Root} aa_{Root}
| | |
$(CH_2)_4$ $(CH_2)_3$ CH_2
| | |
H–N–H   N–H   imidazole = deprotonated
| | imidazolium = protonated
H $\oplus$ C
| $H-N^{\oplus}-N-H$
$pK_a = 10.5$ | |
                          H   H
                          $pK_a = 12.5$ $pK_a = 6$

4.3 Ionization Properties

The following chart shows a series of key *pK$_a$ values* placed on the *pH scale*. Reactions occurring at each relevant pK$_a$ (letters below the scale) are listed. They are average values, which can be very different, depending on their environment within a protein (see Section 4.5).

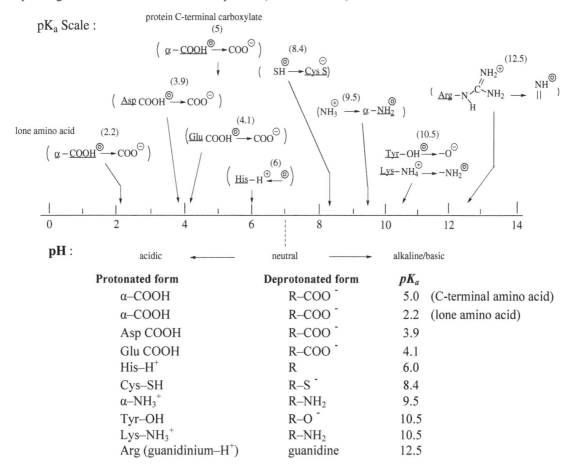

Protonated form	Deprotonated form	*pK$_a$*	
α–COOH	R–COO$^-$	5.0	(C-terminal amino acid)
α–COOH	R–COO$^-$	2.2	(lone amino acid)
Asp COOH	R–COO$^-$	3.9	
Glu COOH	R–COO$^-$	4.1	
His–H$^+$	R	6.0	
Cys–SH	R–S$^-$	8.4	
α–NH$_3^+$	R–NH$_2$	9.5	
Tyr–OH	R–O$^-$	10.5	
Lys–NH$_3^+$	R–NH$_2$	10.5	
Arg (guanidinium–H$^+$)	guanidine	12.5	

4.4 Drawing Peptide Titration Plots

The *isoelectric point* (pI) is the pH at which the *net charge* of a biomolecule is equal to 0. Calculating the charge of a protein at a given pH involves calculating all of their charges, taking into account the percentage of protonation and deprotonation of each species. The following examples demonstrate how to calculate isoelectric points in various situations.

(*1*) For an amino acid, average the pK$_a$ values. For example, with alanine only the following equilibria are involved:

$$\alpha\text{-COO}^- \leftrightarrow \text{COOH}, \qquad \alpha\text{-NH}_2 \leftrightarrow \alpha\text{-NH}_3^+$$

We find that the pI is midway between the 2 pK$_a$ values:

$$pI = [(pK_a(R\text{-}COOH) + pK_a(R\text{—}NH_3^+)] / 2 = (2.2 + 9.5)/2 = 5.85$$

(*2*) When *three* functional groups contribute to the molecular charge in amounts whose charges vary with pH, you must consider the three pK$_a$ values, whether protonation or deprotonation occurs to get to the pH of interest, and whether charge is produced or decreased when the protonation or deprotonation occurs. The following points will help you visualize the calculation by helping to keep the whole set of contributing reaction equilibria in mind and considering whether your answer is consistent with them.

(*i*) Write the structure with all the sites protonated.

(ii) Determine the net charge. If it is 0 or -, there is no pI.

(*iii*) Rank the pK_as numerically from low to high.

(*iv*) N = the charge, pI = $(pK_{a, N} + pK_{a, N+1})/2$

(*3*) If the curves overlap, which they will if two pK_as are within two units of each other, one sketches in the curves for each deprotonation, as if in isolation, then draws the best possible compromise between the isolated curves. Visually, the curves appear to be "merged" into a single transition. The following curve shows them as if the two transitions are clearly resolved. In reality, they will not be.

Case 1: **Tyr—Asp—Ala**

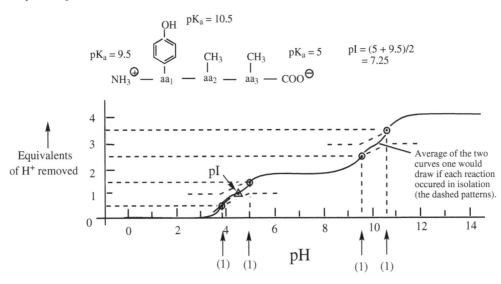

(*1*) Protonate everything … then determine the net charge (= +1 in this case).

(*2*) What is the pI? In other words, at what pH will the charge = 0? Where one negative charge is acquired (where half of the –COOH groups are deprotonated).

$$pI = \text{average of the two of –COOH } pK_as = 4.45$$

Case 2: **Tyr—Ala—Ala**

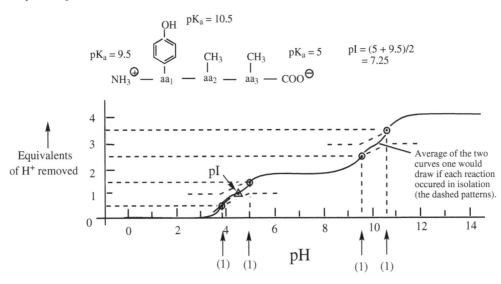

(*1*) The charge of fully protonated peptide = +1.

(*2*) At what pH will the charge be 0?

When all of the –COOH is deprotonated and none of the $–NH_3^+$ is deprotonated.

$$pI = (5 + 9.5)/2 = 7.25$$

Case 3: **Tyr—Glu—Arg**

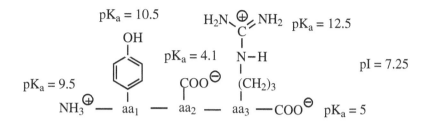

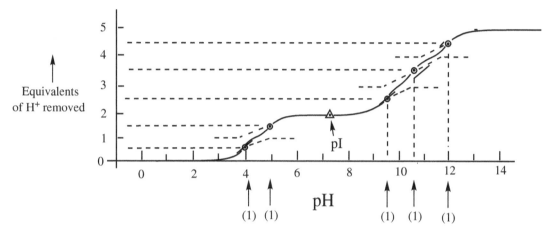

The charge of fully protonated peptide = +2.

$$pI = [(5 + 9.5] / 2 = 7.25$$

4.5 Factors That Influence the pK$_a$ of Protonatable/Deprotonatable Groups in Proteins.

The standard pK$_a$ values can shift, sometimes dramatically, when influenced by solvent factors, as shown in the following examples.

(1) Dehydration (Born Effect): The pK$_a$ changes due to the Born effect when a functional group is buried within the interior of the protein because the dielectric constant is lower than that of water. Lower dielectric constant conditions favor the neutral form. The pK$_a$s of asp, glu, cys, and tyr increase. Those of his, lys, and arg decrease.

-COOH ↔ -COO$^-$ + H$^+$	↑ nonpolar environment	↑ pK$_a$
-NH$_3$$^+$ ↔ -NH$_2$ + H$^+$	↑ nonpolar environment	↓ pK$_a$

When valine 66 in the protein *staphylococcal nuclease* is replaced by asp, the pK$_a$ of the carboxylate is 8.9, *5 units higher* than that of asp in H$_2$O. When val-66 is replaced by lys, the ammonium group has a pK$_a$ of 5.5, *4.9 units lower* than the pK$_a$ in H$_2$O.

(2) Charge-Charge Interactions (Coulombic):

-COOH ↔ -COO$^-$ + H$^+$	↑ positive charge	↓ pK$_a$
-NH$_3$$^+$ ↔ -NH$_2$ + H$^+$	↑ positive charge	↓ pK$_a$

(3) Charge-Dipole Interactions (Hydrogen Bonding):

-COOH ↔ -COO$^-$ + H$^+$	↑ hydrogen bonding to protonated form	↑ pK$_a$
-NH$_3$$^+$ ↔ -NH$_2$ + H$^+$	↑ hydrogen bonding to protonated form	↑ pK$_a$

(Adapted from: Pace, C., N., Grimsley, G. R., and Schultz, J. M. (2009) Protein Ionizable Groups: pK$_a$ Values and Their Contribution to Protein Stability and Solubility, *J. Biol. Chem.*, 284, 13285–13289.)

Chapter 5

Proteins

5.1 Peptide Bonds

Definition: Proteins are polymers composed of α-amino acids linked by sequential peptide bonds. The most basic is the dipeptide:

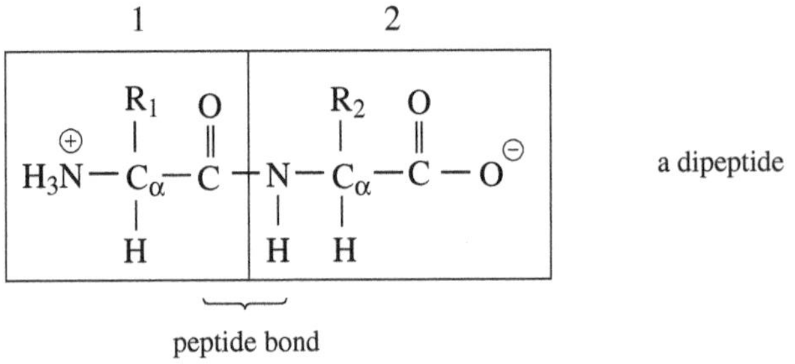

a dipeptide

peptide bond

The *peptide bond* is the fundamental linkage that results in the creation of a protein chain.

In a long polypeptide, the pK_a of the *C-terminus* increases from 2.2 to 5.

Nomenclature:

The '*ine*' suffix is usually replaced by '*yl*'.

e.g.: glycine   →   glycyl

Note that the amino acids are listed in the chain from N-terminus to C terminus, the accepted standard nomenclature.

For example, the substance Aspartame, which is used as an artificial sweetener:

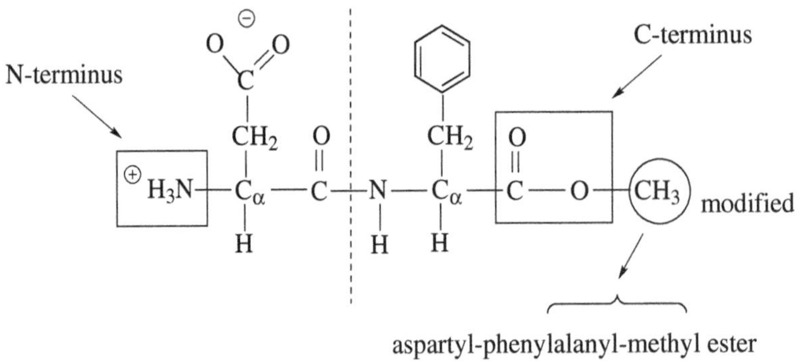

aspartyl-phenylalanyl-methyl ester

Exceptions:	asparagine	→	asparaginyl
	glutamine		glutaminyl
	cysteine	→	cysteinyl
	tryptophan	→	tryptophanyl

The '*in*' is retained, while only the terminal '*e*' is replaced by '*yl*'

5.2 Purification and Characterization of Proteins

Solid-Phase Chemical Synthesis

The key to the synthetic technique is the *solid phase synthesis* approach. This technique involves binding reactants to a ligand, which is bound to an insoluble bead material held inside a pyrex column by a set of "frits," which allow flow of reagents in and out, yet prevent loss of the matrix and bound synthetic intermediate biomolecule. Reactions are carried out by passing reactants into the column, allowing them to react with the column-bound reactant, then flushing them out with suitable solvents. The column is rinsed with the new reaction solvent and new reactant is introduced. The cycle is repeated until the desired length is made. The product is release from the column with a decoupling reagent.

This approach employs similar logic to that used in the chemical synthesis of DNA and RNA.

A key result is that synthetic peptides and protein are as functionally active as "biological isolates." This proved that *synthetic* and *natural* are functionally the same. This was the first clear indication that "living" is a process that is completely the result of interactions between a complex set of coupled chemical reactions.

Isolation from Cells

A typical purification approach involves the following steps:
(*1*) Grow the cells under conditions in which the protein is present in significant amounts, (*2*) disrupt the cellular membranes, (*3*) Remove the large insoluble "debris" by centrifuging the cell homogenate. Larger particles will sediment to the bottom of the tube. One pours off the solution to separate this debris from the dissolved crude sample solution. (*4*) It is important to include additives to preserve catalytic/functional activity: This involves keeping samples cold ("on ice") and using protease inhibitors. Buffers are used to maintain a protective pH. (*5*) Add ammonium sulfate to "salt out" either the desired protein or contaminants. In the former case one collects the precipitate, resuspends it in a planned buffer solution then dialyzes it to remove the salt. In the latter, the desired protein is in the supernatant.

Purification Using Gel Chromatography

This is typically performed using chromatography, the *structural* purity is assessed using gel electrophoresis. The functional activity is determined using a biochemical assay.

Separation is based on the intrinsic molecular differences in affinity of a solute (sample plus impurities) for a moving material (flowing buffer) through a column packed with a "stationary" gel. Binding/unbinding cycles result in differential separation of different materials. This behavior is called *partitioning* and, in gas chromatography, *plating*.

Gel Filtration (Gel Permeation) Chromatography. Smaller molecules are retained longer, while larger molecules *elute* earlier. Larger molecules are less included within and therefore less well retained by the gel network.

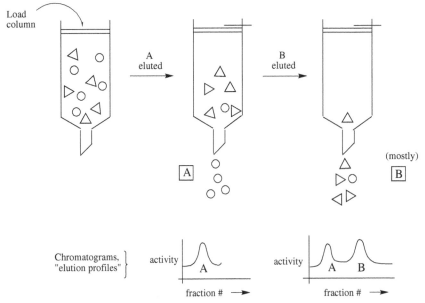

Ion Exchange Chromatography. The column matrix is modified with either cation or anion ligands. For example, negatively charged DNA binds to the cationic diethylaminoethyl ($DEAE^+$) ligand. More negative (and more tightly bound) molecules require higher salt to be dislodged relative to less negative molecules. Separation is based on different mixtures of exposed charges on the biomolecules one separates.

Ligand-dependent Affinity Chromatography. An example involves purifying ATP-binding enzymes by using their tendency to bind to a column containing bound ATP. Free ATP in buffer is used to displace the enzyme from the column matrix, purified from the pre-eluted impurities.

Polyacrylamide Gel Electrophoresis
SDS-Induced Protein Denaturation. The following reaction occurs when SDS binds to a protein:

$$\text{protein (+ or -)} + n\,SDS^- \quad \rightarrow \quad [\text{protein}-(SDS^-)_n]^{n-}$$

$$\textit{native} \qquad\qquad\qquad\qquad \textit{denatured}$$
$$\text{(structurally intact)} \qquad\qquad \text{(little stable internal structure)}$$

Electrophoresis.
In the electrophoresis process, (*1*) the *current* (I) carries the charged proteins, and (*2*) the *voltage* (V) provides the energy that drives their movement. Ohm's Law ($V = I\,R$) captures the relation between the two. The resistance (R) is provided by the gel as the proteins migrate from the upper negatively charge well to the lower positively charged buffer reservoir.

The negative charges of the *n* SDS molecules overwhelm the native charge(s) of the protein. Therefore, the proteins adopt worm-like polyanionic SDS-coated structures, which differ in length. Separation by electrophoresis is predominantly protein size-dependent, and not charge-dependent. Different molecules have different hydrodynamic "sizes," which leads to different mobilities.

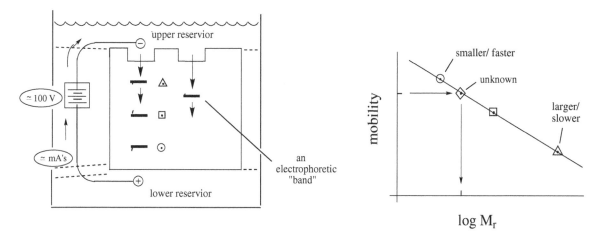

Determining Molecular Weights. One can determine the *molecular weight* (MW) of "unknown" proteins using a *mobility versus log MW plot*. The method involves running suitable calibration proteins along with the sample/unknown protein of interest. Note that the abbreviations MW and M_r are used to indicate molecular weight. The latter actually refers to an experimentally determined "hydrodynamic" value, which also depends on the rotational properties of the molecule.

Assessing Purification. The following figure illustrates an idealized electrophoresis pattern at four different stages during purification of the *glycolytic enzyme* Lactate Dehydrogenase. Contaminant proteins are somewhat randomly distributed prior to the ammonium sulfate stage and become less so after ammonium sulfate fractionation, which selects a specific subclass of proteins. The $(NH_4)_2SO_4^-$ step eliminated proteins based on insolubility. Dialysis removes proteins with molecular weights below the "cutoff" value of the dialysis tubing—typically 12 kDa.

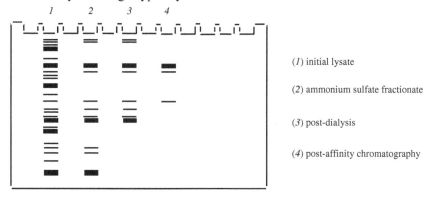

(1) initial lysate

(2) ammonium sulfate fractionate

(3) post-dialysis

(4) post-affinity chromatography

The Differences in the Electrophoresis and Gel Filtration Mechanisms

Note that separation by Size Produces Opposite Migration Rate Trends in Electrophoresis and Gel Filtration Chromatography.

(1) Electrophoresis: Larger molecules move less rapidly in the gel, ending up toward the top of the gel. Larger molecules are held up by the tortuous network of the gel. Smaller molecules move most rapidly, ending up at the bottom of the gel.

(2) Gel Filtration Chromatography: Larger molecules are not impeded as much as smaller molecules as they flow through (and by) the column matrix material. Smaller molecules keep entering the beads, holding them up. Larger molecules emerge most rapidly; smaller molecules emerge later.

Protein Quantification

Absorbance Spectroscopy.

> A_{280}—depends on aromatic amino acid content, which varies with the amino acid content and pH of the protein.

> A_{205}—every peptide bond absorbs in this region. Problem: Interferences due to the buffer and other solution components that typically accompany a semipurified protein.

Chemical Detection Methods.

(1) "*Biuret*" *reaction*:

$$\text{peptide bond} \xrightarrow[\quad (Cu^{1+}) \quad]{\overset{Cu^{2+}}{\overset{\text{(alkalinep H)}}{\curvearrowright}}} \text{pale blue complex} \longrightarrow \overset{\text{read}}{\underset{\text{absorbance}}{}}$$

(2) *Lowry reaction*: dye-coupled Biuret method; depends on comparison with standard protein samples (usually bovine serum albumin, BSA).

(3) *BCA assay*: (Pierce kit) This is an improved version of the dye-coupled Biuret technique. It has fewer interferences. *This is generally considered the preferred technique.* The Cu^{2+} reduction reaction is sensitive to cysteine, cystine, tyrosine and tryptophan residues, in addition to the peptide bonds.

(4) *Bradford assay*: involves binding *Coomassie blue dye* to protein. The dye binds primarily to arginine, as well as weaker interactions with his, lys, trp, tyr and phe. The dye is also typically used to visualize proteins on electrophoretic gels.

Amino Acid Content and Sequence Determination.

Edman Degradation.

This technique involves making PITC-derivatives of amino acids. Because all of the amino acids are labeled with the same reagent, one can compare their contents on a quantitative basis.

Technique 1. Hydrolysis, 6 M HCl, 110 °C, 1-3 days

phenylisothiocyanate (PITC)

amino acids (all of them at the same time)

Technique 2. Reaction with the N-terminus of intact protein (Edman degradation)

trifluoroacetic acid (TFA)

(F_3CCOOH)

another cycle

Methods 1 and *2* both involve the use of *high pressure liquid chromatography* (HPLC) to identify and quantify the amino acid. While this approach has been superseded by the use of mass spectrometry, the "microsequencing" version of it played an especially important role in the 1980s as a way to determine the N-terminal amino acids, thereby allowing one to make a probe DNA suitable to search for the gene in a suitable "DNA library" preparation.

Sanger's Reagent

This technique was developed by Fred Sanger, double Nobel laureate. He won his first Nobel prize for protein primary structure determination.

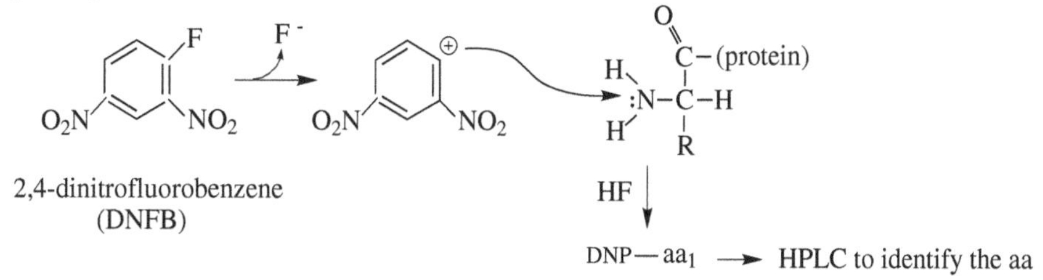

2,4-dinitrofluorobenzene
(DNFB)

DNP—aa_1 $\longrightarrow$ HPLC to identify the aa

DNA-Based Sequence Analysis.

(1) *Dideoxy Sequence Determination.* Fred Sanger won his second Nobel prize for inventing a technique that allows one to determine the sequence of DNA. The method uses DNA Polymerase, primer DNA and special "terminator" nucleotide triphosphates called dideoxy NTPs (ddNTPs). In modern applications, each of the four nucleotides is labeled with a different colored conjugate molecule, allowing one to determine the nucleotide at the terminated end. This is the basis of most modern machine-based determinations.

(2) *Reverse Translation of cDNA.* One can determine the DNA sequence that encodes the protein to indirectly obtain the amino acid sequence of the protein. The sequence of the transcribed mRNA and the 'genetic code' are used to work backwards to *deduce the amino acid sequence* encoded by the DNA. This procedure is called "reverse translation." It is done using reverse-transcribed mRNA sequences called *copy DNAs* (cDNAs). The advantage to this approach versus sequencing the genomic DNA is that the introns have been removed.

Proteins sequence information is also used to get a foot in the door. The sequences of the ends are determined, then the information is used to determine the DNA sequence, which is used to make a probe DNA. This probe is used to isolate the genomic DNA, whose sequence is then determined. This information is also used to make primer DNAs, which allow one to carry out the Polymerase Chain Reaction, as a means to obtain sufficient DNA to clone and manipulate it, if desired.

Sequence-Specific Enzymes. Three examples of sequence-specific digestive enzymes, along with their specificities, are:

(*1*) *Trypsin*:   specific cleavage for Lys and Arg
(*2*) *Chymotrypsin*:   specific for Phe, Tyr, Trp. It also has a "loose specificity" for bulky side groups.
(*3*) *Thermolysin*:   specific for Ile, Leu and Val

By carrying out cleavage reactions with different enzymes, then reconstructing the overlapping patterns, researchers were able to reconstruct the sequence of the entire protein fragment. A similar approach, using DNA fragments, is the typical way to reconstructing the sequence of genomic DNAs. Computers are used to align the fragments and piece together the full sequence. For example, the *protease* Carboxypeptidase B cuts the protein chain, leaving either arg or lys as the C-terminus. Note that the real C-terminus, whatever it is, also remains, allowing one to determine its identity.

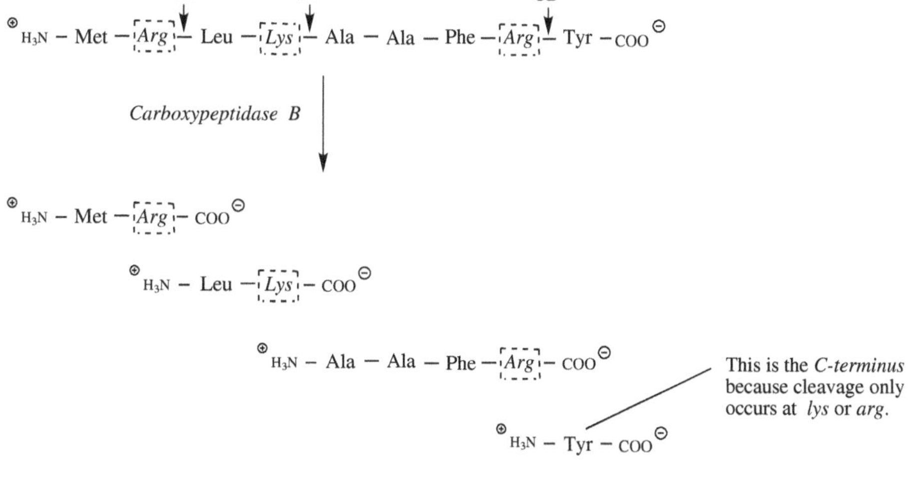

Chemical Modification by Cyanogen Bromide. Cyanogen bromide modifies proteins at *methionine* residues. This provides a way to obtain protein subfragments. Each can then be sequenced using the Edman technique or otherwise manipulated or studied.

Bioconjugates. Many other residue-specific protein modification methods exist. They form the basis of the vast literature on uses of different *bioconjugate* molecules, in which either another type of biomolecule or some artificial molecule is hooked to the initial biomolecule. Two examples include: (*1*) immunoconjugates, in which an antibody is connected to an enzyme, and (*2*) biotinylated DNA, a coenzyme connected to a nucleic acid. Each is described in detail in the *Coenzyme* and *Biotechnology* sections, respectively.

Disulfide Bond Maintenance: β-Mercaptoethanol and Dithiothreitol.
Disulfide bonds: crosslinks between two Cys residues.

β-*mercaptoethanol* (BME).
    This reagent traps reduced sulfhydryl groups. BME and DTT (see below) are used to protect proteins. Most solutions used to prepare proteins contain one of these reagents. At lower concentrations (*e.g.*, 1 mM), both BME and DTT will prevent proteins from forming unintended disulfide bonds. At higher BME or DTT concentrations, the reagent will break disulfide bonds, forming two sulfhydryl groups.
    The cell is generally held under reducing conditions, so relatively few cysteines in most cytosolic proteins are involved in disulfide bonds.

Dithiothreitol (DTT): This reagent is a more efficient reducing agent than BME.

Iodoacetic acid undergoes irreversible reaction with cys –SH groups. It acetylates and protects the cysteine from reforming disulfide bonds. Note that the sulfur is a good *nucleophile* and iodine is a good *leaving group.*

$$R-CH_2-S\text{:}^{H} + I-CH_2-COO^{\ominus} \longrightarrow R-CH_2-S-CH_2-COO^{\ominus} + H^{\oplus} + I^{\ominus}$$

Chapter 6

Protein Structure

6.1 Conformation

A generic *dipeptide* is shown to illustrate the structure of the *peptide bond*, which is formed as a result of a dehydration reaction.

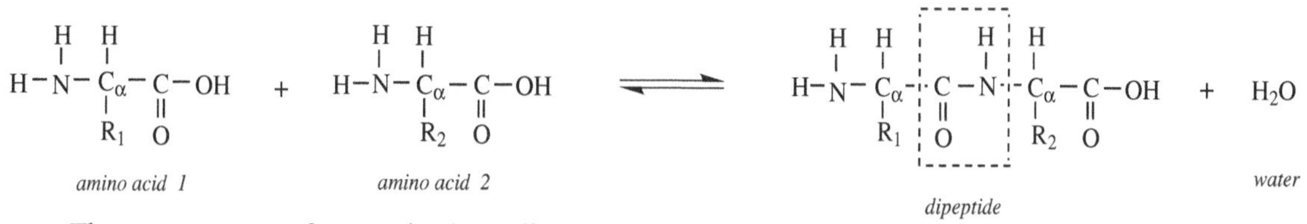

The arrangement of atoms in three dimensions depends on the patterns of connectivity between the atoms and the intrinsic rotational capabilities of the bonds, which is characterized in terms of *torsion angles*.

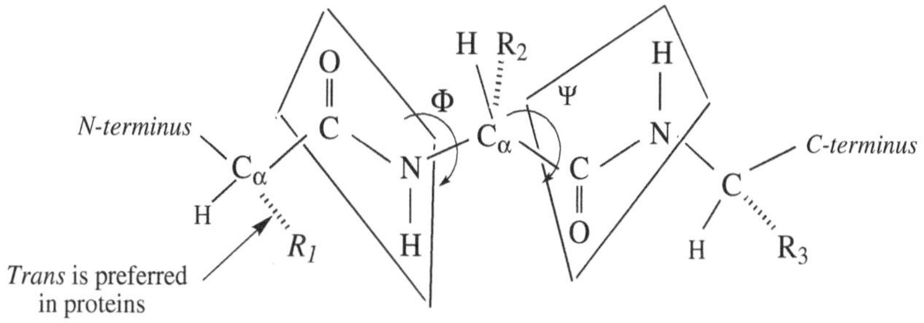

The peptide bond has *double bond character* due to *enamine-ketamine tautomerism*, which is analogous to the more familiar *enol-keto tautomerism*. The atoms shown in the boxes are constrained to being coplanar. The bonds, which are called ϕ and Ψ, can rotate, although the rotation is restricted when the chain adopts a particular structure.

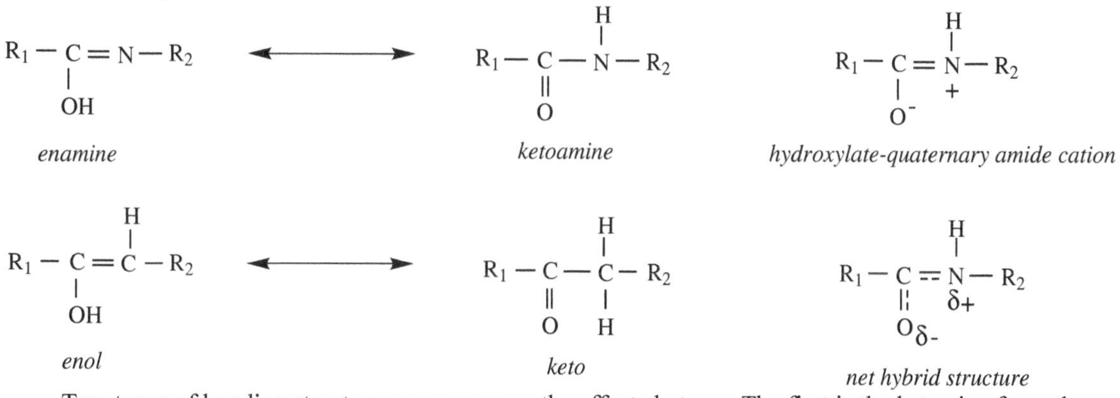

Two types of bonding structures occur among the affected atoms. The first is the ketamine form shown above. The second structure has two changes. (*1*) The carbonyl oxygen has a negative charge and is connected by a single bond to what is normally the carbonyl carbon (as a hydroxylate). (*2*) The peptide bond is a double bond and the tetravalent amide nitrogen has a positive charge. The net hybridized structure due to the two contributing structure has delocalized electrons spanning the bonds between the carbonyl carbon the carbonyl oxygen and the amide nitrogen, with a partially negative charge on the carbonyl oxygen and partially positive charge on the amide. As a result, the peptide bond has a net bond order of 1.5 and does not rotate easily.

6.2 Classification of Substructure

The first level of classification involves the degree of compatibility with water.
(*1*) *Water Soluble*: many enzymes fall into this category. They are typically "globular" in shape.
(*2*) *Water Insoluble*: structural materials, such as Keratin, a primary component in hair and fingernails.

X-ray Diffraction Analysis. The wavelength used is similar to interatomic distances. Light is diffracted by the electron density of an array of proteins embedded within a crystal. This produces a specific pattern of spots on an imaging apparatus that surrounds the mounted crystal. Using computer programs and careful human analysis, one can work backward to calculate the structure of protein.

In an early, highly instructive set of studies, John Kendrew and colleagues solved the structure of *myoglobin*, which carries molecular oxygen in muscle tissues. The structure revealed many general features of protein architecture. (The structure is shown in the *Myoglobin* section.)

Primary Structure. This is the sequence of covalently linked amino acids.
Secondary Structure. These are the regular folded structures found within most proteins, for example, α-helix, antiparallel and parallel β-sheets. Others are the U-turn and extended chain, which are described in detail below.
Tertiary Structure. This refers to the packing and stabilization of regular interconnected substructures from a single chain into the overall globular structure; side chains accommodated. An example of the ATP-binding domain of *hexokinase*, which is composed of an α/β-domain, is shown below.

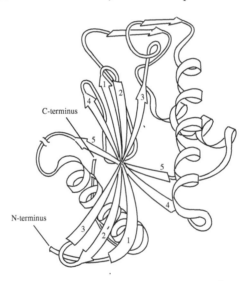

(Adapted from McKee and McKee, *Biochemistry The Molecular Basis of Life* (5th ed.). 2012, p. 149; Fig. 5.21.)

Quaternary Structure. This involves the packing of subunits (different chains) into multisubunit complex. An example is provided by hemoglobin, which forms a tetrameric quaternary structure:

$$2\,\alpha + 2\,\beta \rightarrow \alpha_2\,\beta_2$$

Non-covalent quaternary structure formation often modulates regulatory behavior. Multisubunit proteins exist for three main reasons.
(*1*) Having several subunits is likely to be more efficient than adding more sequence to the length of one polypeptide chain.
(*2*) Replacing depleted or damaged portions in multisubunit complexes, such as collagen fibers, can be more efficiently managed.
(*3*) Interactions within multiple subunit complexes commonly aid in regulating the biological function of a protein monomer.

6.3 Alpha (α) Helices

The α-helix was first described by Linus Pauling. Most α-helices in proteins are right-handed due to steric interference between the carbonyl oxygens and amino acid side chains. The helix conformation is reinforced by hydrogen bonding between each *carbonyl oxygen* and the *amide hydrogen* of the fourth amino acid toward the C-terminus.

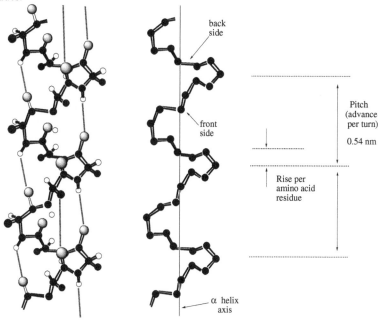

Hydrogen Bonding
(13 inclusive atoms)

The hydrogen bonds that form between the N, O, and H atoms are approximately parallel to the long axis of the helix. The distance between equivalent positions on the α helix is called the pitch and recurs every 0.54 nm. The rise of the helix is referred to as how far each amino acid advances the α helix and is 0.15 nm. Per turn of the helical structure there are 3.6 amino acid residues.

An α-helix is a secondary structure of polypeptides and could be right-handed or left-handed. In the ideal α helix, the distance each amino acid advances the helix, or the rise, is 0.15 nm. For one complete turn of the α helix, 3.6 amino acid residues are required, which consists of approximately one carbonyl group, three N-C$_\alpha$-C units, and one nitrogen. The pitch of the helix is the distance of equivalent positions on the helix is usually occurs every 0.54 nm.

The helical structure contains stabilizing hydrogen bonds between each carbonyl oxygen and the α-amino nitrogen of the fourth residue toward the C-terminus. The atoms circumscribed by the hydrogen bonds form a ring structure containing the 13 atoms, carbonyl oxygen, 11 backbone atoms, and amine hydrogen.

A variant of the α-helix called a *3-10 helix* has a slightly different hydrogen bonding pattern.

6.4 Beta Sheets

Beta sheets form when two or more polypeptide chain segments either fold back upon themselves (antiparallel), or line up side by side in the same direction (parallel), to form an interconnected sheet. These sheets are stabilized by hydrogen bonds between the polypeptide backbone amide hydrogen and carbonyl groups of adjacent chains. The hydrogen bonds are nearly perpendicular to the extended polypeptide chains.

(*1*) In *parallel β sheets*, the polypeptide chains are arranged side-by-side in the same N- to C- terminal direction. The hydrogen bonds are evenly spaced but slanted, as shown in (*A*).

(*2*) *Antiparallel* chains run in opposite directions with respect to N- to C-terminal directions. The hydrogen bonds are perpendicular to the strands and the space between the bonds alternates between wide and narrow, as shown in (*B*). Antiparallel β-sheets are more stable than parallel β-sheets and there can be occasional mixing between the two to form pleated sheets.

A   **B**

β-Barrels. β-Sheet structures are often skewed from planarity. Human retinol-binding protein is characterized by its β-barrel domain. Retinol is a light antennae pigment that is used for vision and bone growth.

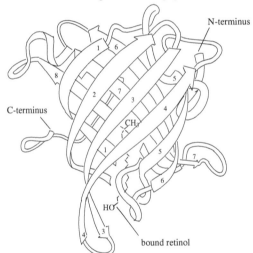

(Adapted from McKee and McKee, *Biochemistry The Molecular Basis of Life* (5th ed.). 2012, p. 149; Fig. 5.21.)

6.5 U-Turns

The structure of *proline* is nearly locked in, but exchanges slowly between *cis* and *trans* structures. The cis form supports formation of U-turns, resulting in a reversal of the direction of the protein chain.

trans peptidyl proline   versus   *cis peptidyl proline*

6.6 Ramachandran Plot

The *Ramachandran Plot* shows the coordinates corresponding to a variety of protein conformations. Solid lines indicate the range of commonly observed φ (phi) and ψ (psi) values (as defined on the first page of this chapter). Large dots correspond to values of φ and ψ that produce recognizable conformations such as the α-helix and the β-sheet. The unenclosed portions of the plot correspond to values of φ and ψ that rarely or never occur.

Each (Φ, ψ) coordinate pair gives the coordinates for one amino acid. The plot consists of all of the data pairs for a given protein. It is constructed after the structure has been determined (by crystallography or NMR analysis). The plot is a fingerprint for the protein *structural topography* present in the protein under analysis. It is not used to determine structure. It is a postdetermination characterization tool. One could, for example, compare the plots of two proteins to get a rapid assessment of the types of secondary structures in each and their relative populations.

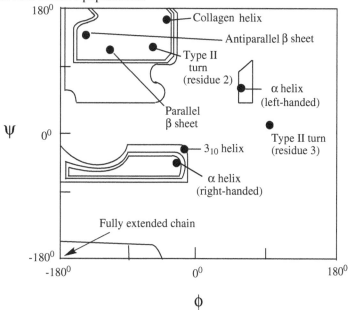

Secondary structures are specific repeating patterns. They occur when consecutive amino acid residues have similar φ and Ψ values. These patterns occur when all the φ *bond angles* (rotation angle about N–C$_\alpha$) in a polypeptide segment are equal and all the Ψ *bond angles* (rotation angle about C$_\alpha$–carbonyl C) are equal. Because peptide bonds are rigid, the α-carbons are swivel points for the polypeptide chain. If these angle values are distorted in any way, the secondary structure can be disrupted. The vacant areas in the Ramachandran plot represent conformations that are rare because atoms would be too close together.

Many amino acid residues fall within the permitted areas on the plot, which are based on alanine as a typical amino acid. Since there are limits on the values of the φ and Ψ angles, certain amino acids are not able to form a secondary structure. Proline is restricted to a φ value of about -60° to -77° because it contains a rigid ring that prevents the N–C$_\alpha$ bond from rotating. By contrast, glycine residues are exempt from many steric restrictions because they lack β-carbons. Thus, they are very flexible and have φ and Ψ values that often fall outside the shaded regions of the plot.

Table 1. Ideal φ and ψ values for some recognizable conformations

Conformation	φ	Ψ
α Helix (right-handed)	-57°	-47°
α Helix (left-handed)	+57°	+47°
3$_{10}$ Helix (right-handed)	-49°	-26°
Antiparallel β sheet	-139°	+135°
Parallel β sheet	-119°	+113°
Collagen helix	-51°	+153°
Type II turn (second residue)	-60°	+120°
Type II turn (third residue)	+90°	0°
Fully extended chain	-180°	-180°

6.7 Stabilizing Factors

(*1*) *The "Hydrophobic Effect."* This is the key driving force. It involves the balance between two huge energies. The first is due to the intrinsic entropy of the chain, which always favors unfolding because that state presents more possible configurations to occupy. The second huge energy is the enthalpy accrued by hydrogen bonding of water molecules surrounding the macromolecule. A third contribution involves "freeing" the water molecules that are "peeled away" from the surfaces that interact when two molecules bind each other and returned to the bulk solvent pool.

(*2*) *Hydrogen Bonding*: This is ubiquitous in biomolecules and provides the major contribution to the enthalpy that drives the "Hydrophobic Effect."

(*3*) *Disulfide Bonds.* They form in an oxidative reaction between the sulfhydryls of two cysteines. The bonded pair is called cystine. They can scramble and thereby inhibit folding. They were a key focus of the proof by Christian Anfinsen that structure begets function in enzymes.

(*4*) *Van der Waals Forces.* Occur between every atom and every other atom; they are significant up to 20 Å. All pair-wise interactions must be calculated to simulate them.

(*5*) *Dipole-dipole Interactions.* Occur between partial charges. In the transient version, the electron density in the two lobes on the barbell-shaped π orbital can switch back and forth in either correlated or anti-correlated synchrony, depending on the attractive or repulsive nature of the interaction.

(*6*) *Ionic Bonds.* Also called a *salt bridge*. This is the interaction between two full charged species, for example, an aspartyl carboxylate group with a lysyl ε-ammonium group. A more relaxed definition includes charge-dipole atom interactions, such as a Na^+ ion with a carbonyl oxygen.

6.8 Thermodynamics of Protein Folding: The Hydrophobic Effect

The primary driving force for protein folding is described as the "hydrophobic effect." Consider the *Gibbs free energy* for protein unfolding, the measurement of the degree of spontaneity (– sign) or lack thereof (+ sign).

$$\Delta G_u = \Delta H_u - T\Delta S_u$$

where ΔS_u is the *entropy*, which is affected by the number of accessible configurations, ΔH_u is the *enthalpy*, which is gained by bond formation, and T is the temperature in Kelvin. The gross numbers under the energy terms are intended to show how the net energies accrue. A more specific plot is described below.

ΔH_u	-	$T\Delta S_u$	$\rightarrow$	ΔG_u	
(+2)		(-1)		(+1)	*Folding* is spontaneous. The protein is predominantly *folded*.

ΔH_u	-	$T\Delta S_u$	$\rightarrow$	ΔG_u	
(+1.5)		(-2)		(-0.5)	*Unfolding* is spontaneous.

Temperature-Dependant Denaturation. Unfolding is more spontaneous due to the increase in temperature, which affects the contribution due to entropy.

        (*1*) At low T (20˚C) the protein is *stable.*
        (*2*) At higher T (> 70˚C) the protein *denatures* (unfolds).

Why? Proteins either fold or unfold depending upon what happens to the *surrounding water*.

    (*1*) At low temperature, hydrogen bonds among surrounding water molecules are intact. They are an example of an enthalpic contribution to the ΔG for folding. Note that the contribution is a positive ΔG (+) for unfolding, since it would be disfavored. The protein always wants to unfold because it becomes more disordered (entropy), but hydrogen bonds overcome that tendency. The encaged protein folds to a small volume to maximize hydrogen bonding.

    (*2*) At high temperature, hydrogen bonds are driven apart by more vibrational energy, and so on. The T is always multiplied by ΔS_u and the ever-present tendency to unfold is enhanced. Also, the many released water molecules gain a lot of entropic energy releasing the folded protein from its cage, allowing it to unfold. The enthalpy is overcome.

Note that the ΔG values are subscripted "f," which means they corresponding to folding, not the subscripted "u" used in the beginning of the section.

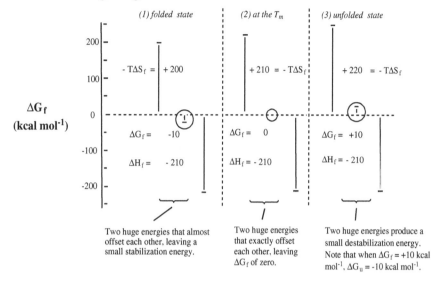

(1) folded state	(2) at the T_m	(3) unfolded state

$-T\Delta S_f = +200$ $+210 = -T\Delta S_f$ $+220 = -T\Delta S_f$

ΔG_f (kcal mol^{-1})

$\Delta G_f = -10$ $\Delta G_f = 0$ $\Delta G_f = +10$

$\Delta H_f = -210$ $\Delta H_f = -210$ $\Delta H_f = -210$

Two huge energies that almost offset each other, leaving a small stabilization energy.

Two huge energies that exactly offset each other, leaving ΔG_f of zero.

Two huge energies produce a small destabilization energy. Note that when $\Delta G_f = +10$ kcal mol^{-1}, $\Delta G_u = -10$ kcal mol^{-1}.

6.9 Chaotropes and the Hofmeister Series

Salts such as sodium perchlorate favor protein denaturation. As a result, they are not used to precipitate proteins, where it is very desirable to maintain the native form. Salts such as $NaClO_4$ and guanidinium chloride are called "chaotropes" because they disturb the hydration cage around the protein. As a result, the protein can unfold to the entropically favored denatured form. The ammonium cation and sulfate dianion actually foster protein folding, so $(NH_4)_2SO_4$ is called an "anti-chaotropic" salt. The chaotropic salt guanidinium chloride (GuHCl) is commonly used to disturb the structure of a protein and determine its relative stability in a quantitative manner. The effect produced by increased [GuHCl] matches the behavior found in spectroscopic thermal denaturation studies.

The relative propensities of salts to induce denaturation or foster folding were first classified by Hofmeister in the early 1900s. The experiments involved placing defined protein and nucleic acid preparations in a broad array of solutions whose cations and anions were varied systematically. Tendencies of salts to induce denaturation were sorted out and placed on a pair of scales, one corresponding to the respective tendencies of a series of anions, and one to those of an array of cations. Some cations, such as sodium and chloride, are only moderately denaturing, Ammonium and sulfate foster folding, so they're the preferred choice in precipitation protocols. (Details are described on pp. 198–200 in Hardin *et al., Cloning, Gene Expression and Protein Purification*, Oxford University Press, 2002.)

6.10 Sodium Dodecyl Sulfate (SDS): Chaotrope Action

Sodium Dodecyl Sulfate is an amphipathic compound that has a polar head group and a nonpolar alkane chain. When SDS is present with a protein in solution the polar head groups encapsulate the protein with the hydrophobic tail positioned to the interior against the folded protein. Because SDS is covering the protein, the *hydrophobic effect* is disrupted and the protein is then able to hydrogen bond with the surrounding solution. This causes the protein to unfold. Note that not all proteins fully denature with SDS.

This reagent is used to convert the proteins to denatured wormlike shapes for analysis of molecular weights using *Polyacrylamide Gel Electrophoresis*, as described in that section.

folded protein   SDS-bound *unfolded* protein

6.11 Visualizing the Energy Landscape

Polypeptides undergo folding to return to their original condition. Some do so by way of forming alternatively folded intermediates while others become trapped in an incorrectly folded manner.

The term "molten globule" is used to refer to a partially folded protein intermediate state found during denaturation within many molecules or their domains. They are collapsed and often have some native, secondary structure features, but have an interior tertiary structure which contains interactions between amino acid side chains that are not yet stabilized.

(*1*) Often with small proteins, folding is cooperative, not involving intermediates.

(*2*) With large proteins, a molten globule is often formed first followed by the return to its native formation.

(*3*) Large proteins with multiple domains often fold separately in each domain, ending with the whole molecule   returning to its native conformation.

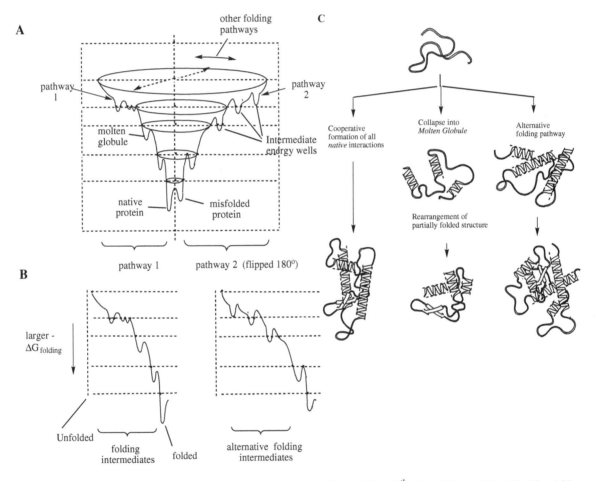

(Adapted from McKee and McKee, *Biochemistry the Molecular Basis of Life* (5th ed.), 2012, pp.158–159; Fig. 5.29–30.)

6.12 Protein Maturation

Making Proteins for Export: The Signal Recognition Process in Translation
The first post-translational processing event occurs *while* the protein is still being synthesized. Proteins that are targeted for processing in the endoplasmic reticulum (ER) contain a hydrophobic set of amino acids in the region located N-terminal to the first amino acid in the mature protein called the *signal peptide*, *propeptide* or *leader sequence*. The following figure shows a schematic of the translation process catalyzed by the ribosome and using transfer RNAs, the messenger RNA and the *signal recognition protein* (SRP). The emerging nascent signal peptide binds SRP, which directs the translating complex to the signal recognition receptor complex, which removes the signal peptide by proteolysis once the "homing" process has been accomplished. The remaining protein is extruded into the lumen of the ER.

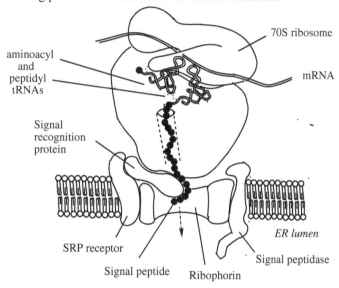

(Adapted from Moran *et al., Principles of Biochemistry* (5th ed.), 2012, p. 693; Fig. 22.32.)

The Molecular Chaperones
Molecular chaperones assist with the folding and assembly of proteins in living cells. Most are classified as *heat shock proteins* (hsp), which are present in plants, animals, and bacteria. They are also found to be active in eukaryotic organelles such as mitochondria, chloroplasts, and the endoplasmic reticulum.

Molecular Chaperones briefly bind to nascent proteins and unfolded proteins that have undergone stressful conditions that have led to denaturation. The chaperone hsp70 can aid in stabilizing nascent proteins (*i.e.*, newly made, yet not fully folded) and reactivating unfolded proteins. Chaperones in the hsp60 family are beneficial in helping proteins reach their developed conformation. If a protein is incapable of folding correctly, the chaperones destroy it.

Note that *arginine* residues line the inside surfaces of the barrel-like multisubunit *chaperonin* complex. Recall that the R group of arg is terminated by a guanidinium group. Both guanidinium and urea disrupt the structure of water. Inside the chaperonin complex, they are thought to "tickle" the water around the folding protein so that it unfolds and refolds correctly. This solves an essential problem in protein folding, recovering misfolded proteins and sending them down the correct pathway.

Model of the E. coli Chaperonin Complex GroES-GroEL
GroES (a co-chaperonin, or hsp10) is a seven-subunit ring chaperonin that often works in conjunction with *GroEL* (a chaperonin, or hsp60). GroES sits on top of GroEL which is made up of two seven-subunit rings that are stacked to form a hollow space where protein folding occurs with the aid of ATP.

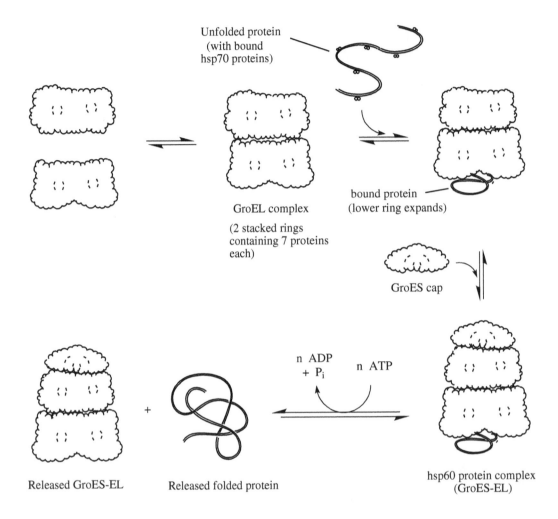

(Adapted from McKee and McKee, *Biochemistry the Molecular Basis of Life* (5th ed.), 2012, p. 161; Fig. 5.32.)

Disordered Protein Binding

Some proteins occur as disordered *coil* forms, which fold into regular secondary structures upon binding a suitable "template" binding partner.

 For example, the transcription regulatory protein, CREB, binds to DNA and increases or decreases the transcription of downstream genes. The CREB protein has a phosphorylated disordered domain called pKID that binds the KIX domain on the transcription co-activator protein CBP. A pair of helices is formed as the two domains bind and undergo folding.

 The following figure shows the process (panel A) and the ordered KIX domain alone in the tube and ribbon formats, and the complex between the peptide and the pKID domain, in ribbon structure format.

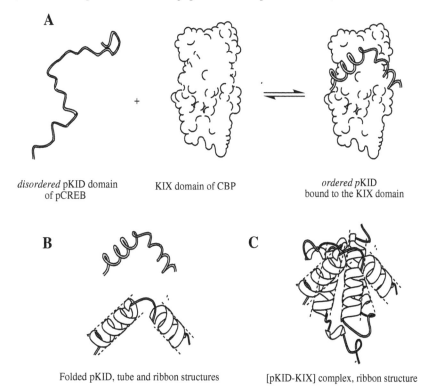

(Adapted from McKee and McKee., *Biochemistry the Molecular Basis of Life* (5th ed.). 2012, p. 155; Fig. 5.27.)

Measuring Reversible Folding and Linking it to Catalytic Activity

Christian Anfinsen demonstrated that the activity of the enzyme Ribonuclease A (RNase A) depends on the integrity of the protein structure. RNase A can be denatured using 8 M urea and a thiol, a sulfur-hydrogen bond (R-SH) which reduces disulfides to sulfhydryl groups. The enzyme can be *renatured* (returned to the *native* structure) by simply removing the urea and R-SH group and using air to oxidize the reduced disulfides.

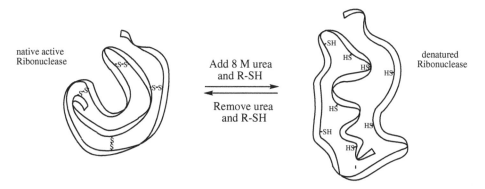

(Adapted from McKee and McKee, *Biochemistry the Molecular Basis of Life* (5th ed.), 2012, p. 156; Fig. 5.28.)

Other Types of Protein Crosslinks. Desmosine is present in *elastin*, where it allows for elasticity and flexibility. Allysine involves cross-linkage between two lysine residues via *Schiff base* formation. It is present in *both elastin* and *collagen,* a triple-helical protein structure that provides the structural rigidity in connective tissue.

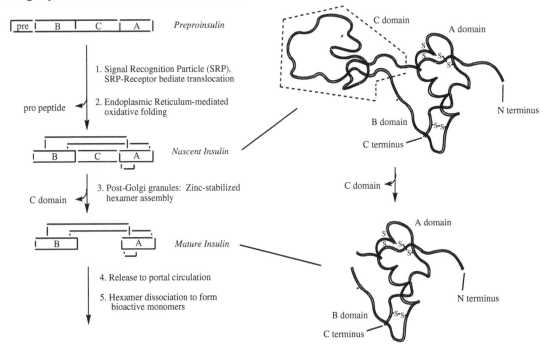

Insulin Maturation. The physiological processing of insulin begins in specialized Islet of Langerhans cells in the pancreas then proceeds through several steps as outlined in the following figure. Blockage of proper folding is predicted to lead to *neonatal-onset Diabetes Mellitus.*

(Adapted from Weiss, M. A. (2009) Proinsulin and the Genetics of Diabetes Mellitus, *J. Biol. Chem.* 19159–19163, Fig. 1.)

The *Signal Recognition* mechanism leads to removal of a hydrophobic *signal peptide*, which directs the protein from the ribosome into the lumen of the endoplasmic reticulum, where further processing occurs. Eventually the protein is packaged into a hexameric zinc-containing complex for transport through the circulatory system. See Weiss, 2009 for details.

Amino Acid Biosynthesis, Catabolic Degradation and Protein Breakdown.
Most biochemistry textbooks contain sections that describe the biosynthesis and breakdown pathways for the common amino acids. While this information is important, as well as relevant to a number of genetic diseases, they are beyond the scope of this one semester treatment. Their relation to the Urea Cycle and protein degradation is described in Section 30.4.

Chapter 7

Ligand Binding and Functional Control

7.1 Oxygen Transport in Blood

Myoglobin and Hemoglobin

Hemoglobin binds molecular oxygen (O_2) and transports molecular oxygen from lungs to tissues within red blood cells. It has a tetrameric quaternary structure with four O_2-binding pockets containing one heme molecule each. *Myoglobin* binds O_2 in muscle tissues.

These proteins provided the basis for our first clear understanding of the relation between protein structure and how their functions are modulated by solution conditions. This "workbench" was used to develop and test many fundamental ideas involving how *cooperative ligand binding* by *multi-subunit ligand-binding proteins* controls their functional activity. It is a great example of how physiological and biochemical levels of function can intercommunicate through *allosteric interactions*.

Heme. Hemoglobin and myoglobin contain the bioinorganic cofactor called *heme* (iron-protoporphyrin IX). Heme consists of a porphyrin macro-ring composed of four pyrrole rings. Reduced iron (Fe^{2+}) is held by chelation in the center of the macro-ring.

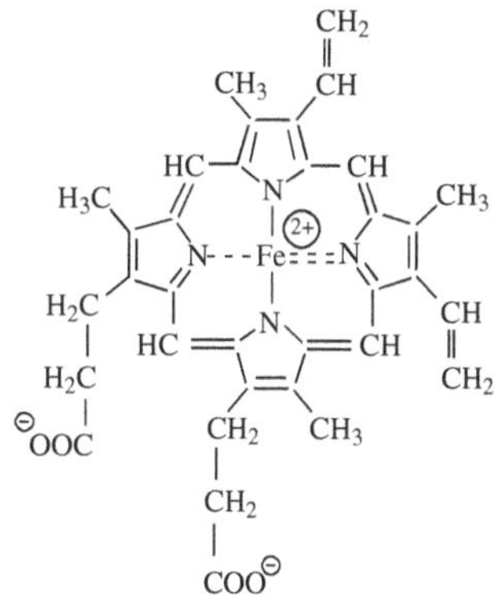

(Adapted from Moran *et al., Biochemistry* (2nd ed.), 1994, p. 5.28; Fig. 5.34.)

Although the figure shows iron covalently bound to only two nitrogen atoms, iron binds equally to all four nitrogen atoms. Both myoglobin and hemoglobin bind to molecular oxygen via this chelated iron.

Cofactor Binding. Globular proteins contain cavities or clefts, whose structural components are complementary to the structure of the specific ligand. These clefts typically contain ionizable residues that perform crucial functions by participating in the chemical mechanisms of ligand binding. When required cofactors are bound within a protein it is called a *holoprotein*. When the cofactors are absent, it is call an *apoprotein*.

$$4 \text{ globins} + 4 \text{ hemes} \rightarrow \text{hemoglobin}$$

apoprotein   *holoprotein*

Myoglobin

Myoglobin is composed of eight helices. The heme group alternately binds iron and oxygen reversibly. Only the α-carbon atoms of the globin polypeptide and the two histidine residues on the side chains are illustrated. In order to make the visual clear, one of the heme acid side chains is moved.

The protein is found at high concentrations in skeletal and cardiac muscle and gives these tissues their characteristic red color. This red coloration resides in the iron atoms of the heme prosthetic groups, where molecular oxygen molecules bind. Heme consists of a tetrapyrrole ring system called protoporphyrin IX with an ionic iron bound in the center. The four pyrrole rings of this system are linked by methylene (-CH=) bridges, so that the entire porphyrin structure is unsaturated, highly conjugated, and planar.

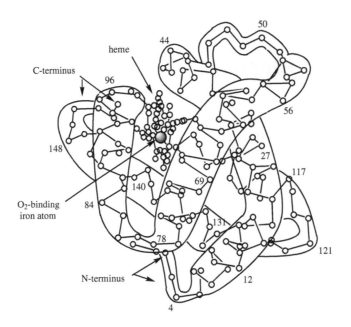

Myoglobin
(Adapted from McKee and McKee, *Biochemistry The Molecular Basis of Life* (5ᵗʰ ed.), 2012, p. 165; Fig. 5.37.)

The protein component of myoglobin is a single polypeptide chain called *globin*, which contains eight sections of α-helix. The folded chain forms a crevice that almost completely encloses the heme group. His-64 forms a hydrogen bond with oxygen, and His-93 is bound covalently to the iron atom within the heme.

Hemoglobin
The protein contains four subunits. Each subunit contains a heme group that binds reversibly with oxygen.

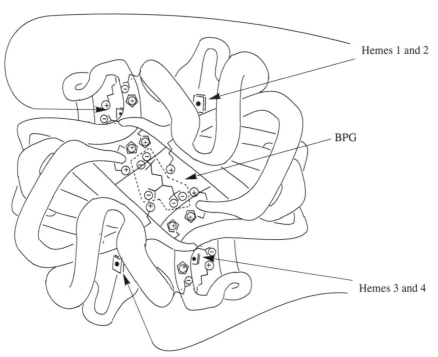

The central cavity of hemoglobin contains six positively charged side chains. The N-terminal α-amino group of each β chain form a cationic binding site. (Adapted from Moran *et al.*, *Biochemistry* (2ⁿᵈ ed.), 1994, p. 5.50; Fig. 5.62.)

7.2 Hemoglobin Oxygen Binding: Cooperativity

Titration Curves Are Ligand-Lattice Binding Curves

So far, we have looked at binding in terms of proton removal from a conjugate base, such as the titratable amino acid functional groups. The pK_a is a benchmark that characterizes the capability of the molecule to bind the proton. A lower pK_a means that a higher H^+ concentration is required to jam the proton onto the molecule.

The standard format for plotting a *binding curve* places the ligand concentration on the *x*-axis and the *percentage of the sites* on the bound (lattice) molecule that are *occupied* on the *y*-axis. This plot is typically a hyperbolic shape, as shown for the binding of O_2 to myoglobin below.

The pH titration plot and the hyperbolic ligand-binding curve contain the same information. Consider the two-proton/conjugate base binding reaction $A^{2-} + 2 H^+ \leftrightarrow H_2A$. The pH-titration plot shows the number of protons removed from the lattice molecule in the two deprotonation steps, with pK_as at 2 and 6. Panel B shows a plot in which the $[H^+]$ corresponding to the pH values in panel A are replotted as $[H^+]$. The number of H^+ bound to the conjugate base (the lattice) are plotted on the *y*-axis. This transformation converts the pH titration curve into a conventional binding curve. Note that a log scale is used. If the scale is not logarithmic, the plot would have the expected hyperbolic shape and all of the lower H^+ concentration information would be pushed into the left side of the plot. The semilog presentation retains the advantage of showing each concentration range with equal emphasis.

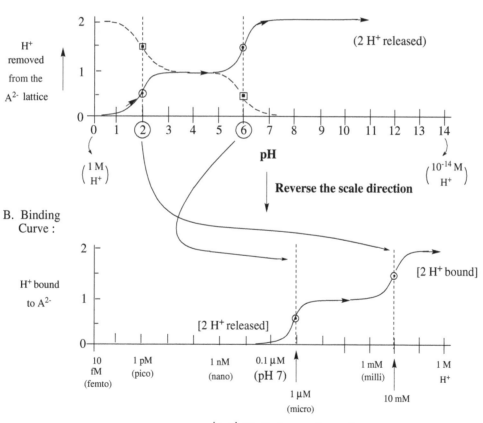

Hemoglobin: Oxygen Transport and Binding Equilibria.

Hemoglobin transports molecular oxygen from lungs to tissues. It is found in red blood cells. The binding curves for binding of oxygen to myoglobin and hemoglobin are shown below.

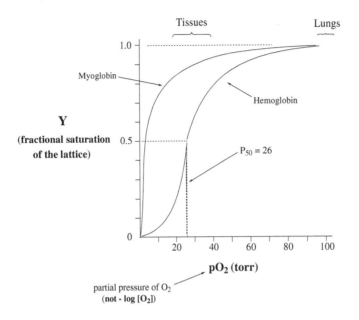

The parameter P_{50} is the *partial pressure* of O_2 required to achieve 50% occupancy of the binding sites on the molecule. Of course, myoglobin only binds one O_2, while hemoglobin binds four O_2s to achieve 100% occupancy. Be clear that we are talking about populations of molecules, not necessarily single species. The P_{50} value for myoglobin-O_2 binding is *2.8 torr*. In contrast, hemoglobin requires a P_{50} of *26 torr*. This shows that hemoglobin binds O_2 with a lower affinity than myoglobin.

Biologically, this means that hemoglobin is nearly saturated with oxygen in the lungs, where the partial pressure of oxygen is high. In contrast, in the tissue capillary beds where exchange between circulatory and cellular systems occurs, the partial pressure of oxygen is low enough that oxygen is released from hemoglobin and transferred to myoglobin, which holds it in the cells. This transfer provides an efficient fuel-supplying system for delivery of oxygen from the lungs to the muscles.

Cooperativity In O_2 Binding

Hemoglobin displays a sigmoidal (S-shaped) curve, which usually indicates the presence of *cooperativity* in the reaction equilibria. The hyperbolic curve for myoglobin does not. A common reason for multimer formation is because it introduces the possibility of controlling the differential binding characteristics of different sites, depending on how much ligand is present.

With hemoglobin, binding the first O_2 makes subsequent molecules bind more easily. This is called *positive cooperativity*. When O_2 binds to the first site, the next binding site changes, making it more susceptible to ligand binding. The third O_2-binding site has approximately the same affinity as the second, and the fourth has a lower affinity. If each subsequent association/binding constant increased relative to that of the previously bound site, the system would show purely *positive cooperativity*. Hemoglobin shows positive cooperativity in the site 1 to site 2 case, but negative cooperativity in the site 3 to site 4 case. At low $[O_2]$, it picks up O_2 fairly efficiently, at higher O_2 pressures it tends to try to release the O_2. This is called *mixed cooperativity*—positive at low $[O_2]$, negative at high $[O_2]$.

BPG Inhibits Binding of O_2 to Hemoglobin. Allosteric Inhibition

Hemoglobin binds the compound *2, 3-bisphosphoglycerate* (BPG), which is derived from the glycolysis intermediate 3-phosphoglycerate.

2, 3-bisphosphoglycerate

BPG is an *allosteric effector* because it binds to the central cavity in the tetrameric complex, a site *other than* the oxygen binding site, the active site of this binding protein complex.

(*1*) When hemoglobin is in the *deoxy conformation*, the positively charged groups bind to the five negative charges of BPG. When the BPG concentration is *low*, hemoglobin has a very *high affinity* for oxygen. As with H^+ and CO_2, binding BPG stabilizes the *deoxygenated* form of hemoglobin.

(*2*) When hemoglobin is *oxygenated*, the β chains are closer together so the allosteric binding site is too small to bind BPG.

(*3*) In the absence of BPG, hemoglobin has a higher affinity for oxygen. When BPG is present, the affinity of hemoglobin for oxygen decreases. BPG increases the proportion of deoxy (T) hemoglobin. The deoxygenated conformation of hemoglobin is often referred to as the *T (taut) state*. Oxygenated hemoglobin is said to be in the *R (relaxed) state*. Oxygen and BPG have opposite effect on the R ↔ T equilibrium.

(*4*) Hemoglobin *transfer oxygen to myoglobin* at the low partial pressures of oxygen in the tissues.

(*5*) At the high partial pressures of oxygen in the lung alveoli, and in the absence of BPG, hemoglobin binds oxygen.

The important physiological consequence is that our breathing is connected to our metabolic status. Oxygen is required as the terminal acceptor of electrons in the Electron Transport process. The allosteric role of BPG provides a direct link between the status of glucose metabolism and the extent of oxygen uptake. When more ATP is around, the glycolytic compound 3-phosphoglycerate is converted to BPG, which limits the amount of O_2 bound by hemoglobin. The BPG concentration is an *indicator* of the status of the metabolic pathway.

Fetal Hemoglobin

Regulation of hemoglobin oxygenation by BPG plays a unique role in the delivery of oxygen to the human fetus from the maternal placenta. Fetuses produce a specific hemoglobin called *hemoglobin F* (Hb F). Like adult hemoglobin, Hb F is tetrameric and contains two α chains. Instead of the two β chains found in adult hemoglobin, Hb F contains two γ globins. The γ globins contain serine instead of the His-143 found in β globins. As a result, Hb F has two fewer positive charges in its central cavity, making fetal hemoglobin have a lower affinity for BPG. In the absence of BPG, oxyhemoglobin forms more easily. Since *fetal hemoglobin* binds BPG poorly, it has a *greater affinity* for oxygen than *adult hemoglobin*.

The Bohr Effect.

Christian Bohr, the father of the famous physicist Neils Bohr, won the Nobel prize for his work on the effect of H^+ on the binding of O_2 to hemoglobin. He demonstrated that H^+ is an allosteric effector, which like BPG favors deoxygenated hemoglobin.

What Do Allosteric Effectors Change?

In the case of allosteric control of a ligand-protein binding equilibrium, the inhibitor or activator modifies the binding protein, which changes the K_a of the binding equilibrium.

We will encounter another case, in which an enzyme is affected by an allosteric activator and an allosteric inhibitor, in the *Allosteric Feedback Inhibition: Aspartate Transcarbamoylase* section. In that case, the allosteric effectors change the Michaelis constant (K_M) of the catalytic reaction.

Many other examples of allosteric effectors have been characterized. See the book *Binding and Linkage* by Wyman, J. and Gill, S. J. (University Science Books, 1990) for a thorough description of this subject and the mathematical approaches used to model the cooperativity.

7.3 Antibodies: Immunological Recognition

Immunology, Protection and "Self." Antibodies circulate in the bloodstream and recognize *non-self* molecules called *antigens*. The immune system is broadly divided into two general types called *cell-mediated* and *humoral*. The latter depends on antibodies; the former involves a complex set of proteins called "complement." Antibody recognition of foreign antigen leads to proliferation of specialized cells that make many more of the antibody of that specific type. Also, the bound antibody tags the antigen as being *non-self*, which leads to elimination of the antigen by another class of circulating cells called macrophages.

Several different *classes* of antibodies exist. The most abundant in blood are of the immunoglobulin G class. This specific antibody is Y-shaped and is composed of four polypeptide chains: two identical light chains and two identical heavy chains. The heavy chains attach to carbohydrates and the light chains attach to antigens. *Disulfide bonds* connect the chains so that the N-termini coincide and create the antigen binding sites. These sites are important for the antibody to recognize specific antigens.

Structure of Immunoglobulin G. The IgG tetramer is composed of two *heavy chains* (H) and two *light chains* (L) that are linked to each other and the opposite chain by disulfide bonds and noncovalent interactions. This forms a tetramer with two identical halves that form a Y-like shape. Each chain contains a constant (C) region and variable (V) region composed of β-barrel domains. Each constant domain gains structure from its disulfide bridges. The two variable domains exist on the end of the fork and contain an identical antigen binding site. A variety of foreign antigen proteins bind to the outer surface of the sites.

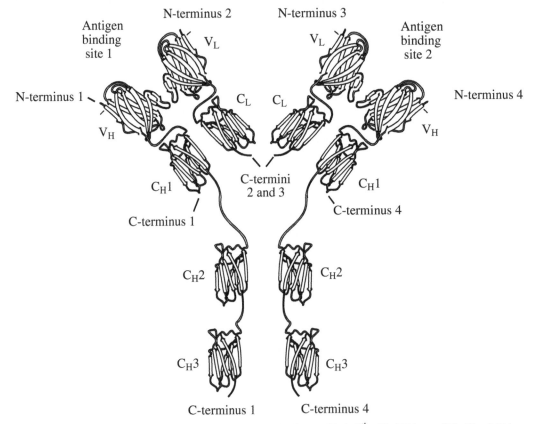

(Adapted from McKee and McKee, *Biochemistry the Molecular Basis of Life* (5th ed.), 2012, pp. 152; Fig. 5.25.)

Antibody Repertoire and Diversity. Genes that encode antibodies undergo a complex rearrangement process called *hypervariable switching*. The resulting antibodies formed after transcription and translation have a wide variety of antigen-binding sites. This process leads to an adapted immune response by allowing recognition and removal of foreign molecules that cover a wider variety of structures. They are a *diversified* form of protection.

Chapter 8

Enzymes

8.1 Enzymes are Biological Catalysts

Purpose and Definitions

(*1*) *Catalysts* increase the rate of reaction by *lowering the activation energy* required to cross the *transition state* barrier.

$$\text{rate} = \text{velocity} = \text{activity}$$

$$\text{rate} = k_1\, c \tag{1}$$

The lower case 'k_1' is the rate constant for a simple one-way (irreversible) reaction and c is the reactant concentration in a simple irreversible reaction. It describes the relation between the rate and reactant concentration. Larger c leads to a larger rate. Enzyme-catalyzed reactions involve a more complicated scheme, which is captured in a generic reaction scheme called the Michaelis-Menten model.

(*2*) Reactants are called *substrates* in biochemistry. The *overall goal* of biochemical kinetics studies is to *characterize the rate of product formation at different substrate concentrations.*

(*3*) Enzymes are very specific and produce high yields. One product is typically made, which is usually stereochemically pure (a single chiral isomer).

Classification

The suffix used to designate an enzyme is -*ase*. The following classes have been defined:

(*1*) *Oxidoreductase*. These enzymes catalyze reactions involving redox changes. $NAD^+/NADH$ and $FAD/FADH_2$ are common cofactors.

(*2*) *Transferase*. A-X + B $\leftrightarrow$ B-X + A   The X group is commonly *phosphate* (P_i), which is often transferred from ATP. Kinase reactions catalyze many crucial regulatory processes. Substrates range from small molecules to proteins. GTP, UTP and CTP are also used as energy sources in the construction of biomolecules, *e.g.*, proteins, carbohydrates and lipids, respectively.

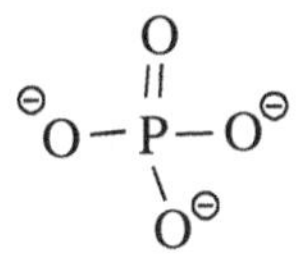

$$\overset{\ominus}{O} - \overset{\displaystyle \overset{O}{\|}}{\underset{\underset{\displaystyle O^{\ominus}}{\diagdown}}{P}} - O^{\ominus}$$

(*3*) *Hydrolase*. These reactions break a bond by adding water. Examples include protease, nucleases and lipases.

(*4*) *Lyase*. This involves adding a molecule to a double bond, without breaking the bond.
The term synthase is used. (Note that ligases are synthetases, not synthases.)

(*5*) *Isomerase* One reactant → one product This involves moving a group to another site within a molecule. They are also called racemases.

(*6*) *Ligase* This class typically require ATP to drive the reaction. They are called synth*etases*. (Lyases are synthases.)

Table 1. Enzyme Class and Subclass Definitions

Class	Name	Reaction	Properties	Enzyme names
1	Oxidoreductases	A(ox) + B(red) ↔ A(red) + B(ox) [1]	Involving oxidation and reduction pairs; usually have NAD/NADH or FAD/FADH$_2$	Dehydrogenase Oxygenase Oxidase Peroxidase Reductase
2	Transferases	AX + B ↔ A + BX	Transfer of one group from one molecule to another	Trans- Transferase Kinase
3	Hydrolases	AB + H$_2$O ↔ A + B	Addition of water to cleave a double or single bond	Many different names
4	Lyases	A (=) X + B ↔ A=B	Addition to a double bond or cleavage to yield a double bond	Synthase
5	Isomerases	A ↔ B	Interconversion of two isomers; always reversible; not limited to stereoisomers	Isomerase Racemase
6	Ligases	A + B + NTP ↔ AB + NDP + P$_i$ or AB + NMP + PP$_i$	Joining two molecules together with input of energy from nucleotide triphosphate	Synthetase

[1] Note that the symbol ↔ is sometimes used in this book to indicate a reversible reaction. The more commonly used symbol involves two half-headed arrows in opposing directions.

$$A_{ox} + B_{red} \xrightleftharpoons[k_{-1}]{k_1} A_{red} + B_{ox} \qquad (2)$$

A *mnemonic* can be used as an aid to remembering these classes:

Example used to remember the enzyme classes: O-T-H-L-I-L

Other **T**eachers **H**ave **L**ied **I**n **L**itigation.

8.2 Enzyme Function: Activity Assays and Enzyme Kinetics

Overview: Kinetics in Three Steps
The purpose of the Michaelis-Menten approach to kinetic analysis of enzyme-catalyzed reactions is to determine how the velocity of the reaction depends on the concentration of added substrate [S]. The rate of the reaction varies from zero to the value of the maximum velocity (V_{max}). The midpoint of the trend is characterized by the Michaelis constant (K_M).

*Curve 1: **The Progress Curve**.* Dependence of product formation on time. One measures the *rate* as the *slope*. The slope changes as the amount of reactant dissipates. The key misconception students have is captured in the following question:

Q: Why does the measured rate/velocity (V) decrease to zero at lower [S]?

The naïve idea many reach is that [S] decreases to *zero*, so the rate becomes 0. This *not* true and highlights a key point that is central to understanding the benchmark-like nature of K_M.

The correct answer is that [S] has decreased to less than ~1/20 the value of K_M. This occurs because too little substrate remains to drive the mass action requirement for catalysis to occur.

If K_M is a large [S], the physiological concentration remaining when the enzyme turns off can be significant, *i.e.* 50 μM for a K_M of 1 mM.

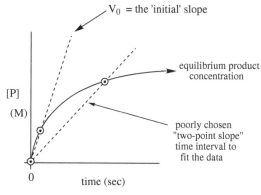

Tangental definition, approaching the maximal initial slope:

$$V_0 = \text{"initial velocity"} = \lim_{t \to 0} \left(\frac{d[P]}{dt} \right) \qquad (3)$$

Two-point approximation method, which must be used with a judicious choice of points to capture the real slope.

$$V_{0,\,two\,point} = \frac{\Delta [P]}{\Delta t}$$

$$= \frac{[P]_2 - [P]_1}{t_2 - t_1}$$

Curve 2: ***The logarithmic Michaelis-Menten Plot***. Dependence of V_0 on the logarithm of the Substrate Concentration.

Note that the non-logarithmic version of this plot has a hyperbolic shape. Do not confuse the sigmoidal nature of the logarithmic plot with the situation in which sigmoidicity appears and the x-axis is [S] not log [S]. In that situation, sigmoidicity is typically an indication that the catalytic mechanism involves significant *cooperativity*.

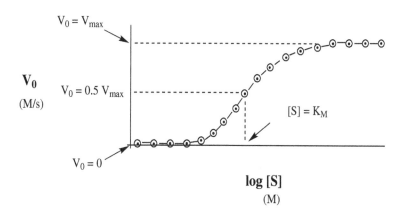

Note: each circle is drawn from the data of a different progress curve.

V_{max} is the "maximum velocity," which is the fastest possible initial velocity under the defined conditions.

V_0 reaches the limiting velocity V_{max}.

$$V_0 = \frac{V_{max}\,[\,S\,]}{K_M + [\,S\,]} \qquad \begin{array}{l}\text{Michaelis-Menten}\\\text{Equation}\end{array} \qquad (4)$$

The equation reproduces the following intuitive behaviors.

(*1*) At low [S], the [S] is << K_M. In this case, $V_0 \sim (V_{max}/K_M)[S]$. Doubling [S] doubles V_0. This is called a first-order kinetic response.

(*2*) At high [S], the [S] is >> K_M, and $V_0 \sim V_{max}$.

Comparing a Michaelis-Menten Plot to a Standard Titration Curve
It is instructive to point out the similarities in the pH-dependence plot and the logarithmic version of the Michaelis-Menten plot. In each case, the x-axis is a function of a concentration and the y-axis is the variable that is affected by the change in concentration. In the H^+ titration plot, the number of protons bound ranges from 0 at low $[H^+]$ to 1 at high $[H^+]$. In the logarithmic version of the Michaelis-Menten plot, one sees the same sigmoidal behavior, but in this case V_0 varies from 0 to V_{max}. If one divided V_0 by V_{max}, the y-axis would also vary from 0 to 1. In each case the benchmark at the center of the transitional action is an equilibrium constant. Note that plot B is an inverted pH titration curve, that is, an H^+-binding curve (see Section 7.2). It plots the number of protons bound, not the number removed, for three representative amino acid R groups.

A. Michaelis Menten Plot

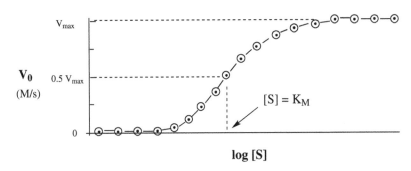

B. Amino Acid R-groups: Representative Binding Curves (*i.e.* Inverted pH Titration Curves).

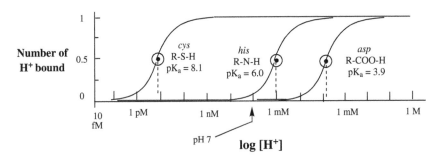

*Curve 3: **The Lineweaver-Burk Plot** (The Double-Reciprocal Plot):*
(*1*) $1/V_0$ is plotted versus the corresponding $1/[S]$ values as data pairs : $(1/V_0, 1/[S])$.
(*2*) This provides an analytical method to calculate the V_{max} (*i.e.* 1 divided by the y-intercept) and the K_M (*i.e.*, V_{max} times the slope).
   One determines the properties of the *substrate* in the enzymatic reaction by observing how changes in [S] affect the rate of product formation. This is like feeling inside a black box to determine the properties of what's inside. Or, more precisely, like putting something into a tube and determining what's inside based on what comes out the other end.
   Consider the following two-step reaction, the generic reaction corresponding to the situation governed by the Michaelis-Menten treatment:

"Catalytic rate constant "

$$E + S \underset{k_{-1}}{\overset{k_1}{\rightleftharpoons}} ES \xrightarrow{k_{cat}} E + P$$

$$k_{cat} = \frac{V_{max}}{[E]_{tot}} \qquad \text{(units: seconds}^{-1}) \qquad (5)$$

The units of the quantity $1/k_{cat}$ correspond to time. This time corresponds to the period required to complete one catalytic event.

The Steady-State Kinetics Reaction Technique

The "steady-state approximation" was developed by Briggs and Haldane. It postulates that the rate of ES formation is equal to the rate of ES consumption. This simulates the situation in which a constant supply of substrate is supplied to the metabolic system. It also assumes that no intermediate compounds are accumulating.

Note that the Briggs-Haldane "steady-state" approach is very different from the Michaelis-Menten method of running the reaction, in which a specified amount of substrate is supplied only at the beginning, then the reaction is allowed to occur until the rate decreases to zero. This occurs because [substrate] is well below the value of K_M. Note that although it is very low it is not zero.

If the [S] is low: $$V_0 = k_1 [E][S]$$

By definition, $V_{max} = k_{cat} [E]_{tot}$, so:

$$V_0 = \frac{k_{cat} [E]_{tot} [S]}{K_M + [S]} \qquad (6)$$

The denominator approximates K_M at low [S].

$$V_0 = \left(\frac{k_{cat}}{K_M}\right) [E]_{tot} [S] \qquad \text{(at low [S])}$$

The ratio k_{cat}/K_M is called the "*specificity constant*." It gives a quantitative way to make the following decisions:

(*1*) Which of a set of different substrates does a given enzyme prefer?
(*2*) Which of a set of enzymes most rapidly produces product from a single target substrate in cross-referenced tests?

Diffusion and Geometric Considerations of Mutual Approach

Catalysis is limited by the maximal rate of possible diffusion of E and S together. The rate constant for diffusion-limited movement is around 10^8 to 10^9 M^{-1} s^{-1} for most biomolecular interactions. This is a second-order rate constant, signifying the fact that two entities must meet to produce a productive encounter.

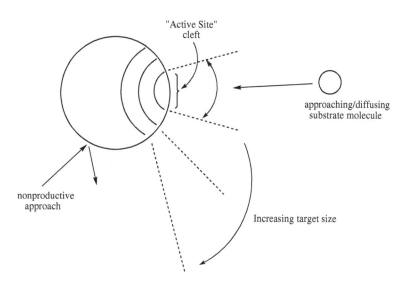

"Active Site" cleft

approaching/diffusing substrate molecule

nonproductive approach

Increasing target size

The Lineweaver-Burk Equation: The Double Reciprocal Plot

$$V_0 = \frac{V_{max}[S]}{K_M + [S]} \qquad \text{Michaelis-Menten Equation}$$

Taking reciprocals:

$$\frac{1}{V_0} = \frac{K_M + [S]}{V_{max}[S]} = \frac{K_M}{V_{max}}\left(\frac{1}{[S]}\right) + \frac{1}{V_{max}}\underbrace{\frac{[S]}{[S]}}_{=1} \qquad (7)$$

The corresponding linear equation is: y   =   m   x   +   b

                                             (slope)   (y-intercept)

Plot the $(1/[S], 1/V_0)$ data pairs as follows to determine K_M and V_{max}:

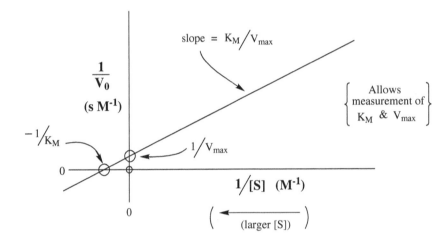

8.3 Requirements for Catalysis

(*1*) The catalyst must be regenerated in the same form following catalysis.

$$E + S \underset{k_{-1}}{\overset{k_1}{\rightleftharpoons}} ES \xrightarrow{k_{cat}} E + P$$

(participation in another
round of catalysis)

(*2*) It does not change the equilibrium free energy difference. It lowers the activation energy (E_a), the energy required to form the activated complex and cross the transition state barrier.
(*3*) A product must form ($E + S \rightarrow ES \rightarrow E + S$ is a "futile cycle").

The Reaction Coordinate. This diagram characterizes the energies corresponding to each progressive state as the reaction proceeds.

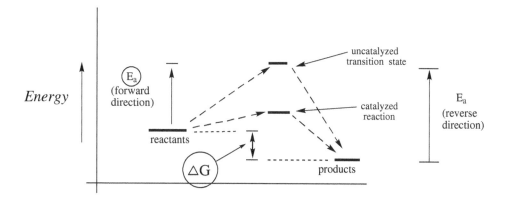

Transition State Activation Energy. The activation energy in the forward direction controls the rate of the forward reaction, which occurs while reactions are also going in the reverse direction. The *ground state* Gibbs free energy (ΔG) characterizes the degree of spontaneity, and poise, of the reaction. As typically defined, *negative* ΔG values correspond to spontaneous reactions in the direction of product(s) formation, however one defines it. *Positive* ΔG values indicate reactions that favor flow in the reverse direction, toward the reactant(s).

$$\Delta G = -R\,T \ln K,$$

where ΔG is the key thermodynamic parameter, R is the universal gas constant, the temperature is in Kelvin, and K is the equilibrium constant.

Table 2. Enzyme Kinetics Plot Summary

Plot Number	Plot	Parameters		Purpose
		x	y	
1	Progress Curve	Time	[P]	Experimental V_0 determined
2	Michaelis-Menten Plot	[S] or log [S]	V_0	K_M & V_{max} intuition
3	Lineweaver-Burk Plot	1/[S]	$1/V_0$	*a*. Calculating K_M & V_{max} *b*. Noncovalent inhibition patterns

8.4 Comparing Enzymes and Relative Efficiency of Use of Substrates

Uses of the Specificity Constant. If K_M is a larger value, for example, in the millimolar range, a large amount of substrate (in biological terms) is required to achieve the same rate.

Two types of comparisons are important to consider when, for example, trying to determine the usefulness of a new pharmaceutical candidate:
(*1*) One can compare two enzymes that can use the same substrate.

(*2*) Alternatively, one can compare how well two variant substrates work with one type of enzyme.

$$S_1 \xrightarrow{\text{E}} P_1$$

$$S_2 \xrightarrow{\text{same E}} P_2$$

Efficacy is a central concept in the pharmaceutical industry. It is a measure of two opposing characteristics of a prospective treatment method (type of molecule). One wants to maximize the effectiveness of action of the drug/approach, yet many very strong drugs have one or more such negative side-effects that the drug is not considered useful. A highly efficacious drug will give an effective result with minimal side effects and downsides.

The comparisons described here encompass a large percentage of the types of problems researchers want to understand when they quantify the relative efficiencies of two drugs with respect to each other. Another scenario occurs when a given pharmaceutical interacts with two alternative enzymes. For example, one might need to compare results obtained with two contrasting patients who have a genetic difference that leads to vastly different amounts of a key enzyme, which is required to metabolize the drug properly.

8.5 Drug Design

Table 3. Design and Discovery of Ligands That May Be Strong Pharmaceutical Lead Compounds.[1,2]

Step	Activity	Purpose
1	*Target selection and validation* Protein (or nucleic acid) identified	Informed by medicinal chemistry experience, genome data, and comparative bioinformatics, chemical validation, gene KO, or RNAi.
2	Target characterization 3D structure determined, drug ability assessed	Obtain recombinant source of target, purify, and develop appropriate assays for biochemical and kinetic analyses, crystallize, and determine the X-ray or NMR structure. Orthologues and construction of homology models are considered.
3	*Ligand/inhibitor studies* Binding and inhibition data (hit discovery)	Exploit the structure in VS, identify ligands from HTS or focused screens, fragment screening by X-ray or NMR methods, and so on utilize *de novo* design, seeking to discover appropriate chemical scaffolds.
4	*Target-ligand structure determination* Elucidate feature that determine affinity and selectivity (Initiate hit-to-lead progression)	Derive accurate structures of target-ligand complexes by X-ray/NMR methods, alternately employ docking calculations
5	*Design chemical modification to ligands* Improve target affinity and develop SAR for compound series (expand to lead series)	Molecular modeling of scaffold modifications seeking to enhance affinity for the target or address potential ADMET issues, bioavailability, and so on, docking calculations can assist modeling information and compounds obtained here can feed back into an iteration of further characterization and design.
6	*Lead series developed, cell-based assay and* ADMET studies commence Investigate biological/pharmacological activity to select preclinical candidates	Testing against whole cells and elucidation of efficacy in a disease model. ADMET investigations are carried out.

[1] *Adapted from Hunter, W. N. (2009) Structure-based ligand design and the promise held for antiprotozoan drug discovery,* J. Biol Chem. *284, 11749–11753.*
[2] *Compounds acquired in Stage 5 go into Stages 3 and 4 in an iterative process. ADMET, adsorption, distribution, metabolism, excretion and toxicity; HTS, high-throughput screening; KO, knock-out; NMR, nuclear magnetic resonance; RNAi, RNA interference; SAR, structure-activity relationships.*

8.6 Enzyme Inhibitors

Reversible Inhibition

Inhibitors are used to investigate the mechanism of an enzyme and decipher metabolic pathways. Natural inhibitors can be used as regulators, and constitute the mechanism of many medicines, toxins and pesticides.

Relative capabilities are characterized by the *inhibition constant* (K_I). For the reaction:

$$E + I \leftrightarrow [E{\cdot}I]$$

$$K_I = ([E]\,[I]) / [EI] \tag{8}$$

Types of Reversible Inhibition.

Competitive Inhibition. (The inhibitor (I) binds free E and competes with S for the same site.)
E is subjected to direct competition between reactions with true and false "substrates" S and I. K_M increases, V_{max} remains the same. Increasing K_M makes sense since more [S] is required to attain the same catalytic rate.

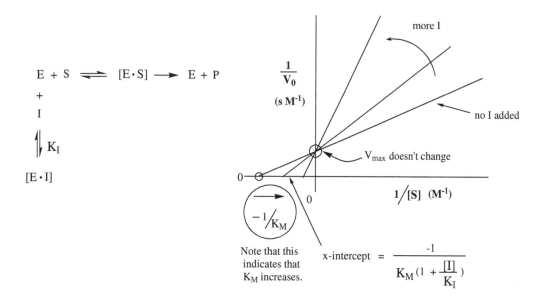

Modified Michaelis-Menten Equation:

$$V_0 = \frac{V_{max}\,[S]}{[S] + K_M\left(1 + \dfrac{[I]}{K_I}\right)} \tag{9}$$

Uncompetitive Inhibition. In this case, I binds the ES complex. Binding of S to E changes E by creating a site for I to bind. I traps S in the enzyme. Both K_M and V_{max} decrease. Note that this makes sense with respect to V_{max}, but the decrease in K_M is anti-intuitive.

$$E + S \; \rightleftharpoons \; [E \cdot S] \; \longrightarrow \; E + P$$

$$+$$

$$I$$

$$\updownarrow$$

$$[E \cdot S \cdot I]$$

$$\text{slope} \; = \; \frac{(1 \; + \; \frac{[I]}{K_I'})}{V_{max}}$$

$$\frac{1}{V_0} \quad (\text{s M}^{-1})$$

more I

no I

V_{max} decreases

$$1/[S] \quad (\text{M}^{-1})$$

$$-1/K_M$$

$$\text{x-intercept} \; = \; \frac{-1}{K_M + [S](1 \; + \; \frac{[I]}{K_I'})}$$

The modified Michaelis-Menten Equation is:

$$V_0 \; = \; \frac{V_{max} \, [S]}{K_M \; + \; [S](1 + \frac{[I]}{K_I'})} \tag{10}$$

where $\; ES + I \; \overset{K_I'}{\rightleftharpoons} \; IES.$

Noncompetitive Inhibition. Here, I binds both E and ES. Note that S binds EI to create a cyclic connected set of equilibria. Two separate sites for I and S exist. Binding of I changes the conformation of E. The K_M does not change, but V_{max} does. Note that the product can be an inhibitor if the catalytic step is reversible (by mass action).

$$E + S \; \rightleftharpoons \; [E \cdot S] \; \longrightarrow \; E + P$$

$$+ \, I \qquad\qquad + \, I$$

$$\updownarrow \qquad\qquad\qquad \updownarrow$$

$$[E \cdot I] \; + S \; \rightleftharpoons \; [E \cdot S \cdot I]$$

Note: S binds [EI] too.

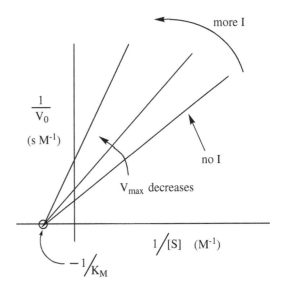

more I

$$\frac{1}{V_0}$$

$$(\text{s M}^{-1})$$

no I

V_{max} decreases

$$1/[S] \quad (\text{M}^{-1})$$

$$-1/K_M$$

$$V_0 \; = \; \frac{V_{max} \, [S]}{K_M \left(1 \; + \; \frac{[I]}{K_I'}\right) + \left(\dfrac{1 + [I]}{K_I' \, + \, [S]}\right)}$$

$$\text{y-intercept} \; = \; \frac{1}{V_{max}} \left(1 \; + \; \frac{[I]}{K_I'}\right)$$

Irreversible Inhibitors. These are generally the result of a displacement reaction, with covalent bond formation.

$$E^{\ominus} \;+\; I-X \quad \xrightarrow[\phantom{k_1 \gg k_{-1}}]{k_1 \gg k_{-1}} \quad E-I \;+\; X^{\ominus} \longleftarrow \text{leaving group}$$

Acetylcholine Esterase. Inhibitor and Antidote Actions

Nerve conduction involves the following reaction, which is susceptible to inhibition by an organic fluorophosphate compound:

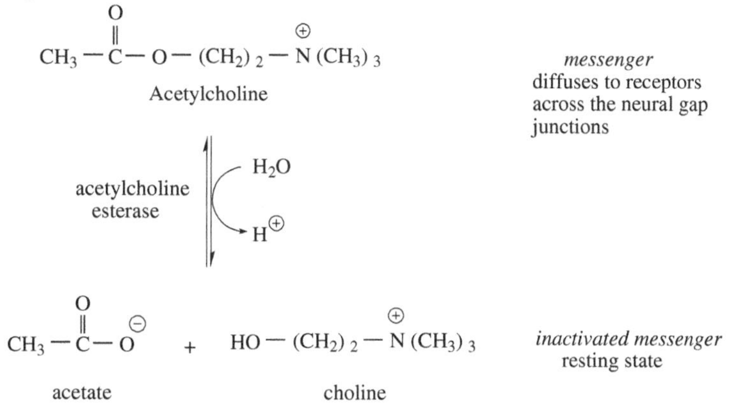

$$CH_3 - \overset{\overset{\textstyle O}{\|}}{C} - O - (CH_2)_2 - \overset{\oplus}{N}(CH_3)_3$$

Acetylcholine

messenger
diffuses to receptors
across the neural gap
junctions

acetylcholine esterase H_2O $H^{\oplus}$

$$CH_3 - \overset{\overset{\textstyle O}{\|}}{C} - O^{\ominus} \;+\; HO - (CH_2)_2 - \overset{\oplus}{N}(CH_3)_3$$

acetate   choline

inactivated messenger
resting state

This reaction is involved in the relay of impulses between nerves, and nerves and muscle cells.

1. *Normal Reaction*

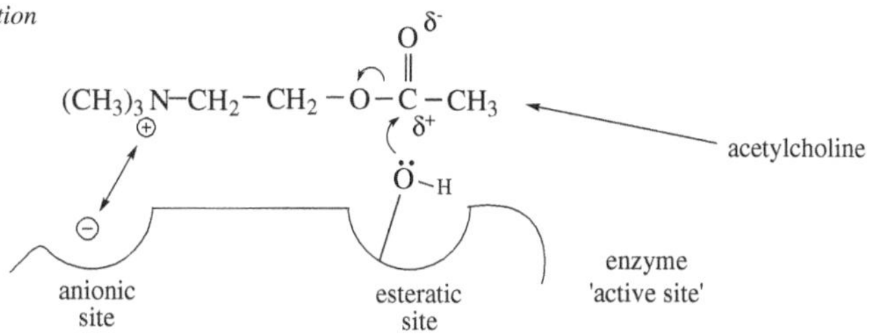

2. *Nerve Gas* - diisopropylfluorophosphate (DFP); a "suicide substrate" or "dead-end inhibitor"

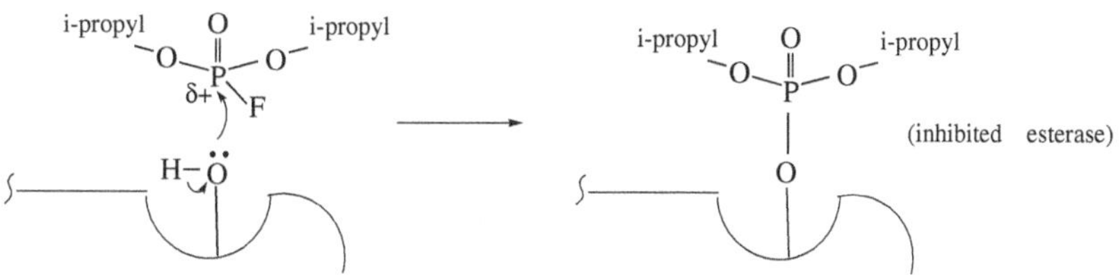

3. *Antidote* - pyridine aldoximine methiodide (PAM)

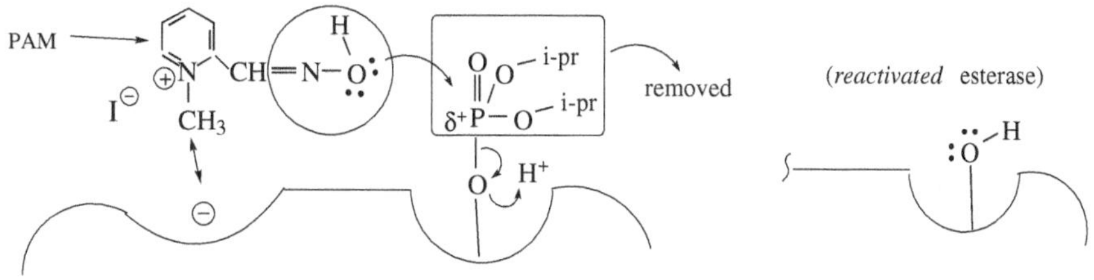

AZT (3'-azido-2', 3'-deoxythymidine). This drug inhibits *reverse transcriptase*, which synthesizes DNA from viral RNA during infection by HIV (see Section 2.4.2). It was the first widely used treatment for AIDS, which has been replaced by protease inhibition-based approaches.

This very reactive *electrophile*, which replaces the hydroxyl group, inhibits the normal active site nucleophile of the HIV reverse transcriptase.

Methotrexate. This drug inhibits dihydrofolate reductase, which is required to make the DNA precursor dTTP. This inhibition selectively kills rapidly dividing cells, such as cancer cells. This technique is used as a successful cancer chemotherapy agent, for example, childhood leukemia and autoimmune diseases such as rheumatoid arthritis and Crohn's disease.

Bisubstrate Enzyme Kinetics

The Michaelis-Menten method can be modified and used to study more complicated enzyme-catalyzed reactions, for example, two- and three-substrate reactions. The following generic reactions show two ways in which the reactants can bind to the enzyme. The horizontal line represents the enzyme surface; the state of the enzyme is indicated.

Sequential Reactions

(*1*) *Ordered*: Reactants bind in a specific order; products dissociate from the enzyme in a specific order.

(*2*) *Random*: Substrates can bind in either order; products can dissociate in a specific order.

One determines K_M values with respect to the concentrations of *each* substrate. For a given substrate, the other substrate is held at a constant concentration. The analysis involves obtaining data then plotting $1/V_0$ versus $1/[A]$, at constant $[B]$, and $1/V_0$ versus $1/[B]$, at constant $[A]$.

The Ping-Pong Reaction
A general scheme is shown below:

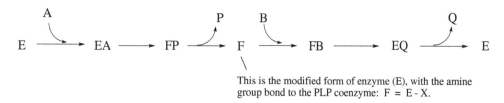

This is the modified form of enzyme (E), with the amine
group bond to the PLP coenzyme: F = E - X.

(*1*) The EA transformation to form FP occurs as follows. The X portion of A (= PX) is attached to E, which creates F
(*i.e.* E-X) and the P portion is released. The A species leaves X behind.

(*2*) *Aspartate Transaminase* creates aspartate from oxaloacetate. The enzyme form designated as F is bound to the
coenzyme pyridoxal phosphate (PLP), which binds and holds the amino functional group that is subsequently transferred
to the proto-C_α atom of oxaloacetate to form aspartate.

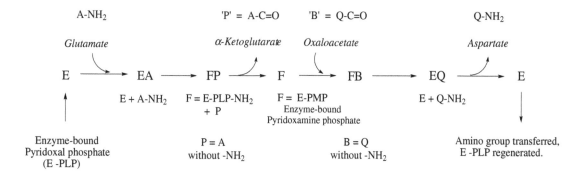

Variations of this mechanism are (*1*) used in two different compartments in the *Malate-Aspartate
Shuttle*, by two different enzyme varients, and (*2*) to produce *pyruvate* from *alanine*, which provides fuel
for the Krebs Cycle.

Kinetic Regulation: Principles

Zymogens and Proteolytic Activation
Proenzymes are inactive precursors that are activated by removal of a peptide, which is an irreversible
covalent modification. Four examples of *pancreatic digestive enzyme* are:

Chymotrypsinogen → Chymotrypsin
Trypsinogen → Trypsin
Proelastase → Elastase
Procarboxypeptidase → Carboxypeptidase

The proenzymes on the left are activated by proteolysis to produce the active enzyme on the right.

Enzymatic Cascades

This involves a catalysis-mediated increase in number of enzymes that occurs at progressive stages in a set of proteolysis reactions. At each step, a new protease *proteolyzes* a new protease precursor. This produces the activated protease with the next required specificity to catalyze the next step in the cascade. The net effect is to produce a very large number of enzymes, greatly amplifying the effect of the initial single proteolysis step. In stage 1, the initiator makes a large number of activated enzyme, each of which activate a large number of enzymes in stage 2. Each stage produces a multiplicative number of enzymes, leading to a huge response, initiated by the single initiator enzyme.

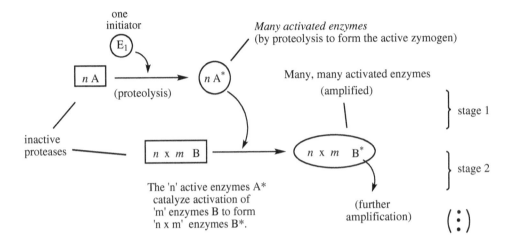

Examples:

(*1*) *Blood Clot Formation* (Coagulation):   The reaction involving an enzyme called Factor IX is missing in hemopheliacs, so their blood will not clot properly.

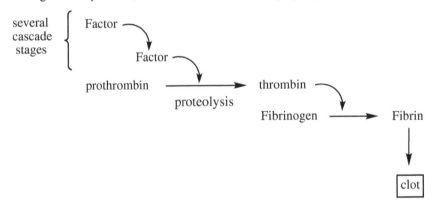

(*2*) *Blood clot dissolution* (breakdown):   Used to dissolve and prevent clots in stroke victims.

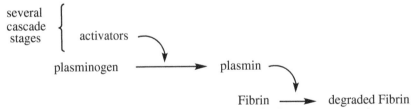

Note that *zymogens need not be enzymes*. For example, fibrinogen is the precursor to the structural protein that forms the "logjam" in the clot, fibrin. Cascades occur in many biochemical processes.

(*3*) A third example of a cascade regulates the *cell-mediated immunity system*. The key reactions are catalyzed by a set of proteases called *complement*.

Feedback and Feed-forward Regulation

Feedback inhibition occurs when the product of a set of linked reactions binds to an enzyme, or other functional protein, in an earlier step in the pathway, thereby inhibiting the downstream reaction it was created by and intervening set of reactions. Regulation is often controlled by *allosteric effector* molecules.

The related concept *feed-forward activation* involves stimulating a later enzyme, or other functional protein, in a pathway by formation of sufficient amounts of a particular activator molecule produced by an earlier reaction in the pathway. This serves to increase the flow by activating the later step and thereby drawing reactants through the linked pathway.

8.7 Allosterism

In this mechanism, a *regulatory effector*, which is usually a small biomolecule, changes the activity of an enzyme or binding protein (or nucleic acid) by *binding to a site other than the active site*. The result is either an increase, or decrease in the capability of the protein to do its job.

(*1*) The effector can either increase or decrease the rate of catalysis. This is done by changing either the K_M (or V_{max}) of the affected enzyme. Recall that the K_M is a measure of the concentration of substrate that must build up and be present to get a reaction to proceed.

(*2*) In the case of the *binding protein*, the effector increases or decreases the binding constant of the protein. This increases the constant itself, which normally does not change.

(*3*) Note that allosterism is a separate concept from feedback inhibition or feed-forward activation. An allosteric effector might lead to these effects; however, it might act on an entirely different metabolic pathway/scenario. In addition, feedback inhibition or feed-forward activation might be produced by a nonallosteric route, such as direct binding of an inhibitor to the active site, or recruitment of a third player, such as a kinase or phosphatase, and so on.

Cooperativity and Induced Fit

Approaches used in this field have been developed in two main directions, involving mathematical models that handle either a *serial* or *parallel* consecutive *buildup* of the allosteric effect. A serial approach is incorporated into the *Monod-Wyman-Changeaux Model* of allosterism. The rate or binding constants increase, or decrease at each consecutive linear *sub-reaction step* in the overall process. The degree of *cooperativity* is quantified as the percent buildup, or decrease, in activity. It is calculated as the *ratio of binding constants, or rate constants, for two consecutive steps*. Positive allosterism means this ratio is greater than 1; negative allosterism means that the ratio is less than 1. The parallel approach, which is incorporated into the *induced fit model* differs in that the buildup is affected by all possible parallel, but linked, reactions.

Both approaches "model" a fit to the data, revealing evidence for cooperativity among the consecutive (and flow-linked) events. For example, controlled switches between the "tense" (T) and "relaxed" (R) states occur when O_2 binds to the four *ligand binding sites* in the tetrameric *lattice protein* hemoglobin. The *R state* occurs has an intrinsically higher affinity for O_2 than the *T state*. This allows for a large range of binding timescales, producing usage rates that vary over a broad range of O_2 requiring activities. The ligand concentration controls its own binding affinity to the lattice.

(*a*) R ↔ T: fast (seconds) changes in activity; (*b*) Covalent changes: slow (minutes to hours).

Allosteric Feedback Inhibition: Aspartate Transcarbamoylase

This reaction regulates the *de novo* synthetic pathway of the pyrimidines, that is, the uridine-, thymidine- and cytosine-containing nucleotides. This highly controlled enzyme consists of 12 subunits. Six are catalytic and six are regulatory, with a number of binding sites for different allosteric effectors.

The enzyme catalyzes the first committed step in the pathway, a common site of regulation. A series of enzyme-catalyzed reactions produces uridine monophosphate (UMP), uridine triphosphate (UTP) and cytidine triphosphate. Since UTP can be used to make thymidine triphosphate, this pathway leads to all of the pyrimidine nucleotides.

Allosteric inhibition occurs when sufficient cytidine triphosphate (CTP) is produced to allow binding to the allosteric inhibitor binding site on ATCase. This regulates against overproduction of CTP. The pyrimidine nucleoside triphosphate binds ATCase. In this case, the enzyme is *allosterically activated*. Thus, if ATP is present, CTP synthesis is promoted. This offsetting pair of effectors serves the grand metabolic purpose of *balancing the production levels* of pyrimidines and purines.

The general effect of allosteric regulators on the K_M is shown for the following cases: (*1*) activator present, (*2*) no effector added, and (*3*) inhibitor present. Note that in some enzyme systems V_{max} changes instead of K_M.

Experimental results demonstrate that aspartate transcarbamoylase by is activated by ATP and inhibited by CTP.

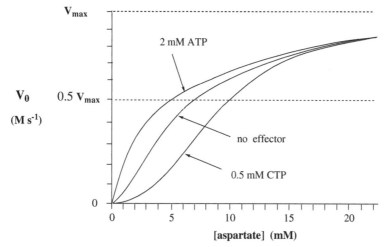

(*1*) The results show how the V_0 of the ATCase-catalyzed reaction depends on aspartate concentration, one of the two substrates. The other substrate, carbamoyl phosphate, was held at a saturating concentration in each set of measurements. Conditions are labeled on the plot.

(*2*) Note that the effect is subtle. The K_M values only change by a few mM, yet a fairly large change in activity occurs. For example, compare the V_0 values at 5 mM aspartate: (*i*) no effector present, 40% of V_{max}, (*ii*) 0.5 mM CTP, 15%, and (*iii*) 2 mM ATP, 50%. Both activation by ATP and inhibition by CTP are verified.

8.8 Phosphorylation and Dephosphorylation

These reactions are catalyzed by *kinase* and *phosphatase* activities, repectively. Kinases are *transferases*, phosphatases are *hydrolases*. This ubiquitous mechanism of metabolic regulation; is very commonly employed as an on-off *switching* technique. The serine hydroxyl group is typically phosphorylated by kinases; threonine and tyrosine are also common substrates. Phosphorylation often leads to *activation* of enzyme activity. However, in some cases the opposite is found. One example is the enzyme *pyruvate dehydrogenase*. To illustrate some aspects of phosphorylation-dephosphorylation, four different situations that lead to regulation of some key aspects of metabolism are described below.

Pyruvate Dehydrogenase
This enzyme complex releases acetyl coenzyme A to fuel the Krebs Cycle. It is controlled by two counteracting enzymes, a *kinase* and *phosphatase* (circles 1 and 2)

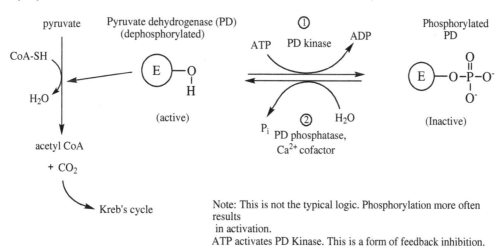

Note: This is not the typical logic. Phosphorylation more often results
 in activation.
ATP activates PD Kinase. This is a form of feedback inhibition.

See the *Pyruvate Dehydrogenase* entry in the *Glycolysis* section for detailed mechanistic features, including the roles of *five different coenzymes* to accomplish the net reaction.

Glycogen Phosphorylase

This reaction controls the release of glucose-1-phosphate from glycogen stores in the liver.

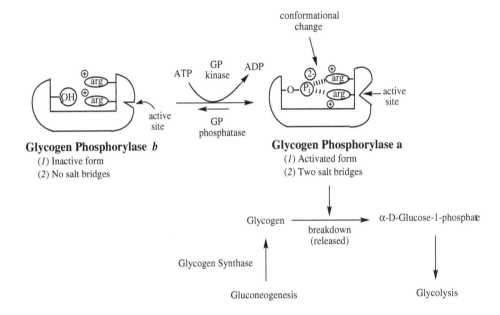

Glycogen Synthase—catalyzes production of glycogen, which offsets the effect produced by GP kinase.

Isocitrate Dehydrogenase

This enzyme is a key step in the regulation of the Krebs Cycle. Catalysis is inhibited by phosphorylation because the dianionic phosphate repels the trianionic isocitrate from the active site.

(See *Regulation* in the *Krebs Cycle* section for the full context and further details.)

Cyclin Kinase

This system controls the "checkpoint" for entry into *mitosis* (and exit from S Phase) in yeast as well as most organisms. It is "very, very, very" complicated according to the cited reference. The following scheme shows the interconnected reactions.

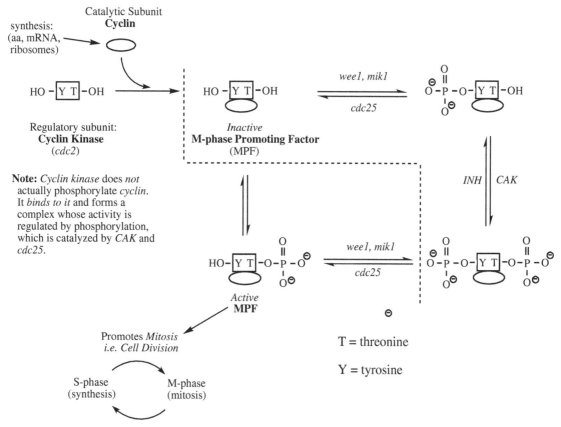

This mechanism leads to the cyclic nature of the cell division process. This is illustrated by the time-dependent progression in *synchronized yeast cell cultures* shown in the following figure, which shows the biochemical changes that occur during the cell cycle in wild-type yeast cells.

The plot shows time-courses for total cyclin (monomer + dimers), active MPF [Cdc25-P], and inactive MPF [phosphorylated Cdc25-P], relative to total Cdc2. The dotted lines indicate the cyclin level and MPF activity that would be measured in a partially synchronous culture. By the author's convention, mitosis is defined as the time period when the tyrosine is more than 50% dephosphorylated.

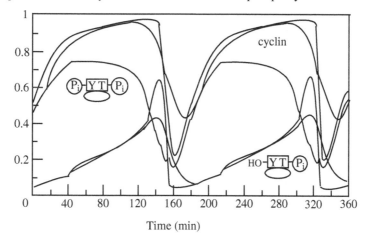

(Adapted from Murray, W. and Kirschner, M., Cyclin Synthesis Drives the Early Embryonic Cell Cycle, ... Mitotic Control in Fission Yeast, *J. Theor. Biol.*, 1995, 173, 283–305.)

Chapter 9

Metabolic Enzyme Action

Catabolic process refer to breakdown of mass to obtain chemically useful energy, as occurs with carbohydrates in *glycolysis* and the *Krebs Cycle*, and lipids in the *β-oxidation pathway*.

The following enzyme classes are common in metabolism:
(*1*) Kinases (a Transferase; class 2); it is not a ligase.
(*2*) Phosphatases (a Hydrolase; class 3); water is the phosphate acceptor.
(*3*) Pyruvate Dehydrogenases (an Oxidoreductase; class 1). Reduced CoA-SH is converted to oxidized CoA-S-R.

A typical logic in regulation: (1) Phosphorylation leads to activation, and (2) dephosphorylation causes inactivation. However, this is not true in general.

9.1 Enzyme Mechanisms

Factors that Contribute to the Control of Enzyme Activities. The following list collects a series of ideas, which were described in previous sections, allowing one a chance to review how each works.
(*1*) The *enzyme concentration*. This is controlled in cells by the levels of protein synthesis, modification, and degradation.

(*2*) The *substrate concentration* produces changes when [S] is less than about 20 times K_M (*i.e.,* when V < V_{max}).
When V = V_{max}, the reaction is said to be *saturated* with substrate. When [S] is less than 1/20 of K_M, the rate of catalysis is 0.

(*3*) *Inhibitors* inactivate E, so the [E] decreases ($E_{act} \rightarrow E_{inact}$). The two general classes are:
    (i) Reversible, which are generally not covalently bound.
    (ii) Irreversible, which form covalent protein-inhibitor complexes.

(*4*) *Allosteric regulators*, which can inhibit or activate. The effect is often *cooperative* in nature.

(*5*) Protein covalent modification. Examples include:
    (*i*) Zymogens cause change in sequence (1° structure) as a result of protein chain cleavage. This is almost never reversible.
    (*ii*) Phosphorylation-dephosphorylation. This reversible on/off switching mechanism involves kinases and phosphatases.
    (*iii*) *Protein modifications*. Examples include glycosylation, myristoylation (lipid addition), ADP-ribosylation (nucleotide addition), ubiquitinylation (attachment of a small protein that tags the bound target protein for controlled destruction), and the many types of cofactors and "natural products" that participate in unique biological niches.
    (*iv*) pH (H^+ on/off); H-bonds, salt bridges

(*6*) *Chaotropes*: (urea, guanidinium, SDS…)

(*7*) *Temperature*: At least two factors affect the kinetics (rate) of a given reaction: (*1*) increasing the numbers of collisions at low to moderate temperatures, and (*2*) increasing denaturation as the temperature is increased above 60 to 70°C. The first idea is captured in the form of the *Arrhenius equation*:

$$k_T = k_0 \exp [-E_a / (k_B T)]$$

where exp $x = e^x$, k_T is the *observed rate constant* (k_1, k_{-1}, k_{cat}…) and k_0 is the *intrinsic rate constant*. E_a is the *activation energy* required to form the transition state configuration. Two E_a values exist correspond to

(1) reactant crossing to products, and (2) the converse crossing. The Boltzmann constant (k_B) is 1.361×10^{-23} J K^{-1} and T is the temperature in Kelvin. The latter idea, protein denaturation, is governed by the *hydrophobic effect*.

(*8*) Variation in the (*i*) *ionic strength* of salt solutions and (ii) *dielectric constant* of the solvent environment. Increasing the value of D of a solvent leads to more shielding of interactions between charges.

Recall that ΔG_{vdW} is proportional to $(-r^{-6} + r^{-12})/D$ and ΔG_{HB} is proportional to $(-r^{-10} + r^{-12})/D$.

Computational Chemistry. Many of these factors can be simulated, in solution environments, using computational methods called molecular mechanics and molecular dynamics analysis.

(*1*) Energy averages are determined by intensive, "all-atom force field" calculations. Vibrations, rotations, translations are accounted for completely, with respect to all types of energies. The theory comes from the field called *Statistical Mechanics*.

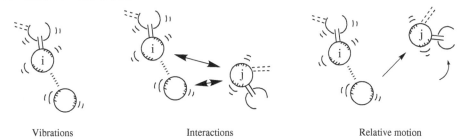

| Vibrations | Interactions | Relative motion |

(*2*) Energies can be calculated and resolved into various contributions due to molecular aspects such as hydrated surface coverage, charge topography, motional rates, conformational accessibility, and so forth.
(*3*) One can simulate reaction mechanism by combining quantum mechanical treatment of the problem, when important, with the dynamics approach, when it is not.

9.2 Modes of Catalysis

Chemical
(*1*) *Acid-Base Catalysis*. This reaction class involves catalytic transfer of an H^+ species. The electron-rich *proton acceptor* species (B) is called the *conjugate base*. Electropositive species, such as H^+ and *carbocations*, often act as *electrophiles* in the role of a *conjugate acid*.

(*2*) *Covalent Group Transfer*. A functional group, obtained from a first substrate, is transferred to a second substrate. Alternatively, it is transferred to an enzyme-held functional group (*i.e.* a coenzyme), which then passes it to the second substrate. Functional groups are classified according to their *group-transfer potentials*. Two typical examples are: (i) amine transfer catalyzed by the pyridoxal phosphate-bearing *aspartate transaminase*, and (ii) phosphate transfer catalyzed by zinc-containing *polynucleotide kinase*.

Binding
(*1*) The *Proximity Effect*. Bringing the reactants together in the active site decreases the effective volume, which increases the effective concentration of the substrate. This drives the reaction by mass action. Details are described below.

(*2*) *Transition-State Stabilization*. The reactant and enzyme favor the transition-state configuration over that of the reactants. The products must be able to dissociate from the active site. Typically, a set of functional groups distort the substrate physically and electronically.

An example occurs twice during the *catalytic triad* mechanism catalyzed by the *serine proteases* (*e.g.*, chymotrypsin). At one point in the mechanism, a hydroxylate nucleophile attacks a carbonyl carbon. Since this would make the carbonyl carbon pentavalent, one bond must go. To make it happen, the carbonyl (C=O) *double* bond is stretched to form a hydroxylate (C-O⁻) *single* bond. This stretch is mediated by two amide hydrogens located in the *oxyanion hole* of the *active site*. (See the *Enzyme Mechanism* section for details.)

9.3 The Reaction Coordinate

This reaction coordinate plot follows the energies of the different reactants, intermediates and transition states and products. It is an energetic map of course of the reaction. They can involve one or several steps, depending on the complexity of the mechanism and amount known. Each step has an activation energy-requiring process in *each direction*. Enzymes are reversible, but their energetics are not the same in each direction. Many reactions form transiently stable intermediates, which exist until the species reacts in the next step of the mechanism.

The following *reaction coordinate diagrams* show what happens when the two primary contributors to catalysis are engaged.

(*1*) The *proximity effect* is shown in going from plot *A* to *B*.

(*2*) *Transition-state stabilization* is added in going from plot *B* to *C*.

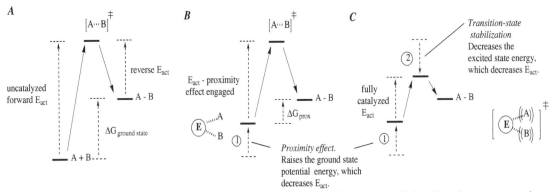

The *Principle of Mass Action*. The poise of a reaction equilibrium is established by the concentrations of reactant(s) and product(s) as a result of the *constraints* imposed by the *equilibrium constant*.

Under *Standard State* conditions, all component concentrations are 1 M. The superscripts in $\Delta G^{o'}$ indicate that the parameter pertains to those conditions. Calculating values at other conditions is addressed in the *Bioenergetics* and *Bioelectrochemistry* chapters.

Enzyme-Substrate Interactions. Weak binding of a substrate to an enzyme is preferred. Strong binding would inhibit catalytic transformations by blocking the release of product(s). As an indie record producer once said, "I do their record and I kick 'em out." Some of the important *forces that mediate mechanisms* are: (*1*) electrostatic interactions (including charge-charge, charge-dipole, and dipole-dipole), (*2*) hydrogen bonds, (*3*) the hydrophobic effect and hydrophobic interactions, (*4*) van der Waals forces, and (*5*) the entropy associated with liberating water molecules from the bound faces of the molecules.

The Proximity Effect. The catalytic groups and substrate(s) are clamped together by the enzyme within the active site. Since the effective volume has been constrained to the active site, it decreases dramatically compared to that of the bulk solvent. Recall that the concentration (c) is the number of moles per unit volume (c = m/V). Because the components are constrained to react within the active site, the local effective concentration increases dramatically. One problem, however, is that trapping the species in one, or a few, configuration(s) costs entropic energy. Nothing is free.

The close juxtaposition of the catalytic groups in the catalytic triad mechanism discussed below is a clear example of the *proximity effect*. As we shall see, the histidine and serine species undergo a complicated series of binding and unbinding reactions in the course of the process required to break the peptide bond.

Reaction Intermediate Formation and Reaction Coordinate Diagrams. The two-step reaction coordinate plot applies the ideas of the one-step reaction to a two-step scenario.

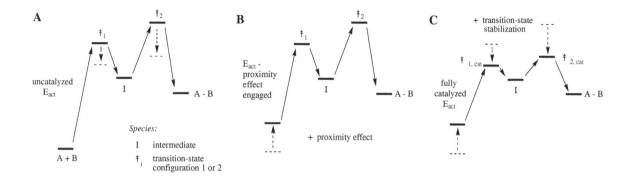

9.4 Induced Fit Revisited

When more than one contact point binds to the substrate, it becomes progressively more inclined to proceed through the reaction. This inclination is called *cooperativity* and regulates a large number of biochemical processes, ranging from *binding of O_2 to hemoglobin*, membrane-induced enhancement of catalysis during *blood clotting*, driving the *binding of repressor* proteins to genetic *operator* sites, and so on.

Transition State Analogs
The induced fit concept has been studied using modified substrates that mimic the substrate. Because they are structurally similar, they fit. Because they contain non-natural functional atoms, they are often potent enzyme inhibitors. This has led to their use as probes of functional groups that contribute to active-site reactivity and in the design of a vast array of pharmaceuticals. Many such compounds are fluorinated. Specific examples can be found in the *abzyme* literature.

Catalytic Antibodies: Abzymes
Consider using a *transition-state analog as* an *antigen* to produce an *antibody*. Since the antigen resembles the *transition-state geometry* of a reacting substrate within an enzyme, it was surmised that they might convert normal substrates to the transition-state, and then on to products. If so, the antibody would be catalytic. A series of these *abzymes* have been produced and do catalyze the intended reaction. The success of this design-driven approach provides strong experimental evidence for the induced fit idea. While abzymes are natural, they are artificial catalysts. The specificity of antibody-antigen recognition is used to accomplish a particular catalyzed reaction. The immunized animal is used to produce the proactively designed catalyst.

Hexokinase: Large-scale Domain Movements
The active site of this enzyme shifts from opened to a closed structures upon binding the substrate glucose and ATP substrates. This closure involves large-scale movement of the surrounding protein domains, which shields the active site from H_2O, a potent competitive inhibitor of the kinase reaction. Incorporating active site closure slows the enzyme rate 10,000-fold, but is required to protect the mechanism, which would otherwise be poisoned by the encroaching solvent.

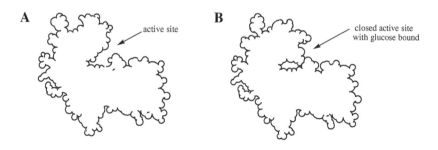

(Adapted from McKee and McKee *Biochemistry the Molecular Basis of Life* (5th ed.), 2012, p. 187; Fig. 6.2.)

Enzyme Specificity. Enzymes can often use more than one substrate. Recall that the *specificity constant* (k_{cat}/K_M) can be used to compare specificities two different substrates with one enzyme. A larger value means that the specified substrate is used more efficiently. This demonstrates that the enzyme can sometimes adapt to different structures, suggesting some degree of *plasticity* in the *recognition process*. Two examples of varied specificities are: (*1*) alkaline phosphatase, which can remove the phosphate from a variety of substrates, and (*2*) alcohol dehydrogenase, which can convert a number of alcohols to aldehydes.

9.5 Acid-Base Catalysis

Acid-base Catalyzed Peptide Bond Cleavage. The following mechanism shows how acid and base species (H^+ and OH^-) each participate in the hydrolysis of an amide bond, such as catalyzed by a protease.

The steps are:
(*1*) H^+ is abstracted from H_2O by the R_4-B: species
(*2*) OH^- attacks the amide C=O
(*3*) carbanion I forms
(*4*) The amide/peptide bond is cleaved and C=O is restored. R_4-B releases H^+ and the secondary amine product forms.

C-H Bond Cleavage. Here, a conjugate base extracts a proton from a substrate, which creates a *carbanion*.

Why Do Enzymes Often Work Best at Neutral pH?
Enzymes depend on pH because the functional groups must typically be in a certain state of protonation/deprotonation to support each of the proton-passage steps required in the mechanism. Determining how an enzyme is affected by pH is a standard first step in characterizing it. It is useful because it provides information on which functional groups are required for catalysis.

The following pH profile shows a situation in which protonation of something destroys the catalytic mechanism below pH 4, and deprotonating something inactivates catalysis at pHs above 9. Why?

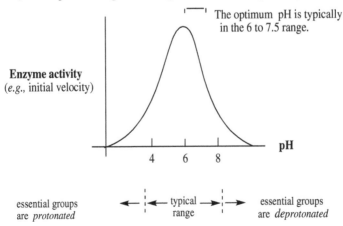

The explanation assumes that the activity of an enzyme depends on the pK_a values of two functional groups in the active site. Let's set one acidic pK_a at 3.9 and one basic value: 10.5 (*i.e.,* asp and lys). These could represent the carboxylate of an aspartate or glutamate residue and the ε-amino group of a lysine, respectively.

$$BH_2^+ \leftrightarrow BH \leftrightarrow B^-$$

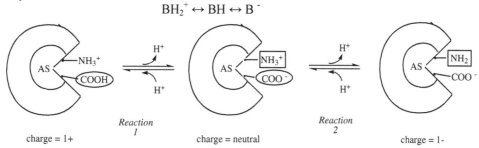

In principle, four types of curves are possible. What are they? What does the enzyme's active site look like, in a simplified sense, in each situation?

(*1*) What does a plot of pH versus V_0 look like when only the BH form is active?

The plot should have a maximal V_0 at a mid-range pH and progressively decreasing values at lower and higher pH values. Where is the pH optimum? In the middle range, around pH 7, with reasonably high activity between pH 6 and 8.

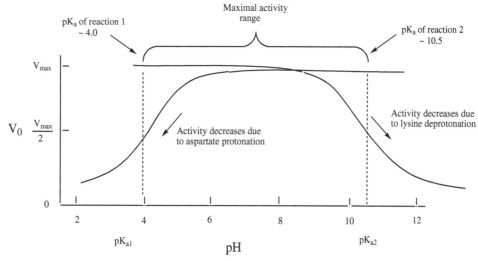

(For mathematical details see Piszkiewicz, D., *Kinetics of Chemical and Enzyme-Catalyzed Reactions*, Oxford Univ. Press, NY, 1977, pp. 61–65.)

(*2*) Which amino acids participate in catalysis?

(*3*) What does the plot look like when only the B^- form is active?

(*4*) What about when only the BH_2^+ form is active?

(*5*) What does the plot look like when the pH titratable residues in the active site don't really affect catalysis?

This can occur when substrates are nonpolar (aromatic, aliphatic) and the attraction between active site and substrate is predominated by van der Waals forces and hydrophobic interactions.

Triose Phosphate Isomerase. Moving a Hydroxyl.

This reaction is discussed in the *Glycolysis* chapter. It converts a compound that cannot be used in glycolysis to one that can. When the six-carbon compound fructose-1, 6-bisphosphate splits into these two 3-carbon compounds, the cell must recover the one that can't be used. The reaction is:

dihydroxyacetone phosphate → *glyceraldehyde-3-phosphate*

(DHAP)   (G3P)

The mechanism is a concerted acid-base reaction.

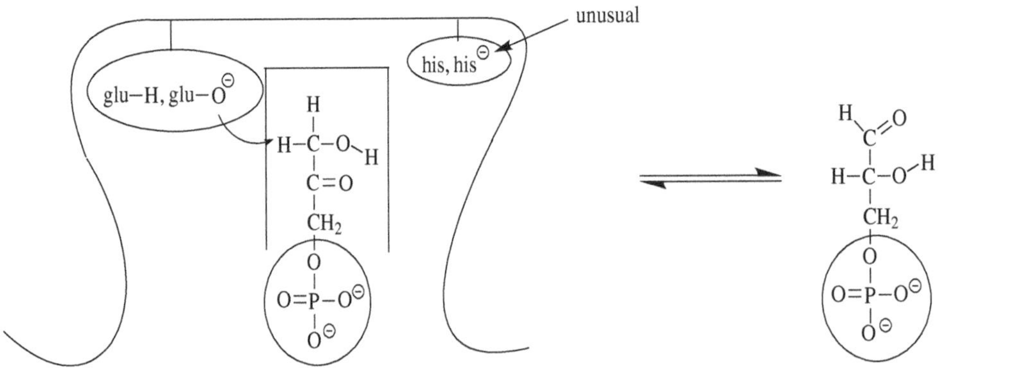

Dihydroxyacetone phosphate
(Substrate)

Glyceraldehyde-3-phosphate
(Product)

The *acid* reaction involves histidine imidazole/imidazolium protonation-deprotonation. The base changes are glutamic acid/glutamate transformations. The following subprocesses occur during the mechanism:

(*1*) C=O is converted to C-O$^-$
(*2*) C-O$^-$ is protonated to form C-OH
(*3*) a C-H is abstracted to form C=C
(*4*) a proton of R=C(H)O-H is abstracted producing R=C(H)O$^-$
(*5*) R=C(H)O$^-$ is converted to R-C(H) =O.

The measured reaction coordinate for triose phosphate isomerase has five resolved steps:

$$E + S \rightarrow [E\text{-}S] \rightarrow [E\text{-}I] \rightarrow [E\text{-}P] \rightarrow E + P$$
$$\quad\;\; (1) \qquad\;\; (2) \qquad (3,\,4) \qquad\quad (5)$$

9.6 Covalent Group Transfer

Group Transfer and Leaving Groups

Many biochemical *group transfer* reactions occur. The good leaving potential of phosphate from ATP is why it is a key carrier of biochemical form of energy. In the first step of a typical mechanism, part or all of S is transferred to E or a bound coenzyme. In a second step, this piece is attached to a second substrate. The product leaves the enzyme with the transferred piece from the initial S.

Group X is transferred from A to B via E. Two steps occur, then E is regenerated.

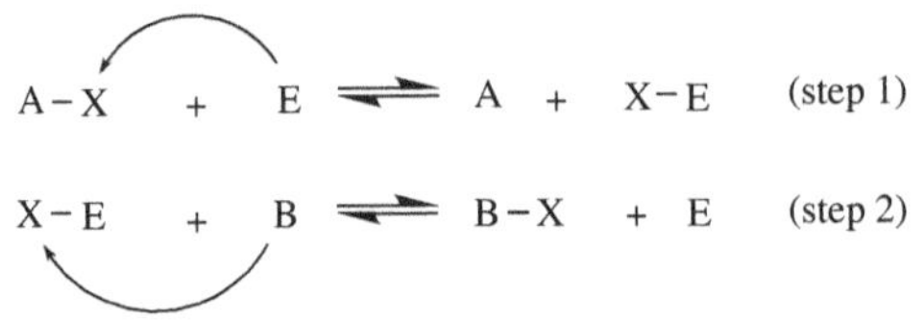

$$A-X \quad + \quad E \;\rightleftharpoons\; A \quad + \quad X-E \quad \text{(step 1)}$$

$$X-E \quad + \quad B \;\rightleftharpoons\; B-X \quad + \quad E \quad \text{(step 2)}$$

Leaving groups (X) have different *group transfer potential* levels, their propensities to leave.

Acetoacetate Decarboxylase. *The Schiff Base.* The next reaction illustrates the use of *Schiff base*. In the acetoacetate decarboxylate mechanism the Schiff base facilitates elimination of a CO_2 from the molecule. Reversal of the Schiff base releases acetone.

In a contrasting case, the benzyl α-carbonyl group of pyridoxal phosphate forms a Schiff base with the ε-amino group of a lysine of *aspartate transaminase*. The *adduct* serves to protect the aldehyde group of bound PLP when it is not engaged in catalysis. Details are described in the *Pyridoxal Phosphate* section.

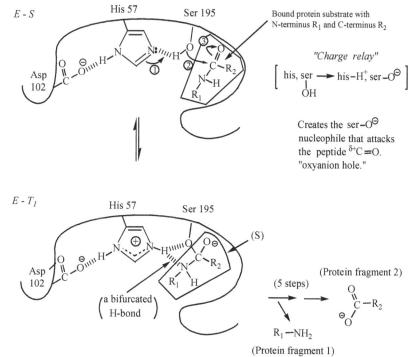

The steps involved in this mechanism are: (*1*) 1 H^+ is used as an acid catalyst and O= leaves, (*2*) the negative charge migrates toward -CO_2, which is eliminated as the –C = C bond forms, (*3*) iminium formation occurs by a *tautomeric shift*, (*4*) acid-base catalysis is reversed: E is regenerated and P forms.

9.7 The Serine Protease Catalytic Triad Mechanism

The Catalytic Triad. The goal of the enzyme is to use the *catalytic triad* to create a serine hydroxylate nucleophile, which attacks the partially positive carbonyl carbon in the peptide. Two protein amide hydrogens stretch the carbonyl carbon-oxygen *double bond* of the peptide substrate, driving it to form a *hydroxylate single bond*. This type of *deformation* of a reactant to get the reaction to proceed is called *transition-state stabilization*. This portion of the enzyme that stretches the bond is called the *oxyanion hole*.

9.7.2 The Charge Relay System. This mechanism *creates the nucleophile* that attacks the protein substrate, in the first step of the reaction. Subsequently, the first product, the C-terminal fragment of the cleaved protein chain (with the newly created amino terminus), is released. The reaction concludes as the

N-terminal fragment (with the newly created carboxylate) is released and the enzyme 'active site' components are returned to their pre-reaction states, in readiness for the next round of catalysis.

The mechanism can be broken into seven steps, as shown below. (Adapted from Moran *et al.*, *Principles of Biochemistry* (5th ed.), 2012, pp. 188-189; Fig. 6.28.)

The noncovalent enzyme-substrate complex is formed, orienting the substrate for reaction. Interactions holding the substrate in place include binding the R_1 group in the specificity pocket.

The binding interactions position the carbonyl carbons of the scissile peptide bond (susceptible to cleavage) next to the oxygen of Ser-195.

Binding of the substrate compresses Asp-102 and His-57. The strain is relieved by formation of a low barrier hydrogen bond. The raised pKa of His-57 enables the imidazole ring to remove a proton from the hydroxyl group of Ser-195. The nucleophilic oxygen of Ser-195 attacks the carbonyl carbon of the peptide bond to form a carbonyl intermediate E-TI$_1$. This species is believed to resemble the *transition state*.

When the tetrahedral intermediate is formed the substrate C - O bond changes from a double bond to a longer single bond. This allows the negatively charged oxygen, the "oxyanion," of the tetrahedral intermediate to remove to a previously vacant position, the "oxyanion hole." It hydrogen bonds to the peptide chain N - H groups of Gly-193 and Ser-195.

The imidazolium ring of His-57 acts as an acid catalyst, donating a proton to the nitrogen of the scissile peptide bond - facilitating its cleavage.

The carbonyl group from the peptide forms a covalent bond with the enzyme, producing an acyl-enzyme intermediate. The initial C-terminal peptide product (P_1), with its new amino terminus is produced. This fragment leaves the active site.

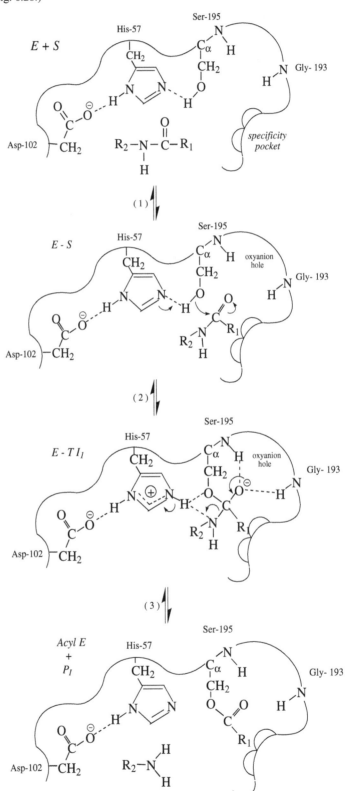

(4)

Acyl E
+
H₂O

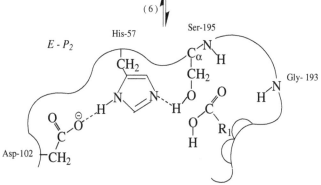

Hydrolysis (deacylation) of the acyl-enzyme intermediate starts when Asp-102 and His-57 again form a low-barrier hydrogen bond and His-57 removes a proton from the H₂O molecule to provide an OH⁻, which attacks the carbonyl group of the ester.

(5)

E - TI₂

A second tetrahedral intermediate (E-TI₂) is formed and stabilized by the oxyanion hole

His-57, once again an imidazolium ion, donates a proton, leading to collapse of the second tetrahedral intermediate.

(6)

E - P₂

The second product (P₂) forms - the amino-terminus of the original protein with a new carboxy terminus.

(7)

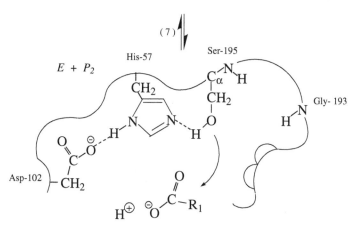

E + P₂

P₂ is released from the active site, regenerating free chymotrypsin.

Amino-terminal product (P₂)

9.8 The Active Site of Tyrosyl-tRNA Synthetase

The following figure shows a reconstruction by *site-directed mutagenesis* of the *transition state* for activation of tyrosine by tyrosyl-tRNA synthetase. A mobile loop envelops the transition state in an *induced-fit mechanism*.

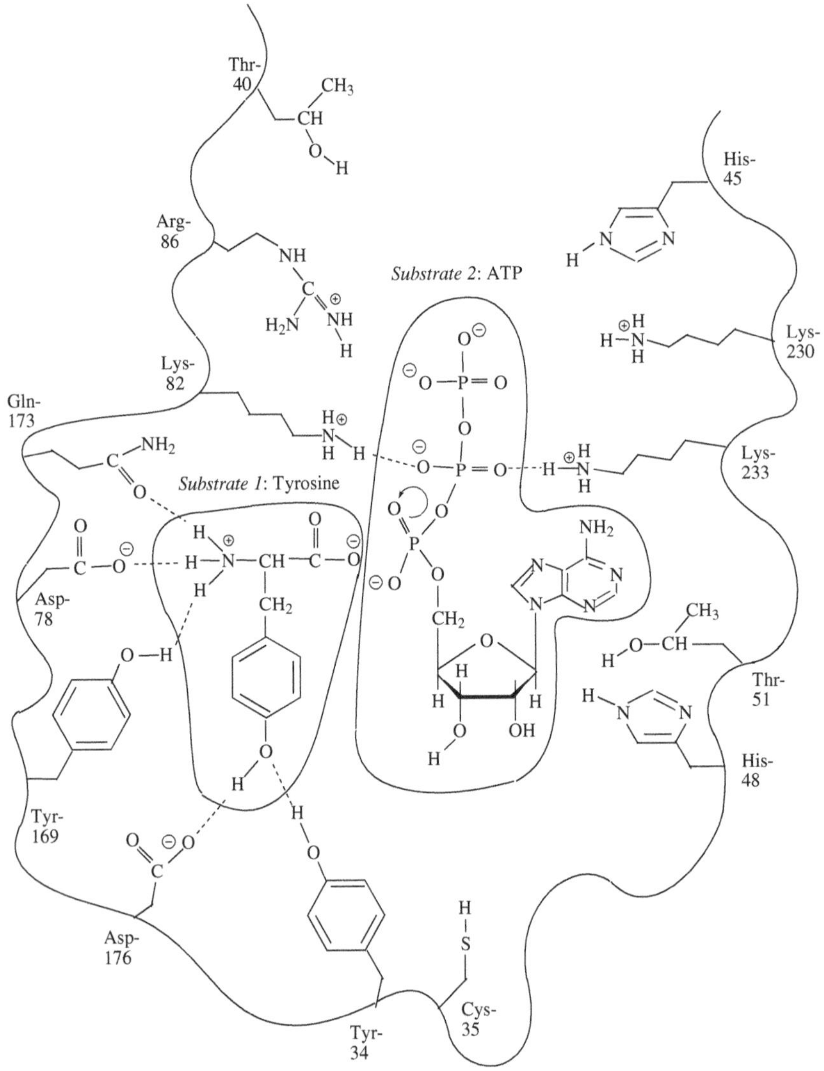

Active Site Protein-Ligand Binding Interactions

This figure shows the large number of hydrogen bonds (at least seven) formed between tyrosyl tRNA synthetase and the two bound substrates, tyrosine (left) and ATP (right). Lysine 82, Arg-85, Lys-230, and Lys-233 all bind to the portion of ATP that becomes pyrophosphate. (Adapted from Fersht, A. R., *et al.* (1988) *Biochemistry* 27, 1581–1587.)

Site-directed Mutagenesis

This technique involves changing the gene sequence in an *expression plasmid vector*. One plans which sequence to replace in order to insert a desired amino acid in a particular position in the protein chain. By making such replacements, one can proactively determine the degree to which a given amino acid contributes to catalysis, binding and other aspects of a mechanism. This approach can be used, along with a technique that determines the energetics of the reactions (*e.g.*, calorimetry, spectroscopy), to determine how much each site contributes *energetically* to binding, structural transformations, and catalysis.

Chapter 10

Coenzymes

10.1 Classification

Cofactors are divided into two subgroups, essential ions and coenzymes.

Essential Ions
(*1*) *Activator Ions* are usually loosely bound. *Examples are:*
    (*i*) Potassium (K^+) is crucial in nerve conduction.
    (*ii*) Calcium (Ca^{2+}) is carried by calmodulin and regulates a huge number of biological processes.
    (*iii*) Magnesium (Mg^{2+}) is present in chlorophyll and activates phosphate transfer from nucleotide triphosphates.
(*2*) *Metal Ions* are usually tightly bound in metalloenzymes. *Examples are:*
    (*i*) Iron (Fe^{2+}) is the redox site in iron-sulfur clusters, which are common in electron transfer redox complexes.
    (*ii*) Zinc (Zn^{2+}) is typically bound to sulfur and histidine ligands, for example, as in *carbonic anhydrase* and *zinc fingers*.

Coenzymes
(*1*) *Cosubstrates* are loosely bound; for example, *ATP* and *UDP-sugars* are used to accomplish a reaction, but are not tightly bound and dissociate after the reaction is complete.
(*2*) *Prosthetic Groups* are tightly bound; vitamins; for example, *pyridoxal phosphate*, a covalently bound coenzyme within *aspartate transaminase*.

10.2 Survey of the Coenzymes

Adenosine Triphosphate (ATP)
Obtaining full activity with enzymes usually requires Mg^{2+}.

Two important examples of *uses of ATP* are:

$$\text{ATP + Glucose} \quad \rightarrow \quad \text{ADP + glucose-6-phosphate} \qquad P_i^{2-} \text{ transferred} \qquad (1)$$

The first reaction is the beginning of glycolysis.

$$\text{ATP + ribose-5-phosphate} \rightarrow \text{AMP + 5-phosphoribosyl-1-pyrophosphate} \quad PP_i^{4-} \text{ transferred} \qquad (2)$$
$$\text{(PRPP)}$$

The product of reaction 2, PRPP, is attached to *nucleic acid bases* to make the *nucleotides*. The pyrophosphate group leaves during nucleoside formation, driving the reaction to completion.

Redox Coenzymes (NAD⁺ and FAD)

Recall the mnemonic "Oil Rig": *Oxidation* is the Loss of electrons; *Reduction* is the Gain of electrons. These two reactions always occur in pairs, leading to the general name *redox reactions*. The generic reaction is illustrated by the alcohol dehydrogenase reaction below. The generic reaction is:

$$A_{red} + B_{ox} \rightleftharpoons A_{ox} + B_{red}$$

where species A passes an electron to B.

Nicotinamide Adenine Dinucleotide (NAD⁺, NADP⁺) (Vitamin precursor: Niacin, that is, nicotinamide)

(*1*) The oxidized form NAD⁺ is used to *oxidize* a substrate. The *reduced* compound NADH is used to reduce a substrate.
(*2*) The coenzyme carries *reducing equivalents* for use in *electron transport*, which fuels *oxidative phosphorylation* by creating a *proton gradient* across the inner mitochondrial membrane.
(*3*) It is typically used in ordered *dehydrogenase* reaction mechanisms:

Example. Alcohol Dehydrogenase. A typical redox enzyme.

(*1*) NAD⁺ receives an electron from ethanol and forms NADH. The electron is transferred attached to hydrogen, as a *hydride*.
(*2*) The kinetics can be measured in either direction, by putting in either the left or right components to drive the intended reaction. The equilibrium and kinetics are *reversible*.

Flavin Adenine Dinucleotide (FAD, FADH$_2$): Made from Vitamin B$_2$

(*1*) Solutions of the oxidized form FAD (which is actually FAD^{2-}) are intensely yellow. The reduced form FADH$_2$ is colorless. This change in absorbance provides an especially good way to follow the kinetics of catalysis and other interactions of flavins in biology.

(*2*) The coenzyme carries reducing equivalents. It functions metabolically like NAD$^+$ (and NADP$^+$) but has a different redox potential, so it can be used to oxidize more difficult functional groups.

(*i*) NAD$^+$: used to oxidize R-OH, C = O, COOH (easier)

(*ii*) FAD is used to oxidize C = C (more difficult) and accept electrons from other carriers: for example, NADH, *ubiquinone.*

Pyridoxal Phosphate: *Contains Vitamin B$_6$*

Schiff Base. The covalent species [E-lys–PLP] is linked by an internal *aldimine* formed by covalent bonding of the coenzyme to the enzyme active site.

The following general linkage is called a *Schiff base.*

Schiff base formation requires several individual reaction steps.

The Ping-Pong Mechanism

An example of this mechanism is catalyzed by *Aspartate Transaminase*. A number of other transaminases (aminotransferases) exist, for example, (*1*) the *isoenzymes* involved in *the malate-aspartate shuttle*, and (*2*) the enzyme that produces *alanine* from lactate, which fuels *gluconeogenesis*.

(*1*) The reactants are the amino acid *glutamate* and Krebs Cycle intermediate *oxaloacetate*. The products are *aspartate* and another Krebs Cycle intermediate *α-ketoglutarate*.

(A) reactant 1	(P) product 1	(B) reactant 2	(Q) product 2
Glutamate (*glu*)	α-Ketoglutarate (α-KG)	Oxaloacetate (OAA)	Aspartate (*asp*)

(E) E-LMP   (F) E-PMP   (E) E-LMP

"F" receives, retains and
transfers the amino group

(2) *PLP and Schiff Base Formation*. PLP has a positive charge and serves as a sink for electrons. The steps in the mechanism are:

Pyridoxal phosphate-lysine Schiff base adduct   *Pyridoxamine phosphate* (PMP)   *Schiff base adduct regenerated*

(*i*) The ε-amino group of a lysine of the enzyme is normally bound to the PLP aldehyde site as a Schiff base. This is like a sword, the aldehyde, being held inert and protected within the scabbard, the Schiff base adduct.

(*ii*) In the first step of catalysis, this Schiff base-bound lysine group is freed from PLP, exposing it for use (*i.e.,* the sword is removed from the scabbard).

(*iii*) in the second, the aldehyde site binds the amino group, which is removed from glutamate. It holds on to it as *a*-keto- glutarate is ejected and the second substrate oxaloacetate (OAA) binds.

(*iv*) In this step, the amine group is transferred to OAA, producing aspartate. *Voila*—but not done yet.

(*v*) To complete the mechanism, the exposed aldehyde, regenerated after the amine transfer, is reconnected to the lysine as the protective Schiff base adduct. The sword has returned to the scabbard.

Another example of a Schiff base-mediated reaction is described in the *Acetoacetate Decarboxylase* section.

Coenzyme A

The free form of coenzyme A (*CoA-SH*) is used to carry the *acetyl* group, which binds the circled sulfhydryl sulfur, forming a thioacetate linkage.

The *vitamin* is *Pantothenic Acid*; the coenzyme includes ADP.

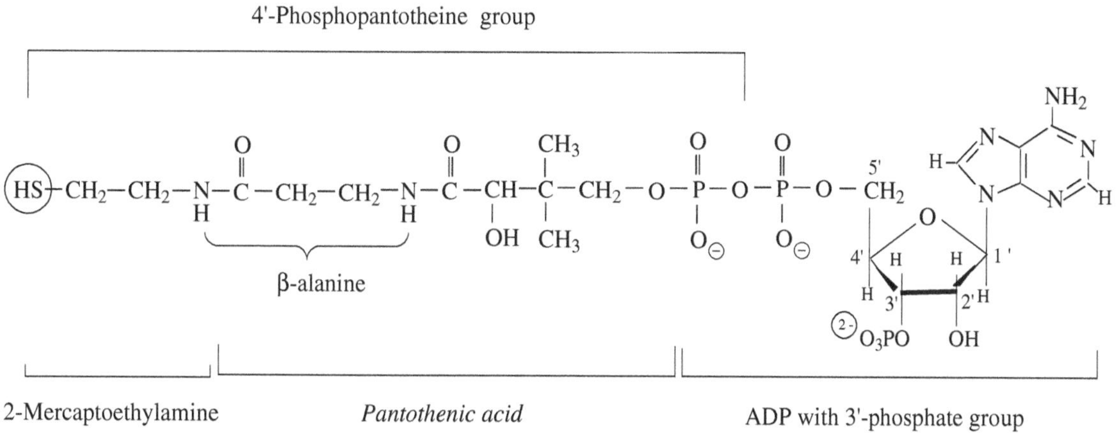

4'-Phosphopantotheine group

$$HS\text{--}CH_2\text{--}CH_2\text{--}N\text{--}C\text{--}CH_2\text{--}CH_2\text{--}N\text{--}C\text{--}CH\text{--}C\text{--}CH_2\text{--}O\text{--}P\text{--}O\text{--}P\text{--}O\text{--}CH_2$$

β-alanine

2-Mercaptoethylamine   *Pantothenic acid*   ADP with 3'-phosphate group

Activated Acetyl. CoA-SH is used in the *Bridge Reaction*, which links glycolysis to the Krebs Cycle. The activated acetyl is transferred to *oxaloacetate* to produce *citrate*. Pyruvate is decarboxylated by Pyruvate Dehydrogenase to yield CO_2 and acetyl-CoA:

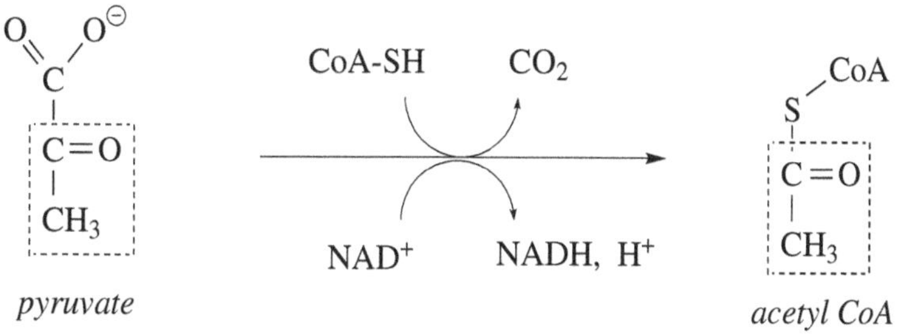

CoA-SH CO_2

NAD^+ NADH, H^+

pyruvate   *acetyl CoA*

CoA-SH is used a second time in the Krebs Cycle to drive the substrate-level phosphorylation reaction catalyzed by *Succinyl CoA Synthetase*.

Acyl Carrier Protein (ACP). This structure is very closely related to CoA. It is required to synthesize fatty acids. Instead of being linked to ADP, the 4'-phosphopantotheine is linked directly to the protein via a serine.

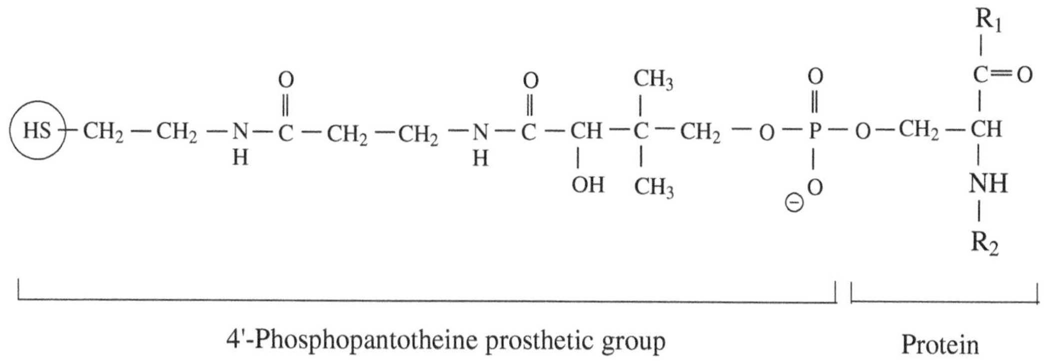

$$HS\text{--}CH_2\text{--}CH_2\text{--}N\text{--}C\text{--}CH_2\text{--}CH_2\text{--}N\text{--}C\text{--}CH\text{--}C\text{--}CH_2\text{--}O\text{--}P\text{--}O\text{--}CH_2\text{--}CH$$

4'-Phosphopantotheine prosthetic group   Protein

Fatty acids are synthesized two carbon units per addition, contributed by ACP-bound acetate molecules.

Biotin (Vitamin H)
Pyruvate Carboxylase.
This enzymatic reaction begins with ATP-dependent carboxylation of the biotin cofactor embedded within the enzyme. Biotin and bicarbonate react to form carboxybiotin. The thioacetate bond is important in metabolism due to the high group-transfer potential of the acetyl group.

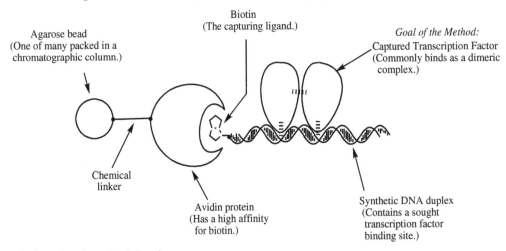

In the second step, the carbanion from enolpyruvate attacks the carbon of carboxybiotin and *transfers a carboxyl group*. This then yields oxaloacetate and regenerates biotin.

Biotin-Avidin Affinity Chromatography.
The protein avidin (from egg white) binds biotin with $K_a = 10^{14}$ M^{-1} (very strong). An important example of the use of this phenomenon in *biotechnology* involves *affinity capture* and *tagging* systems.

Noncovalent Binding: Gel bead "matrix" packed in a chromatographic column.

This technique involves the following steps:
(*1*) synthesize a specific biotinylated DNA,
(*2*) make a nuclear protein preparation,
(*3*) mix the DNA with the protein prep,
(*4*) place the DNA and captured protein(s) through an avidin-agarose affinity chromatography column,
(*5*) elute contaminants, and
(*6*) elute the biotinylated DNA-protein(s) complex with 2-aminobiotin. One recovers proteins that bind the biotinylated DNA sequence.

Folate

Folate is composed of three parts: pterin, *p-amino benzoate* (PABA) and glutamate.

Pteridine

Pterin
(2-Amino-4-oxopteridine)

Folate

PABA

Reduction by transfer of hydride from NADPH occurs in two steps.

Folate

7,8-Dihydrofolate

5,6,7,8-
Tetrahydrofolate

Example: *Dihydrofolate Reductase*

Dihydrofolate reductase converts folic acid to its biologically active form, *tetrahydrofolic acid* (THF). The carbon units carried by THF are bound to N5 and/or N10 of the pteridine ring forming *N5, N10-methylenetetrahydrofolate* (N5, 10-THF). The enzyme *Thymidylate Synthase* uses N5, 10-THF to attach the methyl group to deoxyuridine monophosphate (dUMP), and thereby produce *deoxythymidine monophosphate* (dTMP). As the methyl group is transferred to form dTMP, the folate coenzyme is oxidized to form dihydrofolate. N5, 10-THF is regenerated from dihydrofolate by dihydrofolate reductase and NADPH.

PABA - Glu

$m^{5,10}$-THF   THF

dUMP $\longrightarrow$ dTMP

Occurs by addition of $-CH_3$ at
C5 of the pyrimidine ring.

Dihydrofolate reductase and thymidylate synthase have been studied as targets for *cancer* and Crohn's disease chemotherapy. See *Methotrexate* in the *Enzyme Inhibitor* section. Cells that grow, and replicate, more rapidly, the hallmark of cancerous cells, are more susceptible to growth inhibition by the chemotherapeutic agent than those that are not actively replicating, *i.e.* most normal cells.

Thiamine (Vitamin B$_1$)

Thiamine is used by the "Bridge Reaction" enzyme complex Pyruvate Dehydrogenase to make *hydroxyethyl*thiamine pyrophosphate (HETPP). The circled site is used.

Thiamine
pyrophosphate
(TPP)

hydroxyethyl group
attachment site

TPP is also used by the enzyme *Pyruvate Decarboxylase*, which is required by yeast to synthesize ethanol, and the *transketolases*, which interchange C2 units between sugars to make other sugars, for example, in several *Calvin Cycle* reactions.

10. 3 Metals

Molybdenum.
The structure of the molybdenum cofactor used by *xanthine oxidase* is:

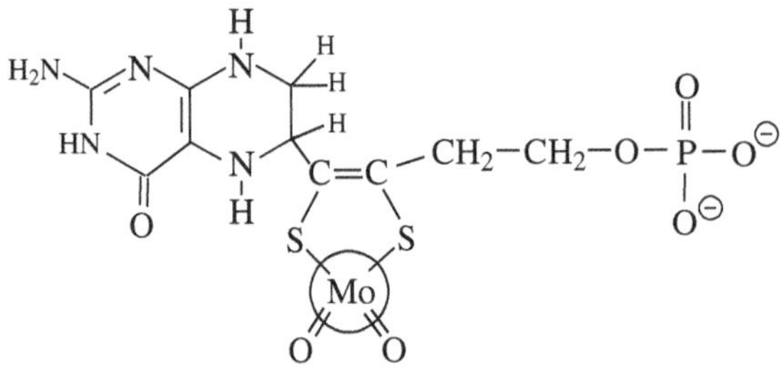

The cofactor contains molybdenum, FAD, and two different Fe-S centers. The enzyme oxidizes hypoxanthine to xanthine, a metabolic intermediate in the *biosynthesis of guanine.*

Vitamin B$_{12}$ (Cobalamin)
Methylated cobalamin transfers a methyl group to *homocysteine* to produce *methionine.* The methyl group is initially transferred to cobalamin from the coenzyme *5-methyltetrahydrofolate.*

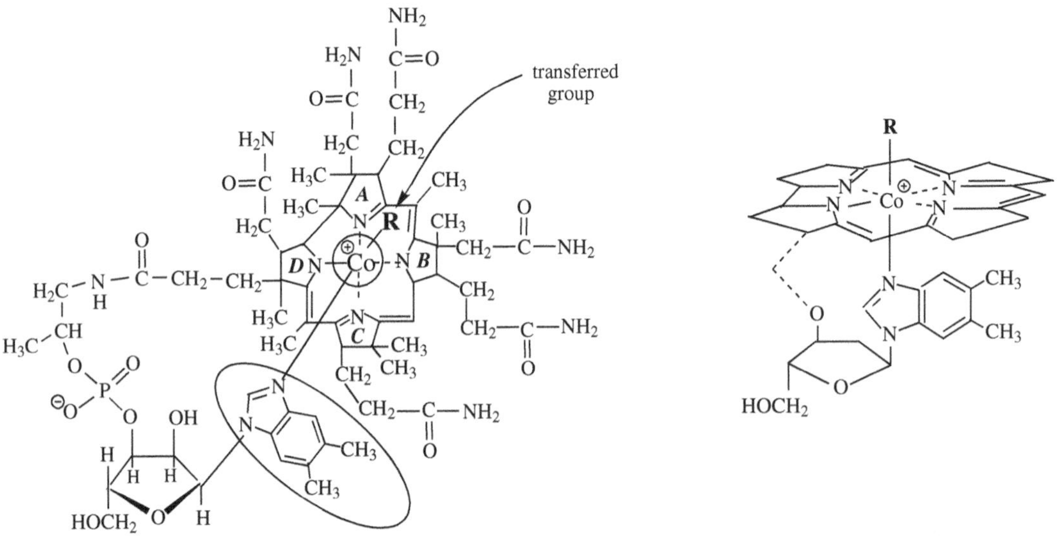

The *corrin* ring system contains four pyrrole rings. The nitrogen of each pyrrole is linked to a central cobalt atom. Another group, 5, 6-dimethylbenzimidazole ribonucleotide, is also attached to cobalt. The structure of corrin is similar to the porphyrin heme group, but lacks one methylene bridge between pyrrole rings A and D.

In another important application, the fatty acid precursor *methylmalonyl coenzyme A* is rearranged by the the enzyme *Methylmalonyl CoA Mutase.* This reaction produces the Krebs Cycle intermediate *succinyl-CoA.* The coenzyme is in the *adenosylcobalamin* form.

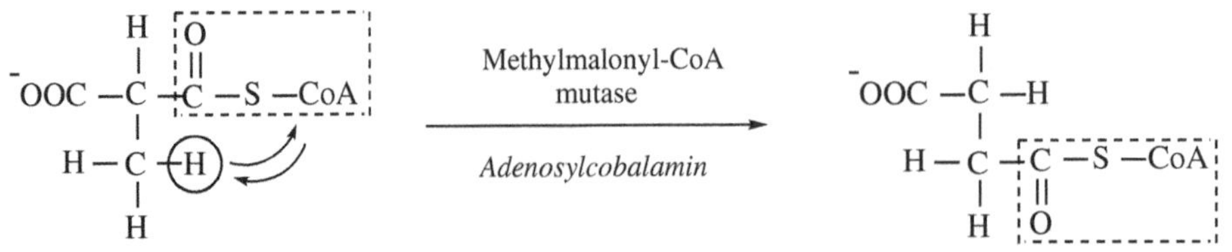

Methylmalonyl CoA   *Succinyl CoA*

Iron-Sulfur Proteins

Iron-sulfur proteins form the core of several redox carrier proteins in mitochondrial *electron transport*. Free sulfurs of the disulfides ($-S^{2-}$) bind the iron atoms in a number of different ligand geometries and stoichiometries. Two examples are:

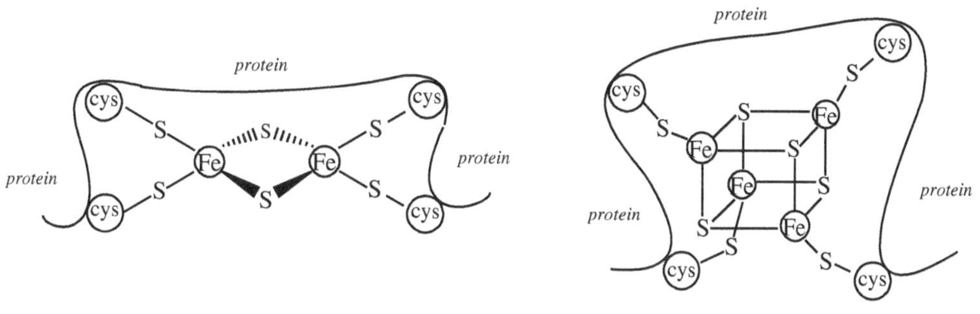

These centers are used in redox reactions in a number of oxidoreductases. In another example, the protein thioredoxin and the small tripeptide compound glutathione are involved in maintenance of disulfide bonds in proteins. Glutathione has access to active sites. Reactions in the Krebs Cycle and photosynthesis depend on thioredoxin.

Transferrin and *Ferritin*. These two protein mediate the transport of the iron atoms, which are intrinsically toxic, through the circulatory system and into cells.

Zinc

Carbonic Anhydrase: Zinc as a Lewis Acid.

This protein catalyzes the interconversion between carbon dioxide and bicarbonate, which leads to the *buffering of* our *blood* (See that section for details.)

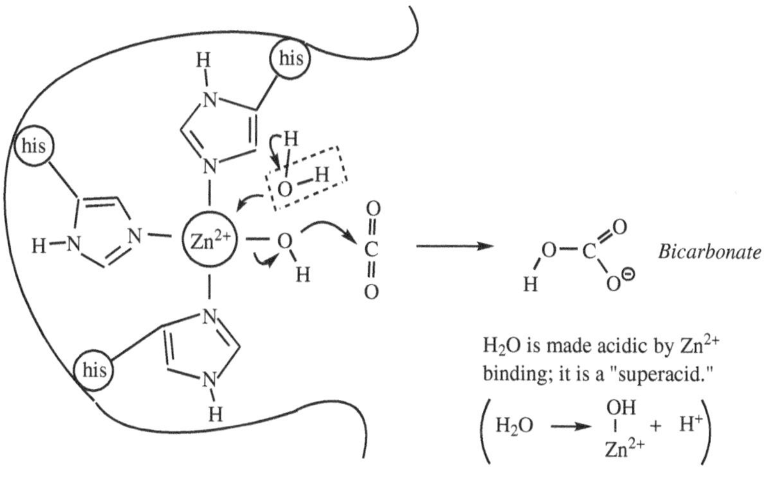

H_2O is made acidic by Zn^{2+} binding; it is a "superacid."

$$\left(H_2O \longrightarrow \underset{Zn^{2+}}{\overset{OH}{\underset{|}{}}} + H^+ \right)$$

Zinc Fingers

Another important role of zinc is its use in *transcription factor* (TF) proteins. Zinc forms a nucleus for four cysteine and/or histidine residues, stabilizing formation of a *zinc finger (ZF) domain*. Different amino acid ligands form unique *ligand spheres* in different fingers, depending on the particular DNA-binding protein.

ZF domains usually occur in several copies in the DNA-binding domain of TF proteins. The fingers fit into the DNA grooves in a sequence-specific manner. Successful recognition leads to *transcriptional activation* of the bound gene sequence. (*See the Nucleic Acids section for details*)

10.4. Carbohydrates-Based Cofactors

3-Phosphoadenosine-5-Phosphosulfate
Sulfate is transferred to a recipient carbohydrate (*e.g.*, heparin) from the activated coenzyme 3-phosphoadenosine-5-phosphosulfate.

Note that phospholipid biosynthesis is analogous in that CDP leaves diacylglycerol when serine or inositol diphosphate are attached to it. (See the *Synthesis of Acidic Phospholipids* section for details.)

Uridylyl Carbohydrates
UDP is a good leaving group. It has a high "group-transfer potential" and is used in the synthesis of lactose.

UDP-galactose

Vitamin C
Vitamin C (*ascorbic acid*) is used as a reducing agent in the *hydroxylation of collagen*.

Upon oxidation, the *diol* functional group, which is composed of two adjacent hydroxyl groups, is converted to a *dione*, which has two adjacent ketones. This molecular system is an efficient *free radical scavenger*. Note the resemblance between ascorbate, ubiquinone and vitamin E (α-tocophorol) and their redox partners. Each supports the dione-to-diol transformation. They all function as both electron carriers and free radical scavengers in a number of biological scenarios.

Ascorbate is used as a reducing agent in a lot of laboratory techniques, often to quench or prevent free radical-mediated processes. It prevents oxidation of carbohydrates, proteins and lipids, yet is nontoxic, so it is a common *food additive*.

10.5 Fat Soluble Vitamins

Vitamin E
Vitamin E (α-tocophorol) is a free radical scavenger. (For details, see the *A Potpourrie of Lipids* section.)

Vitamin A: Cis-Retinal and the Visual Response
The molecule β-carotene undergoes an oxidation reaction, cleaving a double bond and forming two molecules of vitamin A. The alcohol group of vitamin A is then enzymatically oxidized to form *cis-retinal*. *Rhodopsin* is the pigment of the retina that is responsible for forming *photoreceptor cells. Opsin,* a protein within the retina, binds to *cis*-retinal and acts as a visual pigment. When visible light strikes *cis*-retinal in the eye, it is excited and generates a nerve impulse causing *cis*-retinal to convert into *trans*-retinal, which in turn dissociates the opsin. This reaction can be reversed with the enzyme *retinal isomerase* to reform *cis*-retinal.

$\left(\begin{array}{c}\text{orange in}\\\text{carrots}\end{array}\right)$ β-Carotene

Oxidative cleavage to two molecules of Vitamin A.

Enzymatic oxidation of the alcohol at C-15 and isomerization

NADP$^{\oplus}$

NADPH + H$^{\oplus}$

cis- retinal

15 CHO
(aldehyde)

Opsin

Opsin

Retinal isomerase

Light

cis

cis-trans isomerization

trans

trans- retinal

Vision is the result of absorbance of light by the *cis*-isomer, producing *trans*-retinal in the focal center of the eye. This process is called *photo-induced isomerization*. This system is the antenna for our eyesight. Production of the *cis* form *switches* the signal on—*one photon at a time*.

Opsin is the coenzyme carrier protein. In the absence of retinal, opsin is an *apoprotein*. *Rhodopsin* is the *[retinal·opsin]* complex. Transfer of this signal to the optic nerve involves a *G-protein-mediated pathway involving the G-protein Transducin*. (See the *Signal Transduction* chapter for details.)

Coenzyme Q

This coenzyme plays a crucial role in the mitochondrial *electron transport* process, as a *mobile carrier of reducing equivalents*. The second important mobile carrier in that process is a small metalloprotein called *cytochrome c*.

Ubiquinone functions in the oxidation-reduction reactions of cellular respiration. It is a stronger oxidizing agent than both NAD^+ and other flavin coenzymes, and can therefore be reduced by NADH and $FADH_2$. Because it has three oxidation states, the coenzyme can accept or donate either *one or two electrons*.

The three exchangeable redox forms are:
(*1*) the oxidized CoQ (*ubiquinone*)
(*2*) the partially reduced *semiquinone free radical anion*. In electron transport, the free radical exists as the *dianion* CoQ^{2-}.
(*3*) completely reduced $CoQH_2$, (*ubiquinol*).

Vitamin D

Vitamin D is a *steroid* that induces changes in *gene expression* by binding *nuclear receptors*.

Vitamin D₃
(Cholecalciferol)

1, 25-Dihydroxycholecalciferol

The active form is called *1, 25-dihydroxycholecalciferol*. Two separate hydroxylation reactions generate the active form. These compounds regulate the *assimilation of Ca^{2+}* by the body. They work by modulating (*1*) *intestinal absorption,* and (*2*) *deposition* of the cation *into bone.*

Chapter 11

Carbohydrates and Glycoconjugates

11.1 Carbohydrates: Definition (saccharides, sugars)

The simplest sugar is *ethylene glycol*, the primary component of (old-school) antifreeze.

$$
\begin{array}{l}
H_2C-OH \\
| \\
H_2C-OH
\end{array}
\qquad
\left(
\begin{array}{l}
\text{sweet \&} \\
\text{deadly}
\end{array}
\right)
$$

Be careful when storing it outside. It is both sweet and deadly to your children or pets.

11.2 Monosaccharides: Aldoses

Monosaccharides consist of one subunit. The structural designation is derived from L- and D-glyceraldehyde. The letters D and L refer to the experimental direction of rotation of plane polarized light in a polarimeter, dextrorotatory (right) and levorotatory (left), respectively. The designations R and S refer to the absolute attachment of functional groups at the chiral carbon.

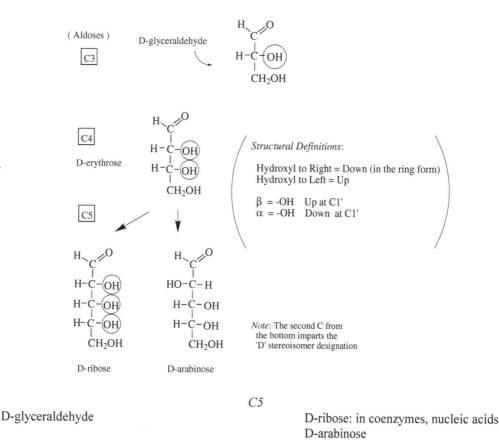

Structural Definitions:

Hydroxyl to Right = Down (in the ring form)
Hydroxyl to Left = Up

β = -OH   Up at C1'
α = -OH   Down at C1'

Note: The second C from the bottom imparts the 'D' stereoisomer designation

C3
  D-glyceraldehyde
C4
  D-erythrose
  D-threose

C5
  D-ribose: in coenzymes, nucleic acids
  D-arabinose
  D-xylose
  D-lyxose

C6
D-allose
D-altrose
D-glucose: fuel for Glycolysis

D-gulose
D-idose
D-galactose: half of Lactose
D-Talose

Aldoses:

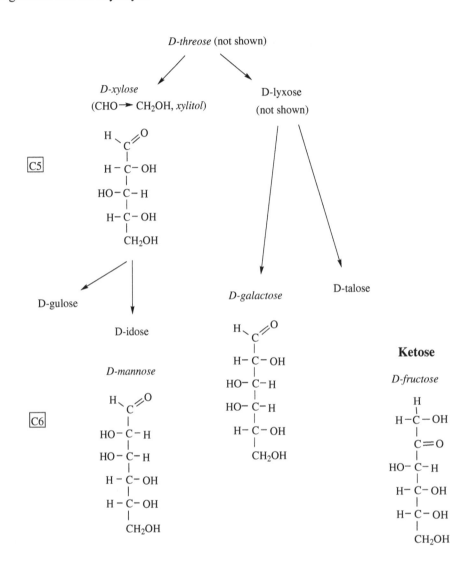

11.3 Monosaccharides: Ketoses

Some important examples of ketoses and some derivatives are:
C3: Dihydroxyacetone is an intermediate formed by *aldolase* in the glycolysis pathway
C4: D- threose. Use of the derivative *dithiothreitol* (DTT) is described in the *Disulfide Bonds* section.
C5: D-xylulose can be converted to the sugar alcohol *xylitol*, a common sweetener in chewing gum.
 D-ribulose is made in plants, where carbon is "fixed" by *Ribulose Bisphosphate Carboxylase*.
C6: D-fructose. The structure, shown above to the right, is a ubiquitous sweetener and food source.

11.4 Structural Features

(*1*) Glucose and galactose are *epimers*, which only differ at carbon 4, where the hydroxyl group is either down and up, respectively.

(2) The L- and D- forms of a chiral compound are called *enantiomers*. These two mirror-imaged compounds rotate plane-polarized light in *levorotatory* (left-handed) and *dextrorotatory* (right-handed) directions in a *polarimeter*.

(3) *Fischer Projections*. The atoms are shown as a vertical chain with C1 at the top. Substituents are shown to the right and left side of the main-chain atoms.

(4) *Haworth projection*s are the standard way to present ring structures formed by carbohydrates. Some defined parameters are shown in the *Glucose Cyclization* figure on the following page.

11.5 Intramolecular Cyclization

Hemiacetals, Hemiketals. The following cyclization reactions form two different structural classes.

aldehyde or ketone + alcohol → hemiacetal or hemiketal

Specific structural transformations are:

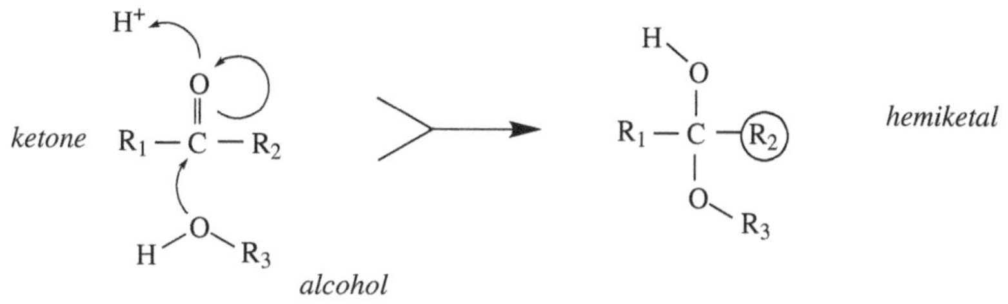

Glucose Cyclization
The following mechanism leads to conversion of linear glucose into the cyclic ring structure.

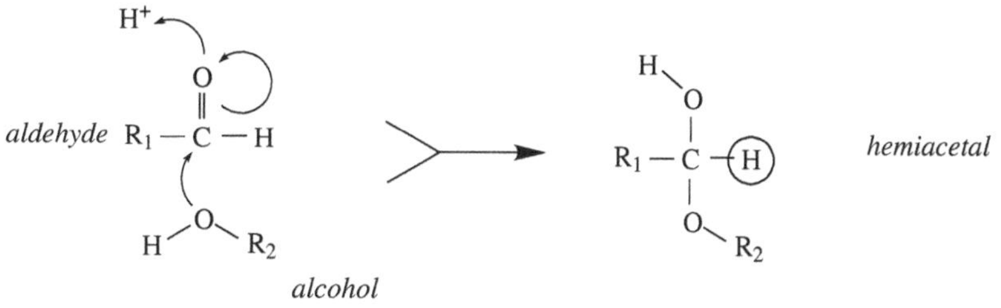

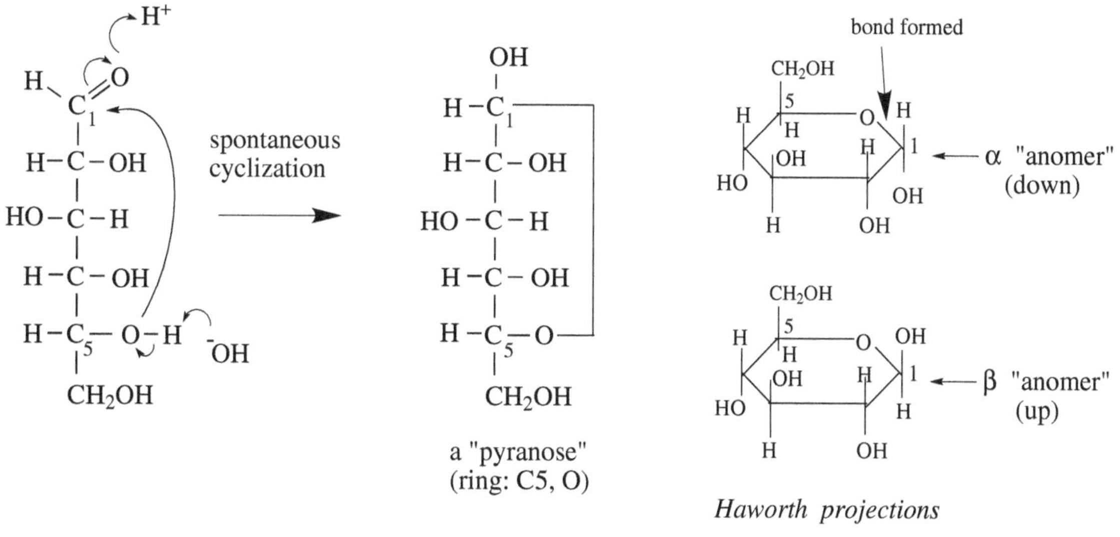

a "pyranose"
(ring: C5, O)

Haworth projections

Ribose forms two structures. The *furanose* form is energetically *less favored* in solution than the pyranose form, yet it is ubiquitous in nucleic acids, ATP, NADH, and so on.

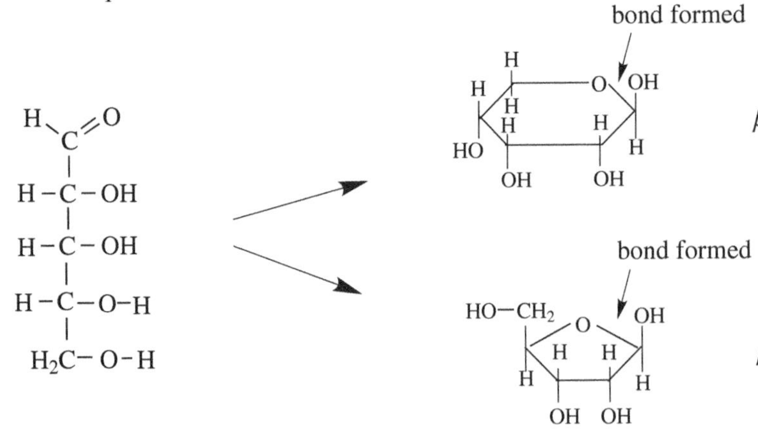

β-*D-ribopyranose*

β-*D-ribofuranose*

11.6 Conformations: Sugar Puckers

Pyranose undergoes a conformational transition from the *chair ring pucker* to the *boat* form.

chair   *boat*

The sugar puckers of a ribose or deoxyribose in RNA and DNA are different. They are referred to as the C-2' endo and C3'-endo sugar conformations. (For details, see the *Nucleic Acids-DNA* chapter.)

One O and three Cs form a plane.

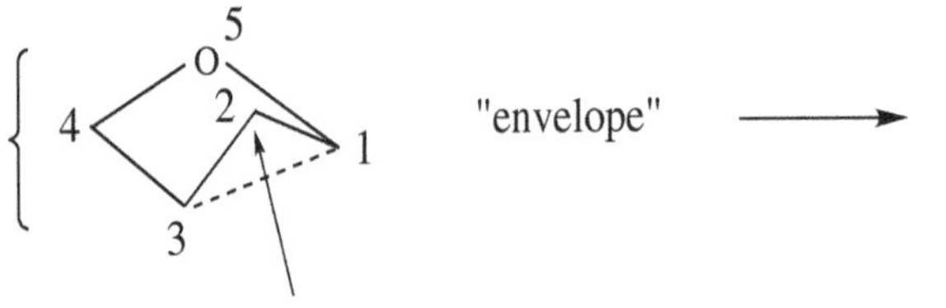

"envelope" ⟶

"Twist," "Pucker" - sugar conformation, torsion angles

This property is important to distinguish different DNA and RNA structures.

One C is "puckered" out of the plane.

11.7 Sugar Derivatives

Sugar Alcohols

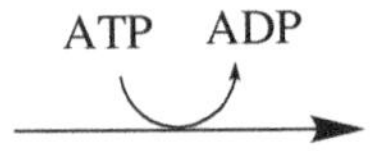

ATP   ADP

sugar alcohol ⟶ sugar phosphate
(*e.g., glycerol*)   (*e.g., glycerol phosphate*)

R-O-H   R-O-P$_i$

Sugar Acids

e.g., glucuronic acid

*Vitamin C
(ascorbic acid)
(in the lactone diol form)*

Amino Sugars

Three major amino sugars exist in cell surface polysaccharides.

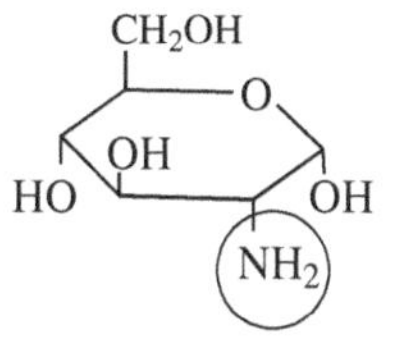

α-D-glucosamine

(GlcN)

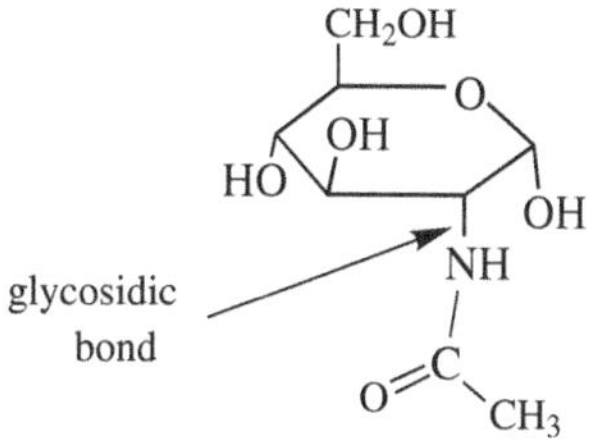

N-acetyl glucosamine

(NAG)

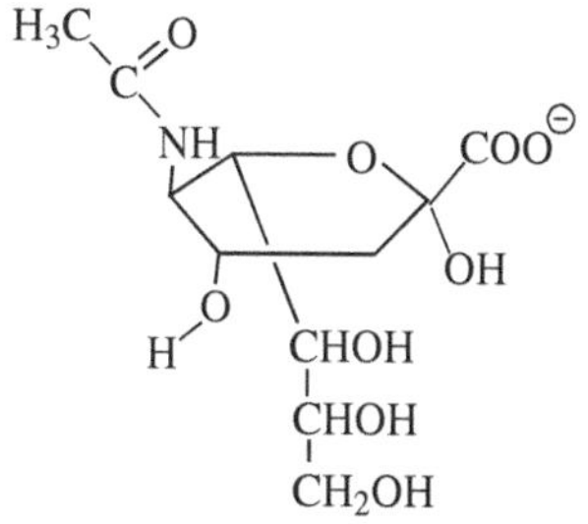

N-acetylneuraminic acid

(NAM)

NAG and NAM are enriched in brain tissue and ubiquitous in protein-linked carbohydrates, such as the *blood group determinant antigens*. Do not confuse NAM with N-acetylmuramic acid (which is shown, for example, in Moran *et al.* on p. 237).

Myo-inositol

This is a monosaccharide that is stabilized because it lacks a ring oxygen. It is released as inositol triphosphate in response to a cellular "second messenger" signal transduction process.

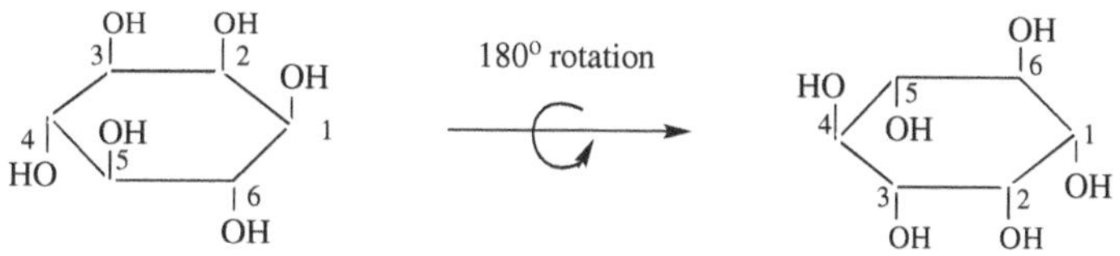

Sugar-X Derivatives

In these cases, X is a "natural product" compound.

Examples:

    (*1*) rose pigment

    (*2*) vanilla flavor

    (*3*) salacin. This derivative of the pain relief medication *aspirin* was found initially in willow bark by caveman-era humans.

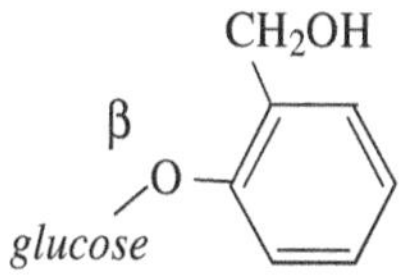

Note the ortho glucosyl and hydroxyl groups on benzene. Lignins in wood are enriched in polymers of such poly phenolic and benzylic structures.

Aspirin (acetyl salicylic acid) is shown and described in the *Eicosanoids* section in the *Lipids* chapter. It also has anti-inflammatory activity and is used to prevent heart attack.

11.8 Disaccharides

Two monomers linked by a glycosidic bond.

    Sucrose. An unusual structural feature is the presence of two anomeric linkages. The formal chemical name of sucrose is: O-α-D-glucopyranosyl-1, 2-β-D-fructose (or fructoside).

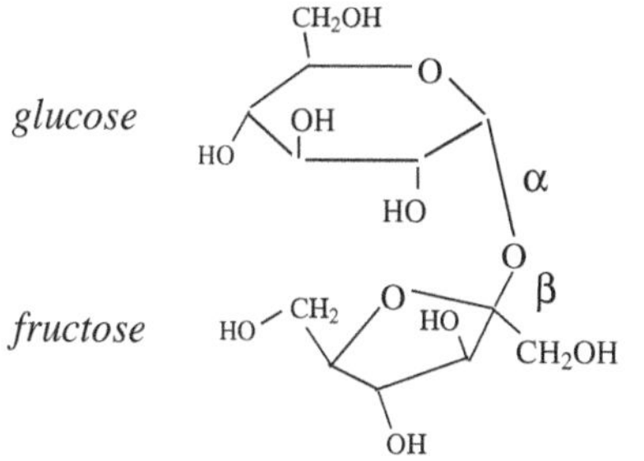

glucosyl (α1→ β2) fructose

Lactose. galactosyl $\beta(1 \rightarrow 4)$ glucose.

11.9 Polysaccharides

Carbohydrate polymers made from the same or mixed monomers.

Purposes

(*1*) Energy precursor storage: compact way to preserve the carbohydrate resources *Example*: Glycogen produces glucose-1-phosphate, which can be converted to glucose-6-phosphate, the fuel for glycolysis.

(*2*) Decreases *diffusion* rates of molecules located near membrane surfaces, which are composed of glycolipids, proteoglycans, glycoproteins and ribose-enriched nucleic acid components.

(*3*) Regulate *osmotic pressure*: High molecular weight carbohydrates attract H_2O because it tries to dilute the carbohydrate enough to establish equilibrium concentrations with that of the bulk fluid (serum, cytosol, organelles, etc.), which is usually much lower. Because the subunits in the polymer are interconnected, they cannot be diluted. As a result, the cell remains wetted and "slippery." This is a key to the structure of many *biocoating* materials.

Starch

This polysaccharide contains both *amylose* and *amylopectin*. Starch is readily degraded to glucose units by the hydrolase Amylase in saliva, the first stage of food digestion.

(*1*) *Amylose* (*unbranched*).

The structure is linear. It only contains $\alpha\,(1 \rightarrow 4)$ linked *glycosidic bonds*.

(*2*) *Amylopectin* is *branched* because it contains both $\alpha\,(1 \rightarrow 6)$ and $\alpha\,(1 \rightarrow 4)$ linkages.

Glycogen

This polysaccharide is essentially the same as amylopectin, except that it contains more branches. The glucose–glucose bonds are cleaved by the enzyme *Glycogen Phosphorylase a*. Branched polymer has more exposed glucose "ends," it can supply them more efficiently than in the linear form.

Cellulose. This polysaccharide constitutes *50% of the carbon in the biosphere.*

(i) All of the linkages are β (1→4). This leads in part to the solid nature of wood.
(ii) Note that the oxygens of every other sugar flips with respect to its neighbors.

Chitin. Present in insect and crustacean *exoskeletons.*

(*i*) Contains only β (1→4) linked NAG subunits.
(*ii*) Note that the sugar oxygens flip with respect to each other.

11.10 Carbohydrate-Protein Conjugates

Peptidoglycans
(*1*) This cross-linked material forms the structural support in *bacterial cell walls.* The peptides are (ala-isoglu-lys-ala) and (gly)$_5$.

(*2*) *Penicillin* inhibits formation of the linkage between *peptides 1* and *2*. Lack of bond formation kills the bacterial cell. The drug, which was discovered by Alexander Fleming, was the first known antibiotic.
(*3*) Lipid-containing examples of peptidoglycans also exist.

Glycoproteins. Three types of linkages exist between protein- and carbohydrate-based species.

(*1*) *O-linked*: Three amino acids are involved: Ser, Thr, hydroxylysine (in collagen).

(*2*) *N-linked*: Generally linked to *asparagine* via an amide linkage.

(*3*) *P-linked*: for example, a phosphatidyl inositol diacylglycerol linked sugar. Note that the lipid anchor is connected to asparagine through the following *sequon* (*i.e.* a small conserved sequence): asn – x – ser/thr.

Proteoglycans

These components of the extracellular matrix contain approximately 5% of the reinforcing structural proteins *collagen* and *elastin*. They also consist of unbranched disaccharides, such as glucuronyl – NAG.

(*1*) *Hyaluronate* and *Keratin Sulfate*. *Hyaluronate* is a disaccharide-repeat unit in a *glycosaminoglycan* polymer found in our joints. The polymer is very viscous and hydrated, so it makes an excellent shock absorber and lubricant. *Keratin sulfate* is a major component of cartilage.

(*2*) *Heparin*, a sulfate-modified carbohydrate, is used to inhibit blood clotting.

Since heparin is highly anionic it competes with DNA-binding proteins and dissociates protein-DNA complexes. *Heparin-Sepharose affinity chromatography* is used to *purify DNA-binding proteins*, such as the *restriction endonucleases*.

(*3*) *Cartilage*. Cushioning material in skin and joints composed of bottle-brush shaped proteoglycans.

(*4*) *Chondroitin sulfate*: High molecular weight carbohydrate ($M_r \sim 2 \times 10^6$ Da) that is an effective cushion between bones, muscle, keratin, skin, *etc*. It *modulates compressibility*. It is a major component of cartilage.

(*5*) *Lysozyme* is an *antibacterial agent* because it hydrolyzes peptidoglycans in bacterial cell walls.

General Physiological Functions of Carbohydrates (CHO)

(*1*) *Attachment*: CHO is added in the Golgi apparatus.

(*2*) *Export*: moving proteins out of the cell. Carbohydrates are typically added when the protein is to be exported (*e.g.*, on antibodies).

(*3*) *Clearance*: destruction of used proteins. The length and degree of oxidation of the carbohydrate are indicator of the status of the structure. These characteristics determine whether the attached protein is preserved in circulation or cleared/destroyed. The components are dismantled and reabsorbed by the cell.

11.10.5 Carbohydrate-Specific Binding Proteins

Lectins are plant proteins that recognize and bind carbohydrates. Recognition depends on the specific monosaccharide or sequence. Purified lectins are used to isolate and characterize carbohydrate chains in CHO-protein structures, CHO-lipid complexes, cells, viruses, and so on.

Example: *Ricin*

Two proteins from castor beans, *Ricin A* and *B*. They form a heterodimer complex in which each subunit has a distinct activity.

Related proteins occur in corn seeds, beans, cucumbers and other plants. The idea is that they are intrinsic pest management complexes, which inactivate translation of an invading mold or bacterium, yet maintain protein synthesis requirements of the seedling.

Ricin B is a *lectin*. *Lectin*s binds specific carbohydrate sequences. They are available as commercial tool to differentiate amongst a variety of carbohydrate analytes. Ricin B attaches to the carbohydrate chains of cell surface antigens, directing the catalytic ricin A subunit to the malignant cells. Once there, the *ribosome inactivating protein* (RIP) Ricin A inactivates protein synthesis in the targeted cell, which kills it. Ricin A *cuts ribosomal RNA* at a specific nucleotide and thereby inactivates catalysis of protein synthesis by the damaged ribosome.

The activities of Ricin A, and other RIPs, have been investigated as possible medical reagents. The premise is that the RIP can be connected covalently to a targeting molecule, such as an antibody or nucleus-directing protein, and the connected protein will escort the RIP into the targeted cell. The RIP will stop all protein synthesis of that targeted cell and thereby stop the cancer.

11.11 Synthesis and Structural Characterization

(*1*) *Synthesis*. It is possible to synthesize carbohydrates by machine, although it is not done as a wide-spread practice. Their intrinsic ability to hydrogen bond inter- and intramolecularly make them stick together. This is useful in their many biological niches, but makes working with them difficult.

(*2*) *Chemical Reagents*. The reagent *periodate* reacts with aldehydes, reducing them to red iodoalcohols. Many other functional group tests exist; for example, the DNS test for 'reducing ends,' the Benedict test, and so on.

(*3*) *Biophysical Characterization*. A lot of analytical work has been done to characterize the structures, motions, interactions and reactions and so forth of carbohydrates. Especially useful information has come from *circular dichroism* (CD), *mass spectrometry* (MS) and *nuclear magnetic resonance* (NMR) studies.

Chapter 12

Lipids

12.1 Structural Overview

Key Property. Water Insolubility. The key feature of lipids is their *solubility in non-polar solvents*, and cellular environments, and *their insolubility in polar solvents such as H_2O*. They have limited or no water solubility. As a result, they block the movement of polar substances.

Structural Diversity. A wide variety of structures exist.
> (*1*) *Fats and Oils*: Amphipathic alkane and/or alkene chain structures with C_n-length tails
> (*2*) *Cholesterol*: an isoprene-derived four-ringed steroid structure
> (*3*) *Eicosanoids*: Prostaglandins thromboxanes and leukotrienes. hormone-like molecules
> (*4*) *Waxes*: C_n - alcohols; derivatives
> (*5*) *Terpenes*: ubiquitous covalent compounds composed of 2 or more *isoprene* subunits
> (*6*) *Cerebrosides, Gangliosides, Sphingomyelins*: all enriched in nerve and brain gray matter

12.2 Saturated and Unsaturated Fatty Acids

Table 1. Correlation Between Micelle Formation and Structure
Selected Saturated and Unsaturated Fats. [1]

Number of Cs	Number of double bonds	Structural Abbreviation	Common Name	T_m (°C)
12	0		*laurate (like in SDS)	44
14	0		*myristate	52
16	0		*palmitate	63
16	1		palmitoleate	- 0.5
18	0		*stearate	70+
18	1	C18:1$^{\Delta 9}$	*oleate	13
18	2	C18:2$^{\Delta 9, 12}$	linoleate	- 9
18	3	C18:3$^{\Delta 9, 12, 15}$	linolenate	- 17
20	0		arachidonate	75+
20	4	C20:4$^{\Delta 5, 8, 11, 14}$	*arachidonate	- 49
22	0		behenate	81+
24	0		lignocerate	84+

[1] See p. 255 in Horton *et al., Principles of Biochemistry* (4th ed.), 2006.

12.2.1 Stability Measurements: Melting Profiles

The *melting temperature* (T_m) is used to quantify the capability of a lipid to form a gel-like complex, called a *liquid crystal*, from a solution containing small aggregates of *dispersed* lipid molecules. The T_m is proportional to the *enthalpy change* ($\Delta H°_d$) for the transformation between dispersed and liquid crystal states. A higher T_m means the complex is more stable; lower T_m means it is less stable.

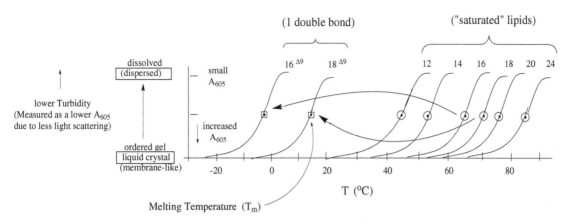

Double Bonds Destabilize Lipid Assembly into Aggregates

The trend of T_m as a function of carbon numbers is plotted below. It indicates that longer chains form more stable complexes. These complexes are called monolayers, bilayers, vesicles, and micelles. They are the types of structures from which membranes are constructed.

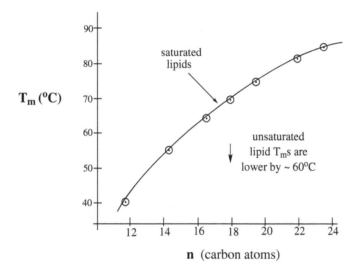

The explanation for this trend is that introducing a double bond produces a kink in the chain. This kink prevents the chain from forming strong surface-surface interactions with the neighboring lipids in a membrane or other *liquid-crystal* type systems, for example, *vesicles, micelles, and lipid rafts*.

The unsaturated chain is less prone to ordered liquid crystal formation. They gel less easily.

12.3 Functions

(1) Lipids are the major structural components of biological *membranes*. Some are not synthesized by humans, so they are essential in our diets. They are a means of *energy storage* and they *form protective barriers* that facilitate sequestration of materials and processes.

(2) Energy Storage. Lipids store about *9* kcal per gram. Carbohydrates only store about *4* kcal/gm.

(3) Vitamins. Examples: *vitamins A* (retinal), *D* (cholecalciferol), *E* (α-tocophorol),
        and *K* (phylloquinone).

(4) Novel functions. Examples include the pain response (prostaglandins); intracellular "hormones" (signal-transduction, phosphoinositides); nerve cell membranes (sphingomyelins), lemon scent (limonene), and many others.

12.4 Diacylglycerol Lipid Derivatives

These structures are the predominant components of biological membranes. *Diacylglycerol* consists of glycerol esterified to two fatty acids. The components are two H_2O-insoluble hydrocarbon chains, a glycerol backbone; and a water-soluble *head group*.

Chain lengths depend on
which fatty acids are present.

Some typical *head group* structures are:

$R - (CH_2)_2 - N^+ (CH_3)_3$ *Phosphatidyl choline*   —causes platelet coagulation

$R - (CH_2)_2 - NH_3^+$ *Phosphatidyl ethanolamine*—involved in lung tissue maintenance; also a major constituent of bacterial membranes

$R - CH_2CH(NH_3^+)(COO^-)$ *Phosphatidyl serine*

$R - CH_2 (CHOH)CH_2OH$ *Phosphatidyl glycerol*   — involved in lung surface maintenance

Other examples (to look up) are:

         Phosphatidyl cardiolipin
         Sphingolipid   — key component of *lipid raft* elements
         Myelins   — enriched in the myelin sheaths of nerves

12.5 Structural Motifs

Monolayers consist of a single ordered layer of lipid molecules, aligned with the *polar head groups* at one end and the *apolar hydrocarbon tails* at the other.

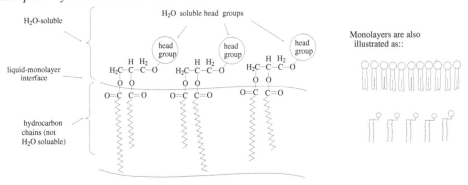

Micelles. Monolayer spherical complex composed of 20 to 100 lipids. The inside chains are dynamic and somewhat disordered. A hydrophobic molecule, for example, greasy dirt, can be trapped inside the central hydrophobic core. This is the concept behind *soap* and *detergent* action. For example, many shampoos contain *sodium laural sulfate* (*aka.* SDS).

Bilayers. These elements are the basis of membrane structure.

outer surface

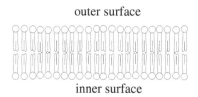

inner surface

Vesicles. These spherical structures can *sequester solutes or solvents* within the central cavity.

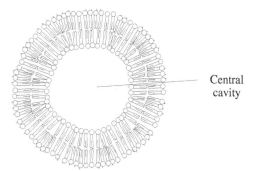

Central cavity

Membranes allow effective *sequestration of charged species*. Charged molecules cannot cross the hydrocarbon layers, which act as a nonpolar barrier to transport between compartments. This phenomenon is critical in the functions of nerve conduction, export of natural products, proper control of transmembrane concentration gradients, signal transduction, mitosis, and so on.

12.6 Assembly

Energetics: Hydrophobic Effect Revisited
As with proteins, the hydrophobic effect stabilizes lipid assemblies. *Enthalpy* accrued by hydrogen bonding between solvent water molecules drives the process, offsetting the large *entropic cost* of trapping the lipids in the organized configuration(s). As with proteins, the assemblies denature. In this case, the lipids *disperse* rather than unfold.

Bilayer stability is affected by the local water concentration, so lipid assembly reactions are strongly subject to reagents that modify the hydrophobic effect. Recall that different lipids impart different stabilities to these liquid crystal-type assemblies, depending on length, degree of saturation, and polar/nonpolar properties. See the *Stability Measurements: Melting Profiles* section for details.

Osmolytes Modify Bilayer Stability
Osmolyte molecules change the *osmotic pressure* of a solution, which can, in turn, affect the stability of adjacent lipid bilayer surfaces. For example, the unusual amino acid *taurine* is made in large concentration within salt-water *clam* tissues. The purpose is to *offset* the *osmotic pressure* induced by the high concentration of salt in sea water. Other osmolytes include glycerol, glucose, guanidinium, urea, and metabolites such as the *polyamines* (*e.g.,* spermidine) which are enriched in cell nuclei, and *neurotransmitter*s (*e.g.,* γ-aminobutyric acid; GABA), which function in nerve conduction.

Induced Assembly In Vivo

Bilayer structures are assembled *in vivo* by *lipid chaperonins* called *snaps, snares* and *annexins*. The tendencies of the lipids to self-assemble to form membranes are enhanced by the proteins.

Natural vesicles are made from cholesterol and sphingolipid-enriched lipid rafts, which initiate the process of packaging proteins intended for delivery from the golgi apparatus to the extracellular space. (For details, see *My Life On A Raft* by Kai Simon in *The Scientist*, vol. 24, pp. 24–29, 2010.)

12.7 Structural and Dynamic Characterization

A few of the tools that have been used extensively to study lipids and their assemblies are:

(*1*) *Freeze-fracture Electron Microscopy.* This approach has been used extensively to produce suitably dissected views of the structure of biological membranes.
(*2*) *Atomic Force Microscopy.* A microscopic stylus is scanned across a molecular sample at molecular dimensions. This reveals the surface topography of the assembly.
(*3*) *Fluorescence Affinity-Labeling and Time-Dependent Tracking.* The *flourescence quenching imaging* techniques allow scientists to follow the movement, assembly, and so forth, of lipids in membrane systems.
(*4*) *Nuclear Magnetic Resonance.* Nuclei such as ^1H, ^{13}C, ^{31}P and ^2H can be attached to lipids in specific positions, allowing researchers to follow them structurally using various NMR techniques. Peak-width and shape measurements and analyses allow characterization of the motional characteristics of the side-chains and head groups, respectively, in the same molecule.
(*5*) *Electron Spin Resonance.* The *nitroxide spin label* can be attached to the head group, allowing a researcher to study the motions of the chain structure, and many other specific details regarding the dynamic nature of the lipid or assembly.

The following dynamic properties have been characterized:
Bilayers Are Charge Impermeable. Charge separation between the inner and outer surfaces of a bilayer leads to a voltage. This potential energy is captured and used to power cells and drive cross-membrane movement of metabolites. When the transmembrane potential is obstructed, the membrane leaks and these functions are lost. Diarrhea is caused by cross-membrane potential decoupling. It is an important example since it is responsible for more deaths worldwide (as a result of dehydration) than any other disease symptom.
Surface Diffusion. Lipid head groups diffuse laterally around the surface of the layer leaflet. The lipid molecules move laterally via two-dimensional planar diffusion.

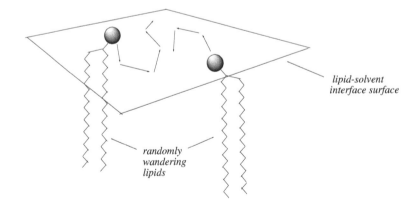

lipid-solvent interface surface

randomly wandering lipids

Cholesterol Solidifies Membranes. Lipid movement is modulated by the stability of liquid crystal structure, which is affected by the lipid properties of the admixture that is present. The well-known lipid *cholesterol* induces solidification of lipid assemblies, a necessary property to retain healthy pliable membranes.

In dietary excess, cholesterol forms *plaques in human arteries*, leading to *atherosclerosis, cardiac arrest* and *strokes*. The medical approach to *controlling the level of cholesterol in human serum* is discussed in the *Regulating Cholesterol Levels* section.

12.8 Eicosanoids

These molecules regulate inflammation, pain, sensitivity, swelling, reproductive processes and are the target of aspirin and similar pain medicines. The precursor fatty acid *arachidonate* is converted into a prostaglandin by a cyclization reaction. The product can be converted to a variety of bioactive structures.

Biosynthetic Pathways

A variety of prostaglandins as well as prostacyclin and thromboxane A_2 can be formed via the *prostaglandin H synthase pathway*. *Arachidonic acid* is oxidized to form *prostaglandin H_2* (PGH_2), which is converted to the other products. The *Prostaglandin H Synthase Cyclooxygenase* (COX) step is inhibited by *aspirin*. Such pharmaceuticals are called *COX inhibitors* or *NSAIDS*.

(Adapted from Horton *et al.*, *Principles of Biochemistry* (4th ed.), 1996, p. 489.)

Biological Activities

These are potent molecules. *Thromboxane A_2* regulates confined changes in blood flow in areas where the eicosanoid is produced. This leads to localized *platelet aggregation*, blood clots, and *constriction of smooth muscle* within the *arterial walls*. In another important role, prostaglandins *induce contractions of the uterus during labor*.

Plants produce *linoleate*, which is essential to humans to fuel arachidonate and eicosanoid biosynthesis. *Linoleate* is required for survival.

| Aspirin | Ibuprophen | Acetaminophen | Rofecoxib (Vioxx®) |

Aspirin (*acetylsalicylic acid*) inhibits the enzyme *Prostaglandin H Synthase Cyclooxygenase* (*COX-1*) activity. Aspirin acetylates a serine hydroxyl near the active site, which irreversibly prevents arachidonate binding. As a result, subsequent reactions in the eicosanoid pathway cannot occur.

COX-1 is expressed constitutively at low levels in a variety of cell types. It regulates *mucin* secretion in the stomach, thereby *protecting* the *gastric (stomach) wall*. The result is that pain, inflammation and stomach health *are* related. The connection really does depend on *how one's tummy feels*.

Two *isozymes* of COX exist. The second one *COX-2* controls the extent of *inflammation, pain* and *fever* in a tissue. The *omega-3 fatty acids,* which are enriched in *fish* and *flaxseed oils*, are used to regulate the onset of *atherosclerosis*. Ingestion of one aspirin per day is advised because it regulates the prostaglandin pathway, and thereby inhibits cardiac arrest.

12.9 Phospholipases

These enzymes catalyze selective removal of fatty acids from the glycerol group of the phosphatidyl group. For example, *Phospholipase C* functions in the *Inositol-Phospholipid Signal Transduction Pathway* (IP-STP) by liberating *inositol-1, 4, 5-triphosphate* from the remaining *diacyl glycerol*. Each component functions as a second messenger. (Details are described in IP-STP section of the *Signal Transduction* chapter.)

12.10 Phosphoinositides

This molecule is a *second messenger* in signal transduction that contains the inositol group in the 1, 4, 5-triphosphate form. *Phosphatidylinositol* (PI) is phosphorylated at C1. The other forms are 4, 5-diphosphoinositol (PIP) and 1, 4, 5-triphosphoinositol (IP$_3$).

12.11 Steroids

The following examples range from (*1*) the much vilified membrane component cholesterol to (*2*) two gender-specific hormones to (*3*) a physiological detergent called a bile salt.

(*1*)   (*2*)   (*3*)

isoprenoid R group

Cholesterol   Testosterone   β-Estradiol   Sodium cholate

Cholesterol
(*1*) Cholesterol makes membranes more rigid, modulating their pliability.
(*2*) Cholesterol is incorporated into chylomicrons in the gut after ingestion. It is stored within cells in plasma lipoproteins.
(*3*) It can suppress transcription of specific genes.
(*4*) By decreasing serum cholesterol levels, one can decrease their risk of coronary heart disease. The *statin* drugs are often used to treat excessive serum cholesterol levels, called *hypercholesterolemia*.

Testosterone and β-Estradiol: These are the male and female sex-determining hormones.
12.10.3 Sodium Cholate: This *bile salt* mediates *lipid absorption* by increasing their solubility.

12.12 A Potpourrié of Lipids

Vitamins (Coenzymes). *Vitamins A, D, E, and K are all isoprenoid* compounds made by plants. As discussed in the *Coenzymes* section, these compounds are important in visual perception, calcium metastasis, and scavenging free radicals.

(*1*) *Vitamin A*
  (*trans*-Retinol)

(*2*) *Vitamin D*
  (Cholecalciferol)

(*3*) *Vitamin E*
  (α -Tocopherol)

The hydroxyl group forms a stable free radical by binding an electron. It's a free radical scavenger.

(*4*) *Vitamin K*
  (Phylloquinone)

Vitamin K and Blood Clotting

This compound is required to convert glutamates into γ-carboxy-glutamate in several blood clotting factors. These modified amino acids line Ca^{2+}-binding grooves that bind a set of Ca^{2+} to create a positive site, which binds strongly to electronegative proteins such as Fibrinogen.

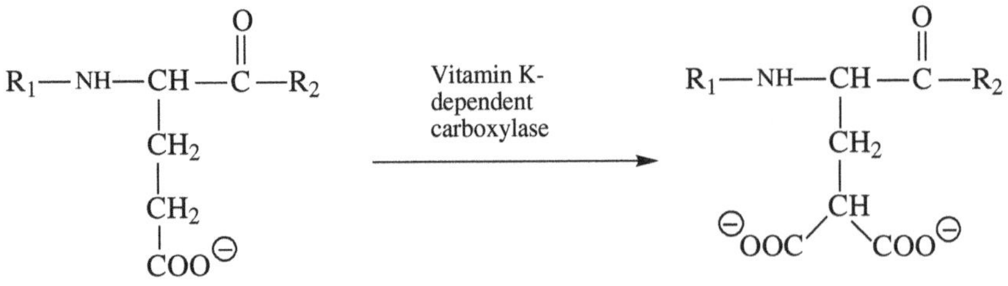

γ-carboxyglutamate

Waxes

Myricyl palmitate is an example of a wax. Formed from a long-chain mono-hydroxylic alcohol, it is a fatty acid ester.

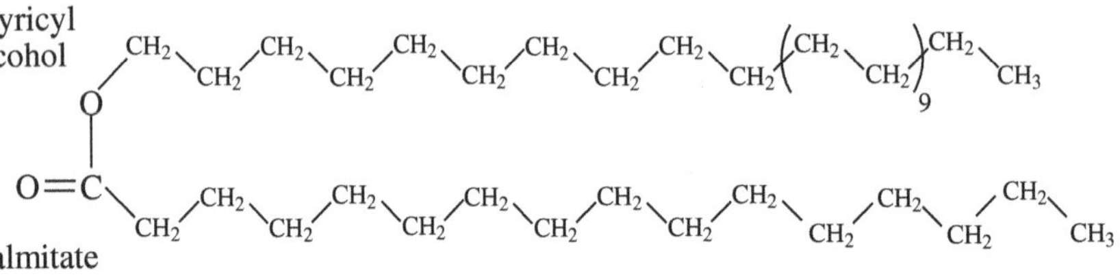

Natural Products. This classification encompasses a huge cross-section of biologically active polyprenyl compounds. Three examples are shown below:

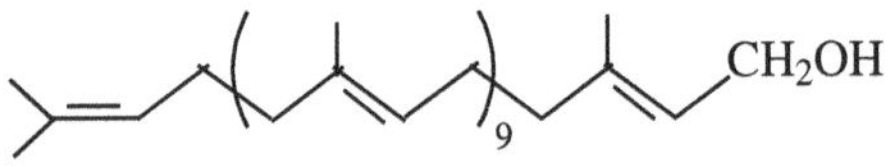

Bactoprenol
(Undecaprenyl alcohol)

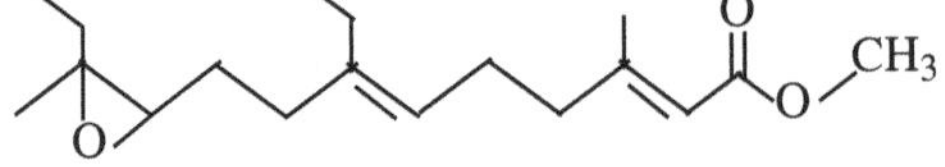

Limonene
(lemon odor, "zest")

Juvenile hormone 1

Limonene is a lipid found in lemons that contributes to their acidic smell.
Bactoprenol is a lipid found in bacteria that is important in cell wall formation.
Juvenile hormone I regulates larval development in insects, that is, molting.

Chapter 13

Membranes

13.1 The Fluid Mosaic Model

Biological membranes contain a variety of components, including lipid-embedded proteins, glycoproteins, proteoglycans, and so on. The lipid and protein content differs on the inside and outside surfaces. The following figure shows the structure of a typical eukaryotic plasma membrane. Typical components are listed.

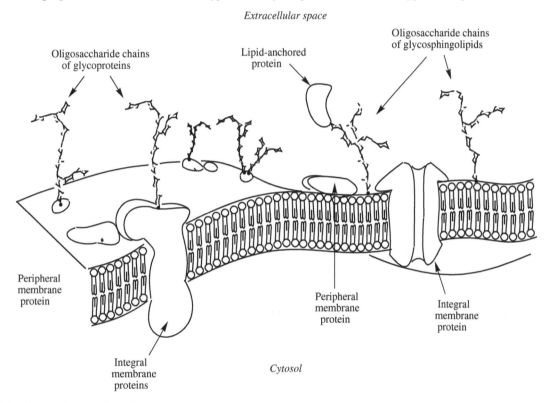

Extracellular space

Oligosaccharide chains of glycoproteins

Lipid-anchored protein

Oligosaccharide chains of glycosphingolipids

Peripheral membrane protein

Integral membrane proteins

Peripheral membrane protein

Integral membrane protein

Cytosol

Membrane-bound Proteins

(*1*) *Anchor proteins*. These are linked to carbohydrates via the N- and P-type linkages described in the *Carbohydrate section*.

(*2*) *Integral proteins*. O-linked; CHO outside; 20 amino acid tail (hydrophobic; membrane-embedded)

13.2 Detergents

These molecules are used to solubilize membrane proteins when purifying a biomolecule or fractionate from cells. Though all of these detergents are not ionic, they are all amphipathic, possessing both polar and nonpolar regions.

Triton X-100
(Polyoxyethylene *p*-t-octyl phenol)

Sodium deoxycholate

Octyl β-D-glucoside

(*1*) *Triton X-100*: This is the most frequently used detergent. 9 or 10 oxyethylene units (n) are attached.
(*2*) *Sodium deoxycholate*: similar to the compound CHAPS. The *bile salt* cholic acid is the protonated species.
(*3*) *Octyl β*-D-glucoside: a lipid-carbohydrate bioconjugate.

13.3 Distribution of Lipids in Biological Membranes

The *phospholipid* contents are *asymmetrically distributed* with respect to the inner and outer leaflets of the human erythrocyte membrane. Patterns found in erythrocyte membranes, as well as that of the bacterium *Micrococcus luteus*, which also shows the asymmetry, are shown below.

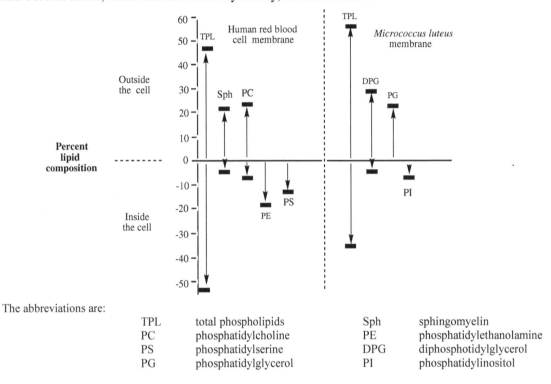

The abbreviations are:

TPL	total phospholipids	Sph	sphingomyelin
PC	phosphatidylcholine	PE	phosphatidylethanolamine
PS	phosphatidylserine	DPG	diphosphotidylglycerol
PG	phosphatidylglycerol	PI	phosphatidylinositol

13.4 The Hydropathicity Scale

Seven-Helix Transmembrane Proteins. *The protein bacteriorhodopsin* from *Halobacterium halobium* is a classic *integral membrane protein*, with seven polypeptide chains that span the membrane. The seven α-helices are bundled and interconnected by consecutive loops that project into the solvent/cytoplasm from both the inner and outer surfaces. The center of the complex contains a channel that transports protons from one side of the membrane to the other.

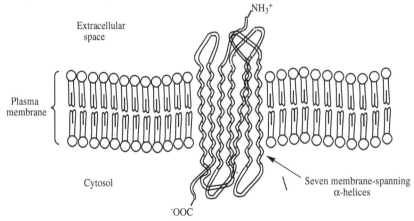

The protein binds the light-harvesting coenzyme *retinal*, which it uses as an energy source to *drive protons through the membrane*. Experimental demonstration of the use of this *proton pump* to drive ATP synthesis is described in the *Oxidative Phosphorylation* section.

Measuring the Hydropathicity of an Amino Acid

One can determine the preference of an amino acid (aa) to *distribute* between *water* and a solvent that mimics the environment within a generic membrane, that is, *octanol*. This involves placing equal amounts water and octanol in one tube. Because they are immicible (do not mix), the added aa will distribute into one or both of the solvents. One determines the *partitioned percentage* of the aa in each layer analytically. This information can then be used to calculate equilibrium constants (and Gibbs free energies) for the partitioning equilibrium $aa_{water} \leftrightarrow aa_{octanol}$ for each of the amino acids.

The results allow one to assess the *hydropathicity* of the molecule, which gives us a sense of their preference for the aqueous environment. More hydrophobic amino acids prefer the interior of the lipid bilayer. The *free-energy changes* for the partitioning process are listed below:

Table 2. Polarity Scale for Amino Acid Residues Based on H_2O-Octanol Partition Coefficients

Amino acid	ΔG for transfer (kJ mol^{-1}) [1]	Amino acid	ΔG for transfer (kJ mol^{-1})
Isoleucine	3.1	Proline	-0.29
Phenylalanine	2.5	Threonine	-0.75
Valine	2.3	Serine	-1.1
Leucine	2.2	Histidine	-1.7
Tryptophan	1.5	Glutamate	-2.6
Methionine	1.1	Asparagine	-2.7
Alanine	1.0	Glutamine	-2.9
Glycine	0.67	Aspartate	-3.0
Cysteine	0.17	Lysine	-4.6
Tyrosine	0.08	Arginine	-7.5

[1] Calculated as $\Delta G = -RT \ln K$,

The Hydropathicity Plot

The hydropathicity plot for *bacteriorhodopsin* is shown below.

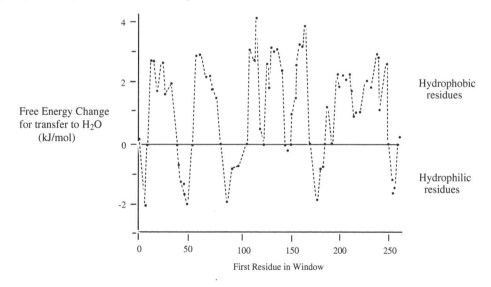

(*1*) *Positive free energies* in the hydropathy plot indicate that the plotted amino acid prefers the *hydrophobic* environment associated with: (*i*) the protein interior, or (*ii*) being embedded within a membrane, in an integral membrane protein.

(*2*) *Negative free energies* indicate that the amino acid will project into the aqueous environment. It is *hydrophilic*.

If one squints, the pattern in the plot has seven positive peaks, indicating the presence of seven membrane-spanning α-helices. The five clear hydrophilic regions (seven are expected) are the intervening loops. Pattern tending toward more hydrophilic sequence are evident in the intervals where they do not cross over into hydrophilic regime *per se*.

13.5 Lipid-Anchored Membrane Proteins

The following figure shows three types of lipid-anchored membrane proteins:

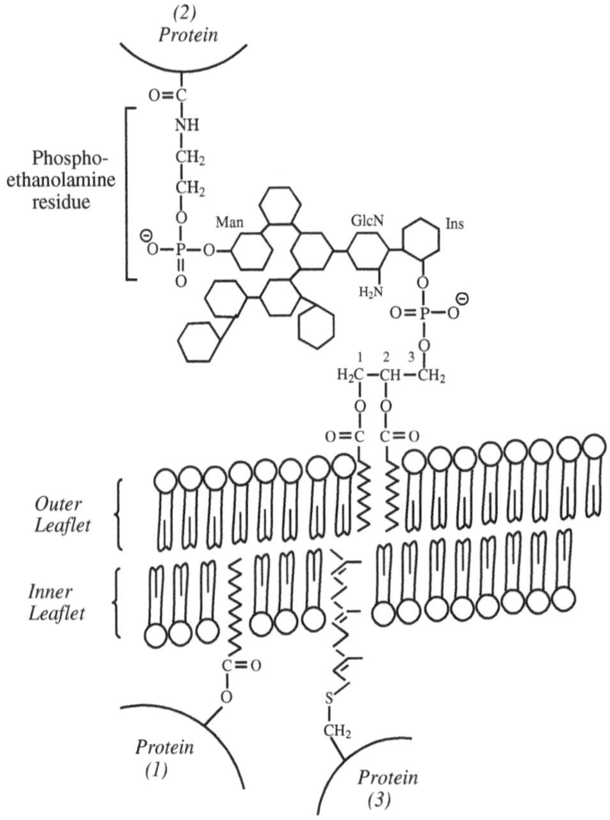

Fatty Acyl Proteins. The inner leaflet of the membrane in anchored to a myristoylated protein.

Glycosylphosphatidyl-Inositol Proteins. The parasitic protozoan *Trypanosoma brucei* has a surface glycoprotein that is anchored by this protein. It is bound covalently to a phosphoethanolamine residue which is also bound to a glycan. Phosphoethanolamine attaches onto the mannose residue of the glycan. The inositol group of phosphatidylinositol is attached to the glucosamine residue of the glycan. The protein is anchored to the membrane by the diacylglycerol portion of the phosphatidylinositol.

Prenyl-Anchored Proteins. This is an example of anchoring via the farnesyl group of a membrane protein that has undergone prenylation. A covalent bond joins the isoprenoid chain to the protein membrane with a thiol group of a cysteine residue close to the protein's C-terminus.

These three anchor types can be found in the same membrane, but it does not resemble the structure illustrated. The following compounds are abbreviated: glucosamine (GlcN), mannose (Man).

13.6 The Erythrocyte Cytoskeleton

Lipid bilayers are essential in the formation of a cell, however, they are very thin and do not contribute much stability to the structural support network. The *cytoskeleton* acts like a protein skeleton within the cell to provide the necessary strength to maintain cell shape, allow cellular movement, aid in intracellular transport and provide protection. The cytoskeleton is distributed throughout the cell, attached to the inside face of the membranes.

The inside face of the erythrocyte membrane is linked to a protein cytoskeleton network composed of *actin* and a fibrous polypeptide, *spectrin*. These components are organized into a meshwork, which is linked to the membrane-bound protein *ankyrin*.

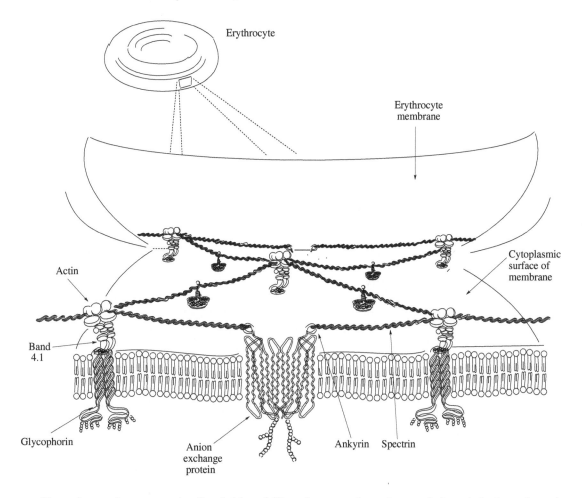

The anion-exchange protein, *Band 4.1*, stabilizes the *spectrin-actin* association. *Ankyrin* anchors the mesh-like pattern by binding the cytoplasm region of band 4.1, which binds to the cytoplasm region of *glycophorin A*. Notice how this molecular framework resembles the re-bar reinforcement in cement structures.

Chapter 14

Transport Through Membranes

14.1 The Transmembrane Potential

An *electric potential* ($\Delta\psi$) is formed by the separation of charges across the membrane of most living cells. The typical standard $\Delta\psi$ across a plasma membrane is -60 mV, with an excess of *negative charge inside* the cell. When a charged solute is moved across the membrane, equation 1 must be altered by adding one term to take into effect the membrane polarization.

$$\Delta G = R \, T \ln (c_2/c_1) + z \, \mathcal{F} \, \Delta\psi \qquad\qquad (1)$$

where z is the unit charge on the solute that is transported, $\mathcal{F}$ is the Faraday constant (96.48 kJ V^{-1} mol^{-1}), and $\Delta\psi$ is the *transmembrane electric potential*, measured in volts. Free energy is released (accrued) as a positively charged solute moves from the positively charged extracellular side of the membrane to the intracellular negative side. In contrast, transporting a negatively charged solute to the intracellular side of a membrane *costs* energy.

14.2 Active Transport

Active transporters are driven by many different types of energy sources, for example, ATP, light, ion gradients. A *direct* source of energy is required to run *primary* active transport. Proton or sodium concentration gradients drive secondary active transport. Some active transporters are fueled by light, for example, *bacteriorhodopsin*, which is described in detail in the *Oxidative Phosphorylation* section.

All organisms have a number of ATP-driven ion transporters, called *ion-transporting ATPases*. Active transporters, such as the Na^+, K^+ and Ca^{2+}, H^+ *ATPases*, create and preserve the ion concentration gradients across cellular membranes.

Table 3. Energy Sources for Active Transport

Type of Transport	Energy source	Transporter	Transported species
Primary	Light	Bacteriorhodopsin	Protons
	ATP	ATPase	Ions
	ATP	P-Glycoprotein	Nonpolar compounds
	Substrate oxidation (electron transfer)	Electron transport proteins	Protons
Secondary	Proton gradient	Lactose permease	Lactose
	Sodium gradient	Active glucose transporter	Glucose

14.3 Ionophores

Valinomycin and *gramicidin D* are potent *antibiotics* that cause cations to leak across membranes, destroying the transmembrane cation gradients. Since these gradients are required to produce ATP and drive secondary active transport, *ionophores kill cells*.

Valinomycin

This is an extremely selective *ionophore*, which binds K^+ 1000-fold more strongly than Na^+. The complex is lipid soluble because the ionophore has nonpolar side chains that interact with the acyl chains of membrane lipids. As a result, the [valinomycin·K^+] complex can diffuse across the membrane and release the ion on the other side.

The secondary structure of valinomycin and the tertiary structure of the [valinomycin·K^+] complex are shown in the following figure.

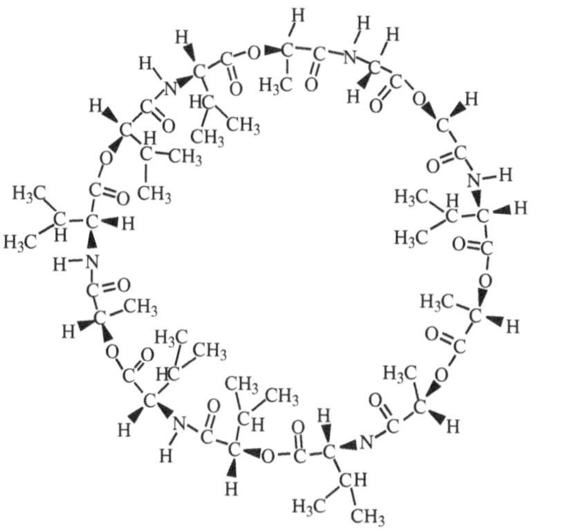

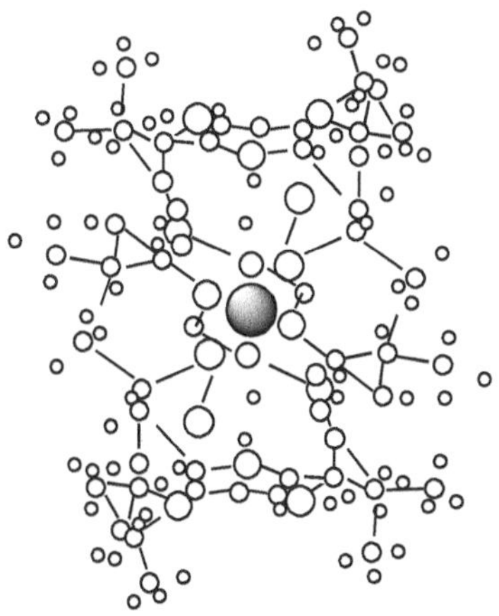

(K$^+$ is in the center)

Gramicidin D

This compound has ion channels that are continuously dissociating and reforming every second, allowing and inhibiting ion conduction with each alteration.

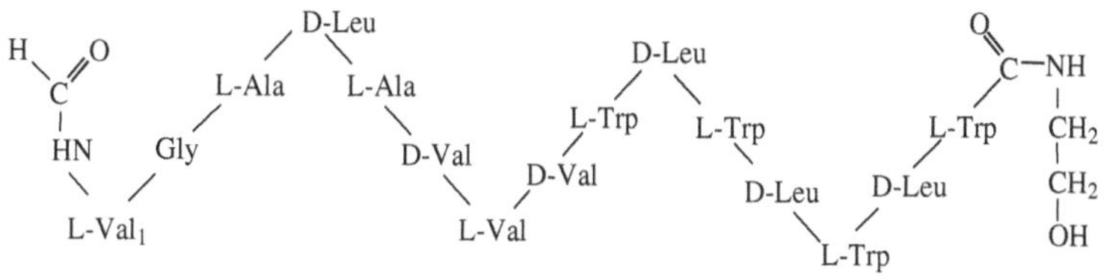

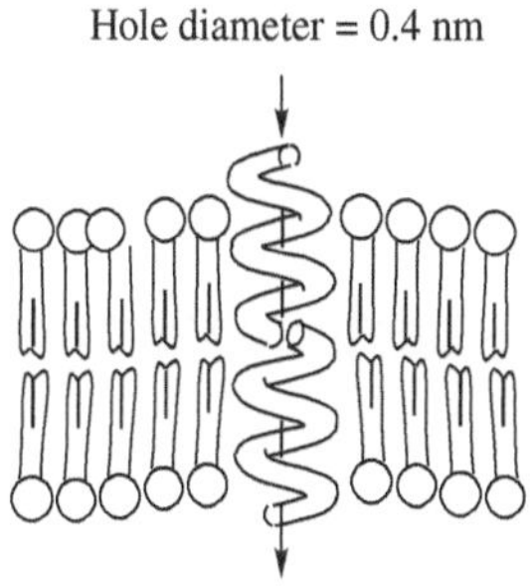

Hole diameter = 0.4 nm

Unlike mobile carriers who need to diffuse through the lipid bilayer, channel-forming proteins do not. Therefore, the rate of transmembrane ion diffusion by gramicidin D is 10^4-fold faster than that of valinomycin.

14.4 The Acetylcholine Receptor Ion Channel

Recalling from the *Enzyme Inhibitor* section, *acetylcholine esterase* degrades the neurotransmitter acetylcholine, which terminates synaptic transmission and resets the open receptor for another nerve impulse. For proper nerve conduction, the appropriate type of *membrane polarization* must be present. The following steps are necessary:

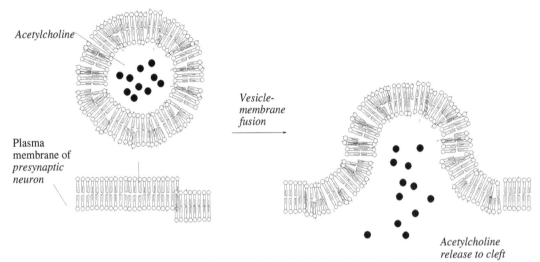

The *plasma membrane* of the postsynaptic neuron is *depolarized* by the released neurotransmitter, which binds the port protein and activates ion channel opening.

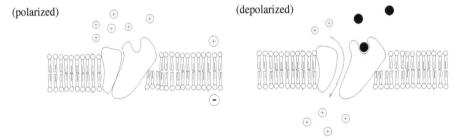

(1) At rest, the channel is closed and the inside of the postsynaptic neuron has more negative charge than the outside, creating a negative membrane potential.

(2) The presynaptic neuron releases *acetylcholine* which then binds to its *receptor*, allowing an influx of sodium ions. The entrance of sodium depolarizes the membrane, reducing the electric potential.

14.5 Lactose Permease and Secondary Active Transport

In the electron transport chain of *E. coli*, oxidation-reduction reactions produce oxidized substrates (S_{ox}) from reduced substrates (S_{red}). During this process, the proteins of the chain *pump a proton across the membrane* with each formation of a redox species, forming a proton concentration gradient.

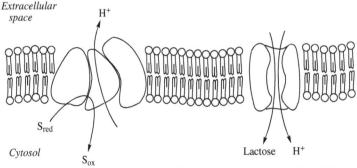

As these protons move down their concentration gradient, energy is released, which causes one lactose to be *translocated* into the cell through a *symport, along* with one H^+, by the enzyme *lactose permease*.

14.6 Mechanism of Transport by Na⁺, K⁺ ATPase

Na⁺, K⁺ adenosine triphosphatase (ATPase) pumps *three* sodium ions *out* of the cell as *two* potassium ions *enter* the cell. Energy must be supplied by the hydrolysis of ATP because both ions are moving up their concentration gradient. The mechanism following mechanism is adapted from Moran *et al. Biochemistry* (2nd ed.), 1994, pp. 12–31; Fig. 12.35.

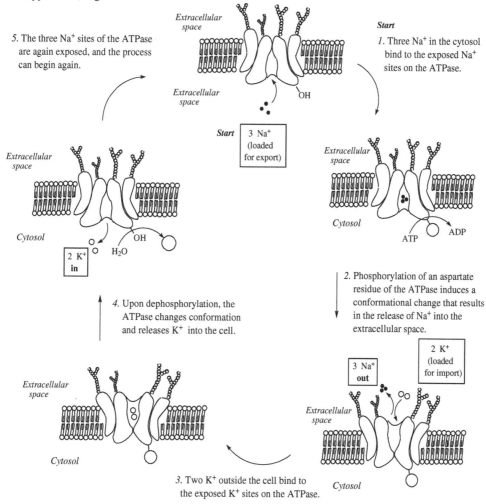

14.7 Ion Channel Blockers

The compounds *ouabain* and *digitoxigenin* are extracted from the purple foxglove. The mixture is used as a heart medication called *digitalis*. These *cardiotonic steroids* are extremely toxic because they *inhibit Na⁺, K⁺ ATPases*. Upon binding to the extracellular domain of the ATPase, the compound freezes the mechanism at the *phosphorylated state*, channel opened outward.

Digitoxigenin

Ouabain

Digitalis works by increasing sodium influx into the heart muscle cells, which effectively activates the *Na⁺, Ca²⁺ antiport* system. Sodium leaves the cell as calcium enters, which increases the strength of the heart muscle contractions

Chapter 15

Signal Transduction

15.1 Signaling Pathways: Hormones, GTPases, Second Messengers and Intracellular Regulation

Signal Transduction occurs when a molecular signal is transferred across the plasma membrane of a cell. The signal is passed to the inside of the cell by the following chain of events:

(*1*) Extracellular *first messenger* molecules (typically *hormones*) bind to a *receptor*, which is closely associated with a *transducer* enzyme, within the membrane. Each receptor is activated or repressed by the signaling/binding molecule, which is designed physiologically to affect its activity in a certain way – to control the intended cellular function.

(*2*) The transducer then transmits the signal that instigates production of the *second messenger* within the cell. Transducers are commonly *Guanine Triphosphatases* (GTPases), which control a variety of biological functions.

(*3*) Second messengers can induce:
　　(*i*) a cytosolic enzyme to modify their substrate protein,
　　(*ii*) a membrane-bound port complex to allows entry of a particular ion, for example, Ca^{2+},
　　(*iii*) and many other responses.

Hormones and *growth factors* are examples of *first messengers* that have specific binding sites on particular *cellular receptors*.

(*4*) A protein within the membrane transports the signal to the cytosol-side *membrane-bound effector* enzyme.

(*5*) The effector enzyme catalyzes formation of the *second messenger*, typically a small molecule (*e.g.,* cAMP, IP_3, DAG).

(*6*) The signal is then transferred to its final destination by the second messenger to produce the *intracellular response*, for example, regulation of some reaction or pathway in the cytosol, nucleus, mitochondria, and so on.

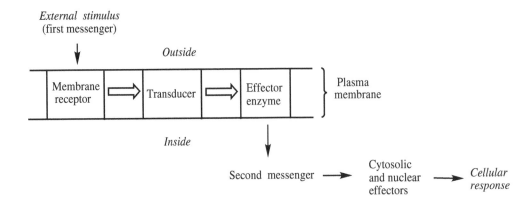

In contrast to the large number of ligands, membrane receptors, and transducers that are known, only a few effector enzymes and second messengers have been found. Therefore, in response, each cell has a specific detecting ability to notify when an extracellular signal resides at the surface but use familiar pathways to carry these signals through to the inside of the cell.

15.2 The Adenylate Cyclase Signaling Pathway

Activation Pathway. When (*1*) an extracellular hormone binds to a *stimulatory transmembrane receptor protein* (R_s), (*2*) a *stimulatory G protein* (G_s) is activated. The G_s protein functions by (*3, 4*) activating the membrane-embedded enzyme *adenylate cyclase*, which (*5*) produces *cyclic AMP*. Cyclic AMP (*6*) activates *protein kinase A* which leads to (*7*) phosphorylation of targeted cellular proteins, and (*8*) the intended intracellular response.

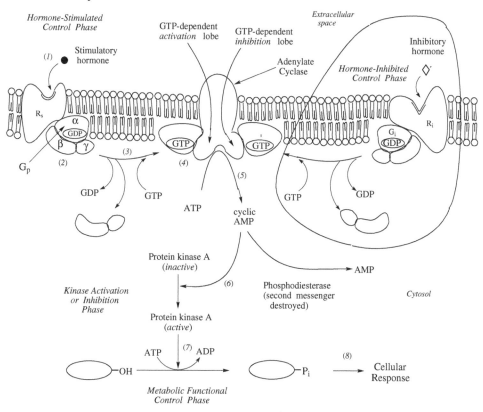

(Adapted from Moran *et al., Principles of Biochemistry* (5th ed.), 2012, p. 288; Fig. 9.46.)

Inhibitory Pathway. This enzyme can also, in specific cases, be inhibited. This is variation on the G-coupling theme is mediated by an *inhibitory G protein* (G_i). When a hormone binds to an *inhibitory receptor* (R_i) in the membrane, it binds adenylate cyclase through the G protein, thereby inhibiting the enzyme. The basic pathway is the same as activation. The specific signal molecules and intracellular targets are different, depending on what is being controlled.

Some hormones or cellular conditions stimulate the stimulatory pathway, others induce the inhibitory route. Cellular development, disease progression and many means of physiological control depend on this type of peripheral membrane signaling system.

15.3 The Inositol-Phospholipid Signaling Pathway

The steps are summarized below:
(*1*) When a hormone binds to its *transmembrane receptor* (R),
(*2, 3*) it activates the *G-protein* G_p.
(*4*) Membrane-bound *Phospholipase C* (PLC) is *stimulated by the activated G_p* to catalyze hydrolysis of the phospholipid PIP_2, which is located on the inside wall of the plasma membrane.

The act of splitting PIP_2 into
(*5*) inositol-1, 4, 5-triphosphate (IP_3) and
(*5'*) diacylglycerol (DAG) begins the process of using second messengers to transport/diffuse the signal to their targets in the cell interior.

(*6*) IP_3 travels and diffuses into the endoplasmic reticulum, acting to open Ca^{2+} channels in the membrane that can only be accessed by IP_3.
(*7*) This releases the stored Ca^{2+} into the cell to produce cellular functions, specifically PKC (see 6').

(*6'*) Diacylglycerol is used in the plasma membrane to activate the *calcium-* and *phospholipid-dependent* enzyme *Protein Kinase C* (PKC).
(*7'*) The phosphorylated target proteins go on to catalyze, regulate, or otherwise affect various metabolic processes, thereby producing a wide variety of cellular responses.

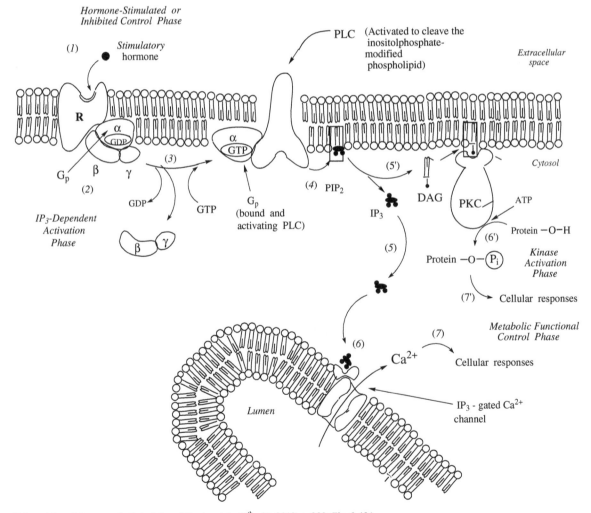

(Adapted from Moran *et al.*, *Principles of Biochemistry* (5th ed.), 2012, p. 289; Fig. 9.48.)

15.4 Phorbol Myristyl Acetate

The plant *Croton tiglium* stores the toxin *phorbol ester* in its leaves. This natural product mimics diacylglycerol (DAG), and thereby *activates* PKC.

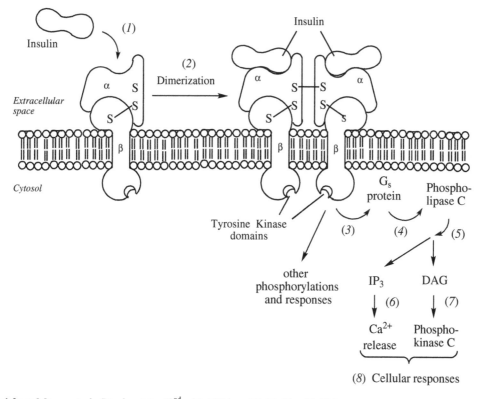

One difference between DAG and phorbol ester is that the toxin is metabolized inefficiently, and therefore *over-activates PKC. Phorbol myristate acetate* is a carcinogen and strong tumor promoter. This may be attributed to the extended activation of PKC, which leads to unrestrained growth of cancerous cells.

15.5 The Insulin Receptor

This *insulin receptor* complex is composed of two parts. The extracellular portion is made of two α chains which both have an insulin-binding site. The transmembrane portion is composed of two β chains, which each contain a cytosolic *Tyrosine Kinase* domain. Glycogen metabolism involves regulation by the activity of both *insulin* and the counteracting protein *glucagon*.

(Adapted from Moran *et al., Biochemistry* (2nd ed.), 1994, p. 12-46; Fig. 12-51.)

Binding of insulin to the α chains causes *autophosphorylation of tyrosine* residues in the intracellular domains of the insulin receptor. Tyrosine kinase also phosphorylates other cellular proteins, creating a cascade of results that produce the overall response to insulin binding.

Prior to use, the insulin monomers are transported through the circulatory system as a *hexameric zinc complex*. For details regarding (*1*) the medical connection between insulin and *diabetes*, and (*2*) the *post-translational* steps in the *maturation of insulin*, see the *Protein Maturation* section.

15.6 Glucagon

The action of glucagon, which ultimately is to stimulate glucose biosynthesis is discussed in the *Glucagon and Fructose-2, 6-bisphosphate* section of the *Gluconeogenesis* chapter. The protein is interesting because it illustrates the multi-faceted binding capability of a single polypeptide. Glucagon is a symmetric dimer that contains: (*1*) a site that binds glycogen or oligosaccharides, (*2*) a catalytic site that binds glucose-1-phosphate or glucose, and (*3*) an allosteric site that binds glucose-6-phosphate, AMP and ATP.

Table 4. Summary of Signals and Target Functions in Signal Transduction

Pathway	Second Messenger(s)	Function Activated
cAMP	Cyclic AMP (cAMP)	Protein Kinase A
Inositol-P-lipid	G-proteins	Phospholipase C (PLC)
Inositol-P-lipid	PIP_2 (IP_3)	IP_3-Gated Ca^{2+} Channel [1]
Inositol-P-lipid	DAG, G-proteins	Protein Kinase C (PKC)
Inositol-P-lipid	Ca^{2+} release	Blood clotting, glucose metabolism, *etc.* [2]
Insulin	tyrosine autophosphorylation	Tyrosine phosphorylation
Glucagon	Fructose-2, 6-bisphosphate	Gluconeogenesis

[1] For example, the released calcium can inhibit glycolysis. Calcium released from the lumen of the ER activates Glycogen Phosphorylase Phosphatase. This traps Glycogen Phosphorylase *a* in the dephosphorylated inhibited form and stops the release of glucose-1-phosphate from glycogen. As a result, the IP_3 signal can turn off the glucose fuel supply.
[2] Calcium also activates Isocitrate Dehydrogenase allosterically, thus stimulating the Krebs Cycle.

15.7 G-Proteins

Range of Occurance. G-proteins mediate the control of cellular responses to external stimuli. They mediate functions such as intermediary metabolism, cell growth and division, secretions, epidermal platelet function, nerve growth, movement toward or away from a chemical stimulus and immune cell (interleukin) function. Some specific examples are:

(*1*) *Transducin* (G_t) mediates visual excitation, after the retinal isomerization reaction occurs.
(*2*) G_k regulates potassium channels in heart muscle.
(*3*) The *β-* and *α-adrenergic receptors* in heart muscle stimulate G_k protein.
(*4*) *Somatostatin* regulates cell growth.
(*5*) *Epinephrine* (adreneline) regulates growth and "fight or flight" responses
(*6*) The insulin response

Targeting Agents. G-proteins are targeted by a variety of malicious and benign effectors. Examples are:
(*1*) *Cholera*. The toxin produced by cholera catalyzes *ADP-ribosylation* of an arginine in GTPases, rendering them inactive in signal transduction.
(*2*) *Whooping cough* (pertussus) toxin operates via the same mechanism.
(*3*) *Caffeine, theophylline* and *theobromine* are various N-methylated analogs of guanine. Since they mimic GTP, they bind and stimulate G-proteins. The result is a set of natural stimulants that are remarkably benign in humans.
(*4*) Phorbol esters promote tumor formation.

Chapter 16

Nucleic Acids: DNA

16.1 DNA and RNA

Discovery. DNA was first isolated and characterized by Frederick Meischer using pus collected from battle wound dressings from the Prussian war.

Structure. Watson and Crick used *x-ray diffraction* data from Rosalind Franklin, and A•T , G•C ratios from Irwin Chargaff, to surmise that DNA is composed of two intertwined strands, the famous *double helix* structure. They proposed that DNA is stabilized by *hydrogen bonding* between A and T, and between G and C in the respective A·T and G·C *base pairs.*

(*1*) These polymers are composed of *ribose* or *deoxyribose*, alternating with *phosphate linkers*. Strand-to-strand binding is mediated by *base pairing,* that is, hydrogen bonding between the centrally located bases.

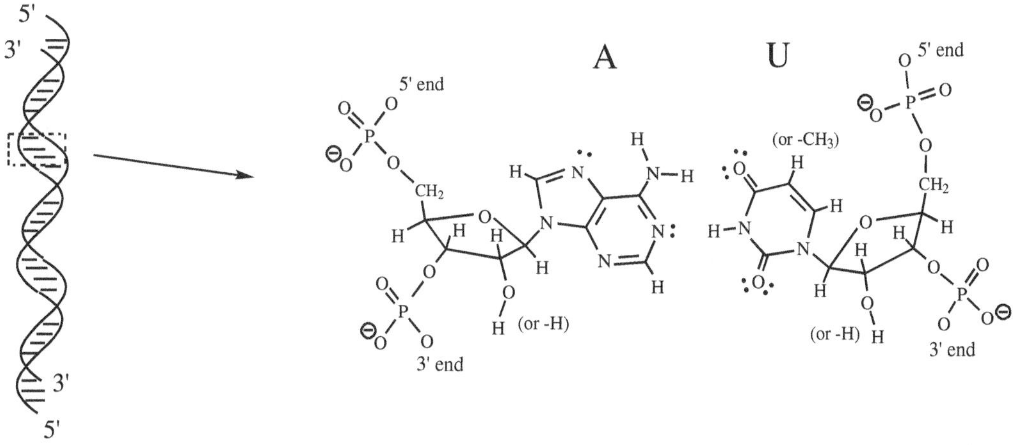
5'
3'
3'
5'
A
U
5' end
3' end
(or -H)
(or -CH3)
5' end
3' end

(*2*) Hydrogen bond interactions occur between the ring substituents to form A•T and G•C base pairs.

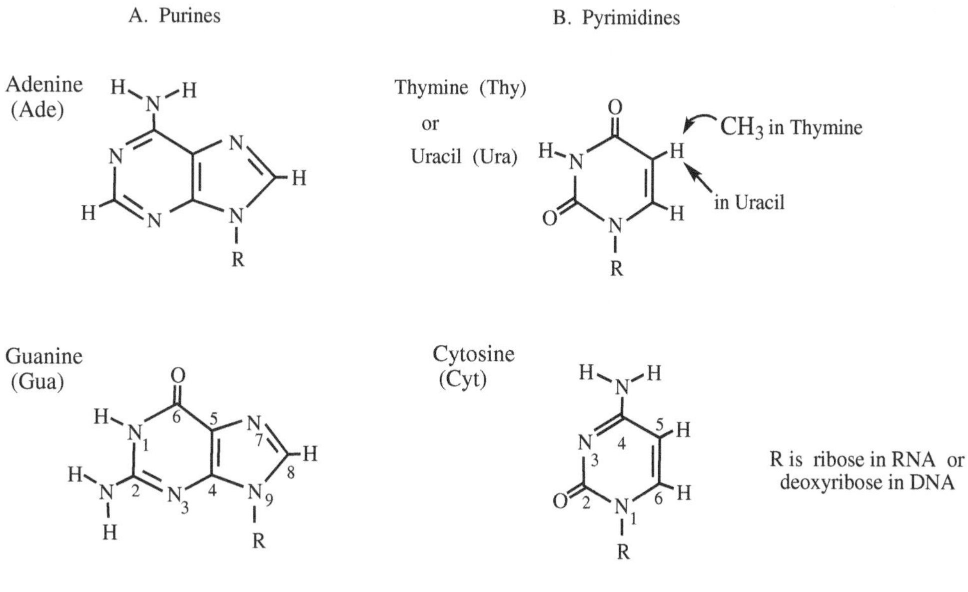
A. Purines
B. Pyrimidines
Adenine
(Ade)
Thymine (Thy)
or
Uracil (Ura)
CH3 in Thymine
in Uracil
Guanine
(Gua)
Cytosine
(Cyt)
R is ribose in RNA or
deoxyribose in DNA

(*3*) The deoxyribose or ribose *C1'* atom is attached to the base ring nitrogen to form the *glycosidic bond.*

Functional Roles

(*1*) *DNA* is a *stable* reservoir of genetic information. Hershey demonstrated that DNA carries the genetic information. Implanting cells with DNA, *transforming* them, can impart the new *genetic phenotype.*

(*2*) In *RNA*, the 2'-hydroxyl group catalyzes *alkaline hydrolysis*, making the chain readily subject to breakage. As a result, RNA *turns over* quickly. It is a transient messenger source derived from the DNA. This allows regulation of mRNA levels. The cell can make a lot of enzyme when the mRNA is initially made then can degrade the mRNA easily when it's served as the template for translation.

16.1.4 DNA Sequence Presentation

Watson-Crick double helices have a *5' to 3' strand polarity.* The best way to show a DNA sequence depends on what type of information is to be conveyed. Five common variations are shown below

DNA sequence	5' - A C G T A T T G G G G T T G G G G T T - 3'
Complementary RNA sequence (lower case letters)	3' - u g c a u a a c c c c a a c c c c a a - 5'

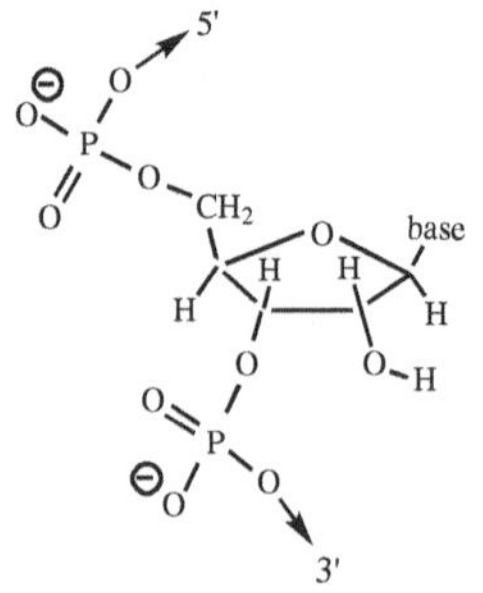

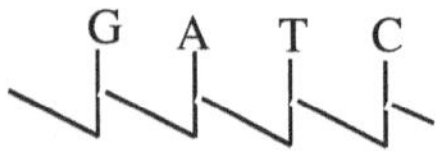

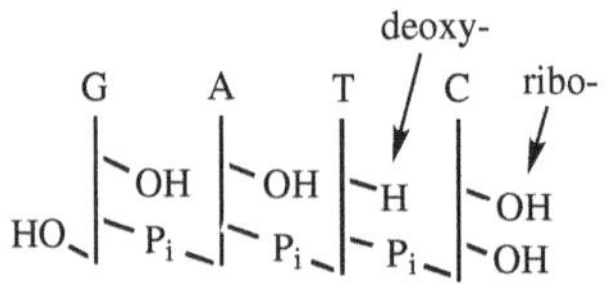

Some others common representations are: (*1*) as letters: GATC, and (2) as the 5'-phosphorylated pGpApC$_{OH}$. (*3*) Purines are abbreviated as Pu or R; pyrimidines are Py or Y. Using these codes, a particular sequence can be written as: PuPuPyPY or RRYY.

Combinatorial DNA/RNA Synthesis

This nomenclature can be used to show a set of *sequences* that *vary at specified positions. Combinatorial permutation* sequences are really more than one DNA. They only vary at a single position. This can be accomplished using synthetic DNA primers in the *Polymerase Chain Reaction* (PCR) technique. When the primers are made, one programs the DNA synthesizer to insert all of the intended nucleotides into a particular step. Since one draws from more than one bottle of reactant, the next position in the synthetic product attached to the solid support matrix is more than one nucleotide (* below).

Coding DNA sequence		5' - A C G T A T T G G G G T T G G G G T T - 3'	
		*	
Product DNAs	*1*	3' - T G C A T A A C C C C A A C C C C A A - 5'	*wild type*
	2	3' - T G C A T A A C C C C T A C C C C A A - 5'	*mutants*
	3	3' - T G C A T A A C C C C G A C C C C A A - 5'	

When the population of molecules is cleaved from the synthetic column, more than one molecule, varying only at the intended position, is recovered. If three bottles were drawn from during one synthetic step, three sequences are made, and so on.

16.2 Physical Properties

Ultraviolet Absorbance of the Bases. Light is *absorbed* strongly by nucleic acid bases at 260 nm. Their *absorbance* occurs with a large *molar extinction coefficient,* so they can be detected with high sensitivity. This property has been used in many ways.

(1) Absorbance approaches provide a convenient way to measure and study the equilibrium poise between single- and double-stranded nucleic acids.

$$\text{2 complementary single-stranded DNAs} \quad \overset{K_a}{\rightleftharpoons} \quad \text{1 duplex DNA}$$

(2 ss DNAs) *[higher A_{260}]*   (ds DNA) *[lower A_{260}]*

Other names for the reversible process are *duplex-to-single-strand equilibrium, denaturation-renaturation* equilibrium and *duplex melting* reaction. *Duplex formation* is called *annealing* and *hybridization* in many molecular biology procedures, for example, PCR. Duplex *dissociation* is typically called *denaturation.*

(2) The Beer-Lambert Law (Beer's Law) allows one to convert measured A_{260} values to molar nucleic acid concentrations.

$$A_{260} = \varepsilon_{260} \, b \, c \tag{1}$$

The parameters are:

A_{260}	absorbance at a wavelength of 260 nm (A_{260})
ε_{260}	the 'molar extinction coefficient' (units: $M^{-1} \, cm^{-1}$)
b	the cuvette pathlength (which is typically 1 cm)
c	the molar concentration

Duplex Stability Measurements are used to study the denaturation of double helix (duplex) structures. This approach can be used to determine *equilibrium association constants* (K_a) for different designed duplex structures. By measuring K_d values at a series of temperatures and DNA strand concentrations, one can determine the $\Delta G°$, $\Delta H°$ and $\Delta S°$ values for duplex denaturation with the specified molecule or molecules.

(1) The stability of a duplex DNA varies with the percent G + C. To do this, chromosomal nucleic acid are *sheared* to 200–1000 base pair lengths for measurements. The stability can be calculated in terms of the *melting temperature* (T_m), the denaturation-renaturation midpoint, as follows.

$$T_m = 69.3°C + 0.41 \, (\% \, G + C) \tag{2}$$

A larger G-C content is correlated with a higher T_m.

(2) Cooperative Melting. When a duplex dissociates, loss of the first base pair makes the next one dissociate more easily. This *cooperativity* makes many duplex dissociation events very rapid. Conversely, formation of the first base pair association nucleates subsequent base pairing. Duplex formation is also a cooperative process.

(*3*) To reinforce the points of similarity, the following plots compare *melting curves* obtained with two different classes of biomolecules:

(*i*) *Nucleic acid*: *duplex-to-single-strands* dissociation-association equilibrium of a simple *DNA*

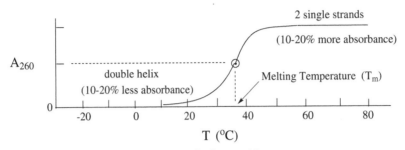

(*ii*) *Lipid*: *liquid crystal-to-dispersed equilibrium* of a fatty acid

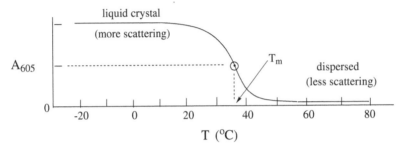

This effect has been used to characterize the energetics of defined nucleic acid structures. Details are described below in the *Secondary Structure Predictions* section.

16.3 Secondary Structure

Secondary Structure Maps
The base pairing pattern, the *secondary structure*, formed by one long RNA strand is shown below as: (*1*) a three-dimensional view, and (*2*) a two-dimensional *secondary structural map*.

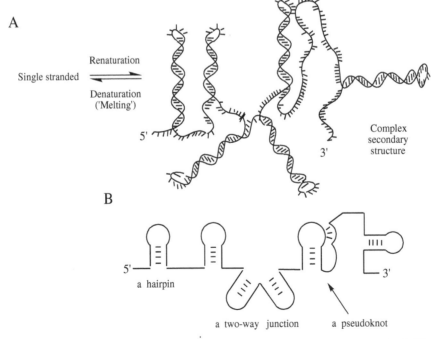

The large secondary structures of, for example, ribosomal RNAs and virus *genomes*, look like complicated city roadmaps.

Empirical Predictions. The data base obtained from duplex melting studies can be used to calculate the best predicted secondary structure, as well as close next-best variants. In mathematical models, a given secondary structure is represented by placing the backbone of the structure on an arc. Base pairs are represented as lines (chords) that connect the two respective nucleotide positions on the arc. Different secondary structural variants have different specific chord patterns, which represent the base paired secondary structure, when folded together. *RNA secondary structure analysis programs* can be downloaded from the Web.

16.4 Backbone Structure

The following tetranucleotide structure shows the hydrogen bonding sites. The phosphodiester phosphates and ribose C1' atoms link the base sequence and backbone together.

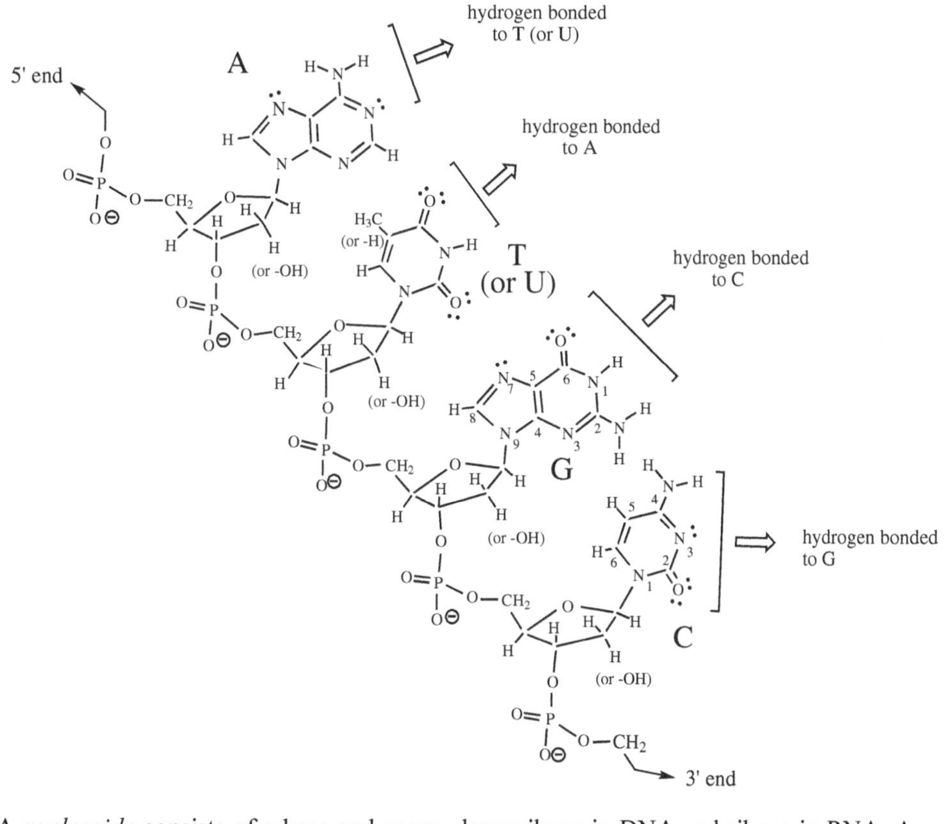

A *nucleoside* consists of a base and sugar, deoxyribose in DNA and ribose in RNA. A *nucleotide* is the base, the sugar and one or more phosphate(s)—connected to any of the *hydroxyl* positions. The following figure compares a *carboxyester* functional group, which is a monoester, with the *phosphodiester* bond.

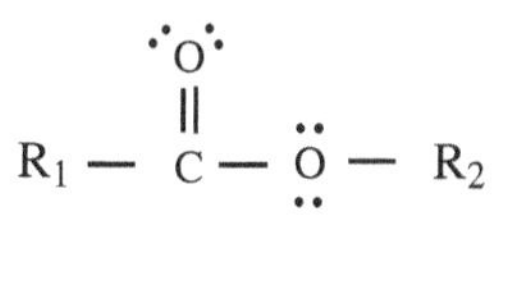

Carboxymonoester 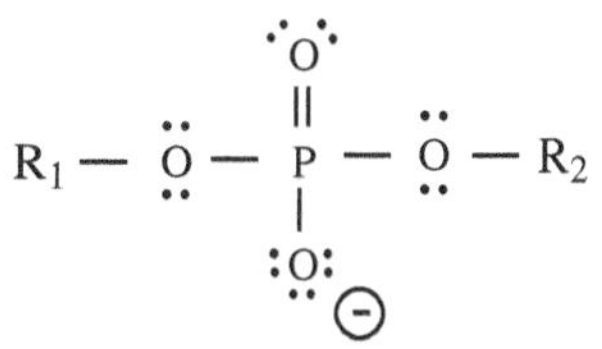   *Phosphodiester*

The phosphate is attached to the sugar at the *3' and 5' oxygen* atoms to produce the interconnected series of *phosphodiester bonds*. In some cases, the *2' oxygen* also binds a phosphate to form a phosphodiester bond. In DNA, the 2' position contains hydrogen, not a hydroxyl group. This makes the chain resistant to breakage via base-catalyzed hydrolysis, which happens easily with RNA (see the *Alkaline Hydrolysis of RNA* section for the mechanism).

16.5 Counterions

Nucleic acids are highly electronegative (*polyanionic*), however the charge is substantially neutralized by bound *counterions*, most commonly Mg^{2+}, Na^+, and K^+. The pK_a for protonation of the phosphodiester oxygen is reduced to below 6 in the polymer, so *one negative charge* is present *per nucleotide* residue. There are two negative charges per base pair. *Counterions* are *atmospherically bound*, exchanging with solvent cations multiple times each millisecond. They are an intrinsic, but transient, part of the nucleic acid structure. *Manning-Record Counterion Condensation* theory is used when considering the charge-dependent properties of nucleic acids.

16.6 Chemical Synthesis

DNAs are synthesized using *phosphoramidite* chemistry via *solid phase synthesis*. Chemical synthesis of RNA is more difficult but can be done, however enzymatic synthesis is usually the method of choice when making specific RNAs.

16.7 Watson-Crick Base Pairs

A·U and *A·T*. The hydrogen bond patterns in the A·U and A·T pairs are shown below. The RNA version is shown. In DNA, the C5 atom of *deoxyuridine* (dU) is methylated to form *deoxythymidine* (dT).

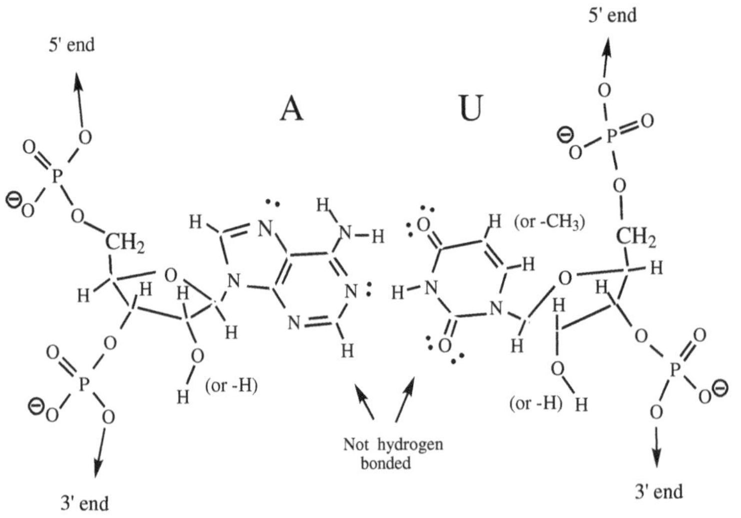

Characteristics are:
(*1*) Two hydrogen bonds form between (*i*) the amino H and carbonyl O, and (*ii*) the imino H and the ring N lone pair.
(*2*) A·T and A·U are more *hydrophilic* than C·G.
(*3*) The A·U in RNA is more *hydrophilic* than A·T in DNA for two reasons:
    (*i*) the 2'-hydroxyl occurs in uridine, while -H occurs in deoxythymidine
    (*ii*) the C5 atom on U is hydrogen; it is a CH_3 in dT.

Drawing Proper Strand Polarities. DNA looks like a ladder. Drawings of base pairs are flawed because they don't convey the true geometry. The *base pairs* shown here are shown in the *plane* of the page. Unlike what is shown, the 5' end of one strand projects directly *into* the page, while the other 5' end projects *out of* the page. This type of strand polarity is called "*antiparallel*." In some special cases, certain DNAs can form *parallel-stranded* structures.

G·C. This base pair is shown as an RNA fragment.

Characteristics are:

(*1*) Three H-bonds form: (*i*) carbonyl O to amino H, (*ii*) imino H to ring N, (*iii*) amino H to carbonyl O.

(*2*) The G·C pair is more *hydrophobic* then A·U.

Molecular Recognition Patterns. Note the differences in H-bonding interfaces in the AT and GC pairs. Specific charge complementarity and proper spatial apposition of the functional groups occurs, and it's different in the two cases. This represents an extended encoded structural message. The details are "written" in terms of charge and shape surface architecture. This is literally what is "read" by proteins in the midst of genetic recognition, decoding, and maintenance in a given biological niche.

16.8 Structural Modifications

Modified Bases in DNA

A number of base modifications occur in DNA. In one remarkable case found in bacteriophage DNA, an entire glucose is connected to the C5-methyl group of thymidines!

The C5 position of cytidine in many eukaryotic chromosomal DNAs is methylated within *GpC Islands* sequences. The presence of this nucleoside, *5-methylcytidine* (m⁵C), is correlated with *epigenetic gene inactivation* in *developmentally specialized cells*.

The Hoogsteen Base Pair: A Rare Tautomer of Adenosine

This base pair contains (*1*) the imino tautomer of adenine and (*2*) the unusual *synclinal* conformation about the *glycosidic bond* (see below).

imino A   *syn* G

The Hoogsteen variant is an unusual, but biochemically significant, base pair. It illustrates how a rare tautomer can lead to the *insertion of a mutation*. Since imino A *decodes in replication* like C, a G would be inserted into newly replicated DNA, instead of the correct T.

The guanosines in *left-handed Z-DNA* also adopt the *syn* G conformation, in which the base is rotated to a position above the sugar. The *syn* G conformation is not as stable as *anti* G due to repulsion between the partially negative charges of N3 and G O1'.

Modified Bases and Base Pairing in RNAs

Many examples of *modified base pairs, triples and other assemblies* occur in *Transfer Ribonucleic Acids* (tRNA), the adapter molecules used to synthesize proteins in translation. Some common examples of modifications are 5, 6-dihydrouridine, pseudouridine, N6-isopentenyladenosine, 7-methylguanosine, and uridylate-5-oxyacetic acid. One particularly complicated modification found in the phenylalanyl tRNA in yeast is called wyebutosine, a fluorescent 3-ringed base.

Ribose Methylation. Ribose is sometimes modified to produce *2'-O-methylribose*, which *short-circuits alkaline hydrolysis* of the RNA backbone. Two examples are the *cap* structure in mRNA and a common modification located adjacent to the anticodon sequence in tRNAs.

16.9 Three-Dimensional Structures

Three canonical duplex DNA structure, the *B, A* and *Z* forms, are shown and described below.

B-DNA. The following figure shows side and top views of B-DNA, in ball-and-stick, space-filling, and line formats. This is the classical structure proposed by Watson and Crick base on Rosalind Franklin's X-ray fiber diffraction data.

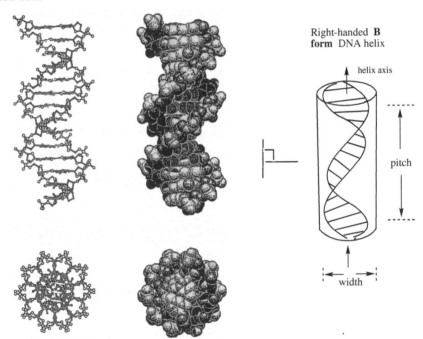

(Adapted from Saenger, W., *Principles of Nucleic Acid Structure,* Springer-Verlag, 1984, p. 262; Fig. 11-3.)

Characteristics:
Most DNA sequences prefer the B form.
Characteristics of B-DNA are:
(*1*) 10 base pairs (bps) per turn; 3.4 Å per base pair (bp) rise.
(*2*) The bps are almost perpendicular to the helix long axis.
(*3*) The B helix is slightly narrower and more stretched out than A-form.

(*4*) Unlike the A-form, the B duplex has *no* central axial cavity.

Sugar Pucker. Deoxyribose has a C-2' *endo* sugar pucker. The *glycosidic torsion angle* between the sugar and base is *anti*.

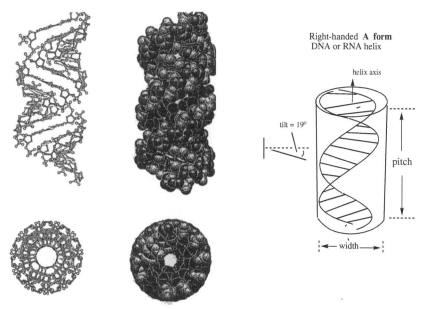

The following figure shows sugar puckers formed by *deoxyadenylate* in B- and A-form DNA, respectively.

(Adapted from *Structure and Conformation of Nucleic Acids and Protein-Nucleic Acid Interactions,* (M. Sundaralingam and S. Rao, Eds.), 1980, University Park Press, p. 487.)

A-DNA. The canonical A-form structure appears below. Note the shapes of the major and minor grooves; the major groove is more deep and narrow.

(Adapted from Saenger, W., *Principles of Nucleic Acid Structure,* Springer-Verlag, 1984, p. 257; Fig. 11-2.)

Characteristics: The A form is preferred in RNA, which does not adopt the B form.
   (*1*) There are 11 bps per turn; with a 2.3 Å per bp rise.
   (*2*) The helix is wider than B form; not as stretched out (slightly more stout).
   (*3*) The bps are tilted 19° from perpendicular to the helical long axis.

Sugar pucker. The ribose has a C-3' *endo* sugar pucker.

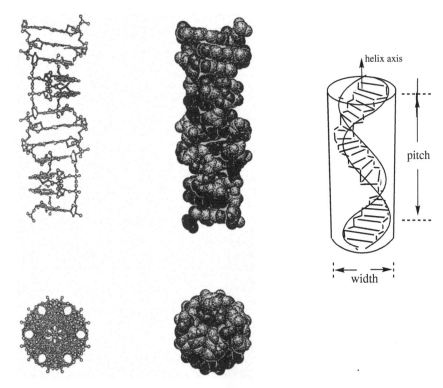

Z-DNA. The molecular structure of left-handed poly(dGdC)·poly(dGdC) is based on coordinates for Z_1-DNA derived on the basis of the hexanucleotide d(CGCGCG) in side and top view. The *zig-zagged* alternating phosphodiester backbone conformations give the molecule its name.

(Adapted from Saenger, W., *Principles of Nucleic Acid Structure*, Springer-Verlag, 1984, p. 286; Fig. 12-2.)

Characteristics:
    (*1*) The duplex is left-handed.
    (*2*) Z-DNA has 12 bp per turn and a 3.8 Å per bp rise. The helix is more stretched out than B-DNA.
    (*3*) It is longer and narrower than B- and A-DNAs. The width is 18.4 Å. Those of the B and A forms
        are 23.7 Å and 25.5 Å.
    (*4*) The base pair tilt angle (9°) is half-way between those of A- (19°) and B- (1°) DNAs

Sugar Pucker. The Gs adopt *syn* glycosidic torsion angles, while the cytidines are in the *anti* conformation. The alternating base-sugar juxtapositions are accommodated by alternating sugar puckers and the Z-formed backbone.

Zig-zagged Backbone

Two different phosphodiester conformations exist. The Z-DNA is stabilized by > 2.7 M NaCl and >10 mM $MgCl_2$, both of which relieve charge repulsion between closely juxtaposed backbone phosphates.

Structure of a G•C base pair in Z-DNA is shown below. The sugar adopts a C-2' *endo* conformation, and the base is in the *anti* structure, as in B-DNA.

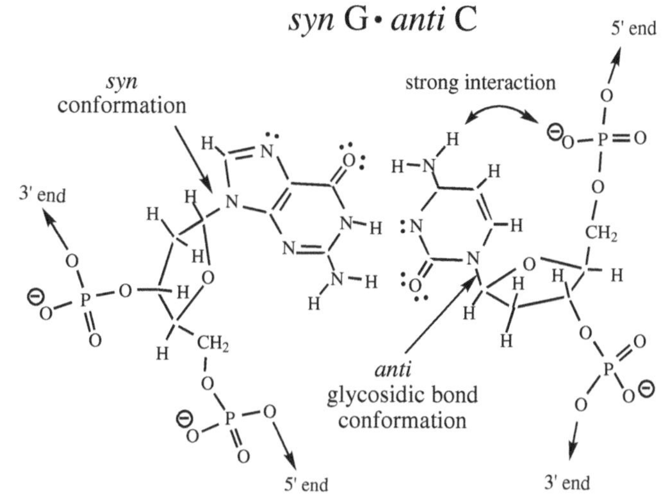

syn G • *anti* C

(Adapted from Saenger, W., *Principles of Nucleic Acid Structure,* New York, Springer-Verlag, 1984; p. 287.)

(*1*) In solution, the *anti* conformation usually predominates in free nucleosides. Guanine nucleotides can adopt the *syn* conformation because favorable electrostatic attraction occurs between the C-2 amino group and the 5'-phosphate.

(*2*) Synthetic alternating purine-pyrimidine sequences form Z-DNA at elevated salt concentrations. In contrast, the presence of negative supercoils in plasmid DNA that contains an alternating CG motif can induce Z-DNA formation at *in vivo* salt concentrations.

The Structural Parameters of A-, B- and Z-DNA are summarized below.

Property	A-DNA	B-DNA	Z-DNA
Helix handedness	Right	Right	Left
Repeating unit	1 base pair	1 base pair	2 base pairs
Rotation per base pair	32.7°	34.6°	30°
Base pairs per turn	+ 11	+ 10.4	- 12
Inclination of base pair to the helix axis	19°	1.2°	9°
Rise per base pair along the helix axis	0.23 nm	0.33 nm	0.38 nm
Pitch	2.46 nm	3.40 nm	4.56 nm
Diameter	2.55 nm	2.37 nm	1.84 nm
Glycosidic bond Conformation	*Anti*	*Anti*	*syn*
Sugar Pucker (Ring Torsional Conformation)	C-3' *endo*	C-2' *endo*	C-2' *endo* at C C-3' *endo* at G

(Adapted from Moran, *et al.*, *Biochemistry* (2nd ed.), 1996, p. 24-22; Table 24-3.)

16.10 Recognition of Sequences

Sequence Microheterogeneity. Structural studies have shown that DNA is really a mixture of B and A forms. An example is the dodecamer sequence from the *lac repressor binding site* characterized by Richard Dickerson and colleagues using x-ray crystallography.

Sequence Microheterogeneity refers to the presence of many subtle structural handles on one DNA sequence. The grooves contain different functional groups, which are read when proteins bind to their recognition sequences. The surfaces and charge characteristic encode the recognition information.

To recognize a sequence, a DNA-binding protein must be able to tell how it differs from the structure of another incorrect potential binding site sequence. This means it must be able to differentiate correct and incorrect DNA surface architectures in terms of charge, shape, hydrophobicity, and so forth.

Bent DNA. The detailed structure depends strongly on local sequence details. For example, specific *adenine-enriched* sequences can make DNA form a structure called *Bent DNA.* Adopting the bent form involves formation of a bound *spine of waters* in the *grooves.*

16.11 Genetic Mutations and Antisense Nucleic Acids

Genetic Mutations
Incorrect nucleic acids are rarely made by replication. Despite the *proofreading* ability *of DNA Polymerase,* mutations occur regularly in human cells. Rare tautomers, *e.g.,* the Hoogsteen base pair described above, occur at a rate of ~ $10^8 s^{-1}$. We have extensive *DNA repair* systems to remove and replace incorrect or structurally damaged nucleotides.

Some mutation is useful. Mutations can lead to three outcomes, (1) evolutionary improvement, (2) no effect (a *silent mutation*), and (3) genetic damage that is functionally apparent (*e.g.,* disease, *temperature-sensitive mutation*). The latter can be inherited. Cancer produces the disease after sufficient time has passed for the essential mutations also accumulate and produce their effect. For example, a series of specific mutations occurs in the pathway that leads to *colorectal cancer.*

Antisense Duplex RNA
The typical mRNA is translated to form a protein. The strand produced for this purpose has as its encoding progenitor the sense DNA strand, which produced the sense mRNA strand.

Antisense RNA is the sequence that is complementary to the normally transcribed mRNA. If antisense RNA' is made, it can bind the mRNA and form a [mRNA·antisense RNA] duplex. This renders the mRNA unusable in translation, which prevents formation of the encoded protein. The *antisense RNA inhibits gene expression.* The antisense approach is used naturally by bacteria to limit the number of copies of *plasmid DNA* it makes, which is called *copy number control.*

Antisense Triplex DNA
This idea is used in techniques designed to bind a *third* DNA (or RNA) strand to a duplex DNA, thereby rendering the duplex sequence incapable of use in transcription. *Triple helix* (triplex) structures can form in three ways:

Gene Therapy involves using designed nucleic acids to targeted a specified gene function for control by binding the antisense nucleic acid. The advent of *Small Interfering RNA* technology is an extension of this idea. The latter approach is described in the named section in the RNA chapter.

16.12 Unusual DNAs

Triplex DNA

Duplex DNAs can bind a third strand via base pairing to form a triplex DNA. Sequences that form "triplexes" are common in the upstream control regions of eukaryotic genes. Triple helices always have both parallel and antiparallel strand-strand pairing interactions.

Base pairing schemes are shown in the *Antisense Triplex DNA* section. Complexes can form from three strands, two strands, where one is a hairpin duplex structure, and from a single strand that has two hairpin turns. The three base triples [T-A~T], [C-G~C$^+$] and [C-G~G] have been studied in triplex DNAs. The ability of a given triplex to form depends on sequence, cation type, and temperature.

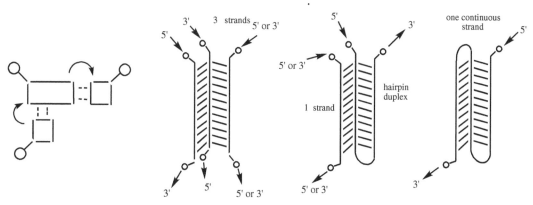

Quadruplex DNA

These structures contain strands arranged in 4-fold symmetry about the central axis.

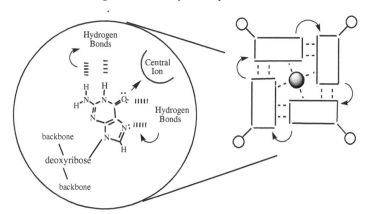

Variants. Synthetic *quadruplex* DNAs can adopt either parallel or antiparallel structures, depending on the sequence, cation conditions, temperature, and the presence of certain proteins.

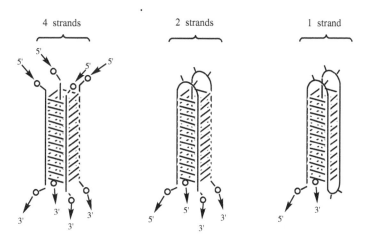

Certain G-rich strands can form both triplex and quadruplex structure, depending on the sequence and ionic conditions. (For details, see Hardin *et al.*, Biopolymers, *Nucleic Acid Sciences*, 2001, 56, 147–194.)

Biological Importance
Quadruplex DNA has received attention in its role as a protein-binding *aptamer*. A specific quadruplex binds the *blood clotting factor Thrombin*. This interaction inhibits binding of thrombin to *fibrinogen*, thereby *inhibiting blood clotting*. The structure of the protein-quadruplex complex has been determined by x-ray crystallography.

Guanine-rich sequences are very common in the *telomeres* of chromosomes (see below), the *GpC Islands* in transcriptional promoters, and immunoglobulin gene-switching sequences. An HIV coat protein binds HIV RNA and makes it form a quadruplex.

Cruciform DNA
This structure is formed from an extended duplex DNA fragment composed of a *palindromic DNA*, which is also called an *inverted repeat*. The two-step *extrusion* process is shown in the figure below.

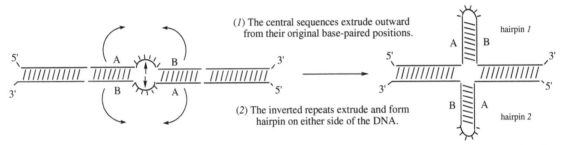

(1) The central sequences extrude outward from their original base-paired positions.

(2) The inverted repeats extrude and form hairpin on either side of the DNA.

This sequence is cut by a *cruciform DNA-dependent nuclease*. More complicated extra-duplex bonding arrangements called *Gierer trees* can also form.

16.13 Stabilization of Nucleic Acids

Contributions. The following factors contribute to the stability of nucleic acids:
(1) The *hydrophobic effect* forces bases inward. It constitutes about two-thirds of the duplex stability. A large gain in stability occurs when the H_2Os surrounding the strands maximize hydrogen bonding, which squeezes the strands together.
(2) Base pairs provide $\leq 1/3$ of the stability.
(3) The order of hydrophilicity of the bases is: U (T) > C > A > G; the latter is the most hydrophobic.
(4) About 10–20% of stability is due to van der Waals forces (London dispersion forces). These forces are the result of dipole/induced dipole interactions between the nearby atoms.

Adjacent base pairs undergo *stacking* interactions, like coins in a roll.

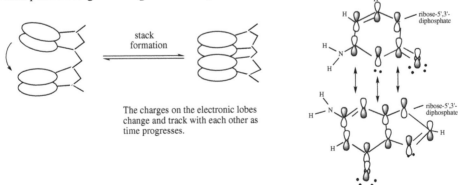

stack formation

The charges on the electronic lobes change and track with each other as time progresses.

*London dispersion force*s occur between π electron orbitals on the lobes of atoms in *adjacent stacked bases*. The electron densities in the orbitals switch back and forth, correlated in time with each other, to produce a bond that is negotiated transiently. They are *transient dipole-dipole interactions*.

Phosphate Groups. Metastability

(*1*) Buckminster Fuller developed the concept *tensegrity*, in which stability is gained at the expense of a tense structural standoff. For example, many modern mountain tents are held up by the tense tentpoles, held taut against the reinforced pockets they fit into.

The phosphodiester phosphates on each strand are *polyanionic,* so they naturally repel each other. This repulsion is held in check by counterions, as we'll see below, so this stability of the DNA is a tense standoff. DNA has tensegrity. This is also a case of *metastability*. The helix is stable, but given the chance to break, it might dissociate.

(*2*) The strands are fairly easily separated. If that were not true, transcription and replication would be more difficult because the polymerases could not separate the strands easily. Strand-strand repulsion in DNA supports transcriptional gene expression. Specialized single-stranded binding proteins actually help the duplex separate to single-stranded form.

(*3*) *Counterions* offset this *electrostatic repulsion* between the strands. Physiological salt concentrations nearly neutralize the negative charges on the strands, but not completely. About 10% remains per phosphate. This result is predicted using *Manning-Record counterion-condensation* theory.

Hydration

(*1*) The grooves of DNAs are filled with water molecules, bound to the atoms of the backbone and base functional groups. The minor grooves in B-DNA are densely packed with water, which form bonds with adjacent water molecules as well as the exposed atoms of the bases. Water molecules in the major grooves hydrogen bond to phosphate oxygen atoms and the keto and amino groups on the bases. Approximately three water molecules bind to each phosphate group and about 20 total water molecules are present per nucleotide in DNA. B-DNA is stabilized as a result of the rigidity induced by these hydrogen bonds.

(*2*) Double-stranded DNAs are stabilized by hydrophobic interactions, base stacking, hydrogen bonding, and electrostatic repulsion.

(*3*) The *hydrophobic effect* is especially important. The purine and pyrimidine rings are typically buried inside of the double helix, inaccessible to the aqueous solvent. Similar to proteins, the entropy of DNA is increased by the decreased order (larger number of combinations) of water molecules in the vicinity of exposed hydrophobic functional groups of the bases. This high entropy is counteracted by binding of highly ordered water molecules, which hydrogen bond with base functional groups in the grooves.

(*4*) The structure of nucleic acids is not rigid. DNA is constantly *breathing*. a process in which the hydrogen bonds between base pairs constantly break and reform, allowing small degrees of "conformational heterogeneity" to occur in the chain.

16.14 Secondary Structure Predictions

Duplex Melt Measurements

Thermodynamics for association of DNA duplexes and RNA secondary structures have been assessed by methods that involve determining the melting temperature as a function of [strand]. The ΔG_a values calculated for the neighboring base pairs of a specific RNA secondary structure are shown below. Such calculations are used to predict the most stable structures.

Example: The following figure shows an application of data to evaluate a possible secondary structure for a 55 nucleotide fragment from R17 virus. The net calculated stabilizing ΔG is -21.8 kcal/mole at 25°C. All of the energies are in kcal mol⁻¹.

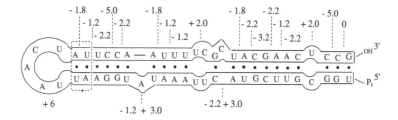

Verification of Predicted Structures

The predictive approach involves using a computer to look at how well evolutionary conservation occurs at specific sequences.

(*1*) A large number of *chemical modification,* and *chemical-* and *nuclease-cleavage methods,* have been used to study nucleic acids. The techniques allow one to verify the sequences and structures involved in particular secondary and tertiary structures.

(*2*) Gel electrophoresis-based *chemical protection* techniques and *DNA footprinting* have been used to determine which functional groups bind to binding proteins. Chemical modification and nuclease cutting experiments allow one to determine protected atoms on the nucleic acid, and thus the protein binding surface.

(*3*) *Alignment of sequences* is used on a daily basis to look for matches between one sequence and a set of sequences in some data base. Many applications have been developed in crime analysis, lineage determination, and so on.

Example: The DNA left in fecal deposits of Artic wolves has been used to track the territories of individuals and populations. Single wolves range over hundreds of miles, within days, looking for food.

G·U and U·U Base Pairs

The *mismatched G·U base pair* is a sturdy base pair. It is special and occurs as a *wobble base pair* in codon·anticodon interactions between mRNA and tRNAs. The ΔG_a of G·U is negative; it is a legitimate base pair. When evolutionarily conserved sequences are compared in the context of the known secondary structure the pair occur within, G·U is found to be conserved. Evolution knows and uses the energetic fact that G·U forms a stable base pair to accomplish biologically important pairing interactions.

Stacking Energies for the Ten Possible Dimers in B-DNA

DNA has different ΔG_a values than the analogous RNA. The following table gives numerical values for *all possible ways* in which two bases can pair in the context of defined *nearest neighbor* base pairs.

Stacked Dimers [a]	Stacking Energies (kJ mol^{-1})	Stacked Dimers	Stacking Energies (kJ mol^{-1})
↑ C•G / G•C ↓	-61.0	↑ T•A / A•T ↓	-27.5
↑ C•G / A•T ↓ ↑ T•A / G•C ↓	-44.0	↑ G•C / T•A ↓ ↑ A•T / C•G ↓	-27.5
↑ C•G / T•A ↓ ↑ A•T / G•C ↓	-41.0	↑ G•C / A•T ↓ ↑ T•A / C•G ↓	-28.4
↑ G•C / C•G ↓	-40.5	↑ A•T / A•T ↓ ↑ T•A / T•A ↓	-22.5
↑ G•C / G•C ↓ ↑ C•G / C•G ↓	-34.6	↑ A•T / T•A ↓	-16.0

[a] Arrows designate the direction of the sugar-phosphate backbone and point from C-3' of one sugar to C-5' of the next.

[Adapted from Ornstein, R. L., *et al.* (1978), An Optimized Potential Function for the Calculation of Nucleic Acid Interaction Energies: I. Base Stacking, *Biopolymers* 17: 2341-2360.]

16.15 Chromosomes

The genetic state of a chromosome and levels of gene expression are maintained by many *factor* proteins. These proteins are regulated by modification of amino acids, binding of specific proteins and nucleic acids, and many other moderating influences. The basic ideas are described below. For details, see books such as *Molecular Biology* by Weaver and *The Cell* by Alberts *et al.*

Required Features
Chromosomes must contain the following elements:
(*1*) *Histones* are protein complexes that bind chromosomal DNA like beads on a string.
(i) They are enriched in lysine and arginine.
(ii) The four types of proteins form an octameric core around which about 200 base pairs of native DNA wraps within chromosomes.
(iii) Control of the transcriptional activities of particular genes involves enzyme-catalyzed reactions that cause histone methylation, acetylation, phosphorylation, and others modifications. This is called the *Histone Code.* (Details are described below.)

(*2*) *Centromeres*. These structures are the site of spindle fiber attachment and contain specific unique sequence motifs.

(*3*) *Telomeres*. These are complex protein DNA complexes at the chromosomal ends.
(*i*) *Telomerase*. This RNA-protein replicase (a reverse transcriptase) carries its own C-enriched template RNA to make the complementary G-rich strand of telomeric DNA. Details are described below.
(*ii*) The enzyme maintains the ends of the chromosomes, which are required for long-term survival.
(*iii*) Telomerase is required for telomere maintenance, and is involved with cell aging and cancer.

(*4*) *Autonomous Replication Sequences* (ARS) are the internal start sites for the discontinuous *replication bubble* form of DNA synthesis found in eukaryotes.

(*5*) *Replicases*. The DNA polymerase complex makes DNA in a template-requiring process that produces two daughter duplexes from one maternal duplex sequence.
(*6*) *Transcription Promoters*. These are sequences such as the classic TATA box in prokaryotes; and typical GC-enriched sequences in eukaryotes.
(*i*) *Transcription Repressors*. In the *operon model*, they bind the *operator DNA* and thereby repress transcription
(*ii*) *Transcription Activators*. Also called *transcription factors* and *transactivator* proteins.
(*7*) *Translation*. *Protein synthesis* involves three types of RNA: ribosomal, messenger and transfer. Details are discussed in the RNA chapter.

Propagating DNA
The following example show a cross-section of the contexts in which DNA is propagated.

Yeast Artificial Chromosomes
YACs contain the necessary information to operate as chromosomes in yeast cells, providing a minimal model for chromosomal structure-function studies.

Plasmid DNA and Supercoils
    A plasmid is a closed-circular DNA that can be replicated within a host cell. It typically encodes an antibiotic resistance gene and has an "insertion site" in which one inserts specifically designed DNA fragments, such as gene sequences of interest.
    When the DNA is internally connected end-to-end, the possibility of forming different *topological isomers* exists. These species can have a different number of twists in the overall ring structure, called *supercoils*. A plasmid with more twists is considered to have a more *negative helical density*. A *relaxed plasmid* has a more *positive* helical density.

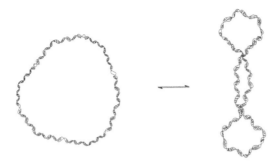

Supercoils have different intrinsic energies. The number of supercoils can be changed by enzymes called *Topoisomerases*. Relaxing superhelical stress is accomplished by type 1 Topoisomerases. The other type, called *DNA Gyrase*, uses ATP as an energy source to impart negative superhelical density, that is, add more supercoils.

The mechanism of the Type 1 enzymes involves making a nick, winding or unwinding of the duplex strand, then resealing the nick.

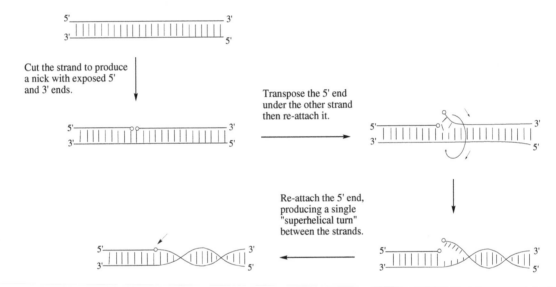

The type 2 enzyme (*1*) wraps a strand around itself, (*2*) breaks one of the two adjacent strand segments, (*3*) passes the other strand between the broken stranded gap, and then (*4*) reseals the gap.

Viruses

These species consist of a nucleic acid genome, which is circular or linear DNA or RNA, depending upon the virus. Some viruses require two circular species, for example, the plant Geminiviruses. Some require *helper viruses* to carry out infection. Some have several different functional forms. Details the complicated propagation system of HIV are discussed in the *Virus* section of the *Cell Biology Review* chapter.

"Rolling-Circle Replication" involves production of multiple connected genomes, which are later cut apart. Viruses usually have very few genes. They encode sufficient RNA or DNA, coat proteins, control factors, and so forth to integrate themselves into the host cell and occupy their genetic niche.

Telomeres

Telomerase Catalyzes Chromosome End Extension. The steps in the telomere buildup process catalyzed by telomerase are shown below.

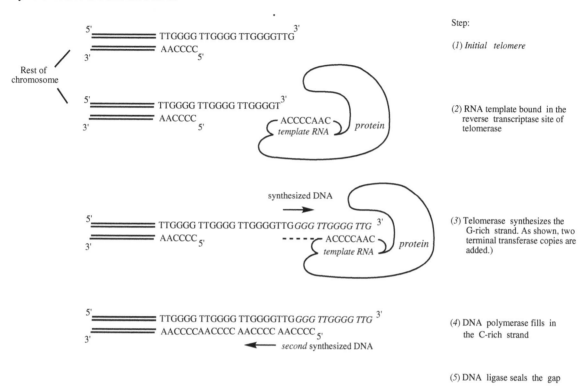

Step:

(1) Initial telomere

(2) RNA template bound in the reverse transcriptase site of telomerase

(3) Telomerase synthesizes the G-rich strand. As shown, two terminal transferase copies are added.)

(4) DNA polymerase fills in the C-rich strand

(5) DNA ligase seals the gap

For details see Osterhage, J. and Friedman, K. (2009) Chromosome End Maintenance by Telomerase, *J. Biol. Chem.* 284, 16061–16065.

The Telomerase [Protein·RNA·DNA] Complex. Chromosomal DNA is bound by the telomerase RNA template within the protein, which is analogous to a protein *hand* grasping the duplex RNA·DNA complex.

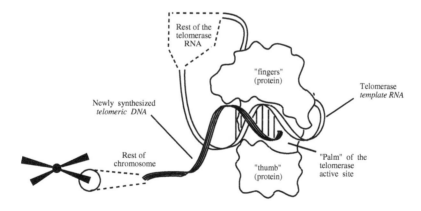

(Adapted from Alberts *et. al. Molecular Biology of the Cell* (4th ed.), 2002, p. 264; Fig. 5-43.)

Nucleosomes

Two molecules of each of the four *histone* proteins, H2A, H2B, H3, and H4, make up an organized octameric complex called the *nucleosome*. Each one is wrapped with about 200 base pairs of DNA for about 1.75 turns. This coiling of DNA around the core nucleosome causes positive supercoils to form in the remaining DNA which are relieved by eukaryotic topoisomerase II, an enzyme that binds to chromatin. Of the wrapped DNA, about 146 base pairs are closely associated with the histone octamer and make up the nucleosome core particle. After the removal of the nucleosides, the naked eukaryotic DNA has about one negative supercoil per original nucleosome.

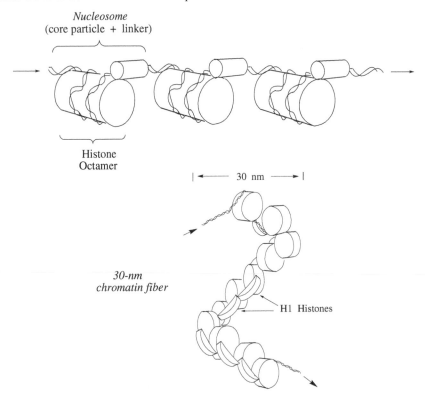

Linker DNA is found between each core particle and is usually about 54 base pairs in length. In higher-order chromatin structures, illustrated in the beads on a string structure, the *fifth histone, H1*, binds to both the *linker DNA* and the nucleosome *core particle*.

16.16 Some Protein Nucleic Acid Binding Motifs

Zinc Finger Proteins

These proteins bind the grooves of DNA. Two or more fingers are typically found per protein. Each finger has a central Zn^{2+} and four ligands composed of His and/or Cys residues.

Leucine Zippers

Protein-protein binding can occur between two *Leucine Zipper Proteins*. This allows two DNA-binding proteins to use their back sides to bind to each other while binding two separate DNA strands with the DNA-binding site.

Two different proteins could bind through their zipper domains to a single protein. The two dimers differ by which DNA sequence the interchanged protein binds. The net effect is that the dimer can select two different genetic activation situations. Only three proteins are required instead of four. The common protein plays the role of two proteins, depending on which partner it binds. This allows DNA binding proteins with different genetic functions to mix and match in order to bind different genes. This is an example of a *combinatorial binding* system.

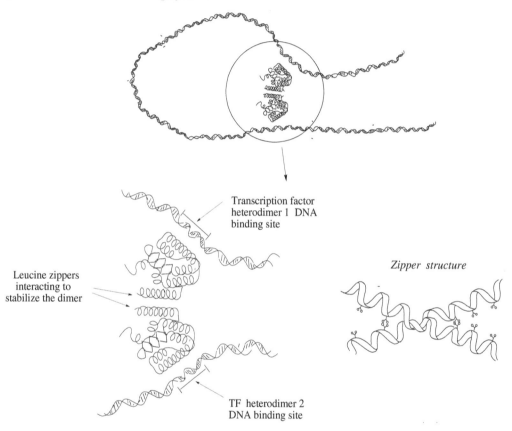

Transcription Factors (TFs) bind via hydrophobic interactions to each other to form heterodimers. This is a necessary step to activate transcription of the receptive promoter sequence. Different combinations of proteins can bind together: TF-1 to TF-2, TF-3; TF-4; and many others.

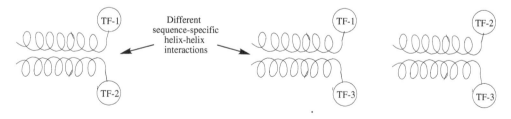

16.17 Recombination

This process is mediated by protein complexes. A well-studied example is the *RecA* complex. The process occurs at a four-stranded structure called a *Holliday Junction*. Such structures have been modeled using synthetic DNA fragments. They are similar to cruciform DNAs.

The *V(D)J Recombinase* mechanism is used to produce the *diverse array of antibodies* made by the *humoral immune system*. See Alberts *et al.* for detailed descriptions of each of these areas.

Chapter 17

RNA

17.1 Cells Contain a Variety of Types of RNA

DNA molecules act as the storehouse for genetic information. RNA molecules, in contrast, operate in several different forms to express the genetic information. Inside of a cell, RNA can be found in many copies for a particular purpose specific to its type. There are four major classes of RNA:

(*1*) *Ribosomal RNA* (rRNA) is incorporated into the ribosome. Ribosomes are made up of protein and RNA and are the site of protein synthesis. Ribosomal RNA makes up the majority of the RNA of a living cell, accounting for 80% of the total cellular RNA. It is the active catalytic focus of the ribosomal reactions. Most of the proteins can be removed yet the individual activities are retained.

(*2*) *Transfer RNA* (tRNA) transport activated amino acids to the ribosome to add to the growing peptide chain of the protein being made. Transfer RNA molecules are short in length, usually about 73 to 95 nucleotides, and make up about 15% of the total cellular RNA.

(*3*) *Messenger RNA* (mRNA) are responsible for coding the DNA into amino acid sequences to be transported to the ribosome for protein synthesis. They are the least stable of the RNA molecules and account for about 3% of the total cellular RNA.

(*4*) *Noncoding Small RNA* molecules also exist in all living cells. Most are found in the nucleolus portion of the nucleus. Some have catalytic functions dealing with proteins. After RNA is synthesizes, some small RNA molecules are active in the processing for modification of these. Some examples are small nuclear (sn) RNAs, small nucleolar (sno) RNAs, micro (mi) RNAs and silencer (si) RNAs.

17.2 RNAs Have Stable Secondary Structure

The difference between an RNA polynucleotide and a DNA polynucleotide is the presence of 2'-hydroxyl group and the replacement of the base uracil for thymine in DNA. These bases only differ slightly, with replacement of the methyl group at the 5 position of thymine by a hydrogen in uracil. The backbone of both RNA and DNA linked by 3' to 5' phosphodiester bonds.

RNA polynucleotides fold back on themselves to form *hairpin* structures composed of complementary base pairs in duplex regions and intervening loops. This is the essence of RNA *secondary structure*. One of these is a *hairpin*, or *stem-loop*, which is made when the folding causes short regions of complementary bases pair with one another. This is displayed in the *cloverleaf* secondary structure of transfer RNA.

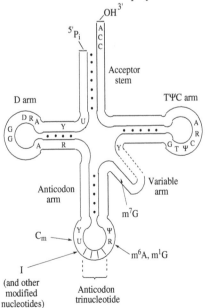

The dashed lines represent the Watson-Crick base pairs between nucleotides. Some nonstandard nucleotides are present and listed. R is a purine nucleotide and Y is a pyrimidine nucleotide.

m^{1}A, 1-methyl-adenylate;
m^{6}A, N^6-methyladenylate;
Cm, 2'-O-methylcytidylate;
D, dihydrouridylate;
Gm, 2'-O-methylguanylate;
m^{1}G, 1-methyl-guanylate;
m^{7}G, 7-methylguanylate;
I, inosinate; Ψ, pseudouridylate; and
T, thymidylate.

17.3 Tertiary Structure: Transfer RNA

X-ray crystallography of tRNA^{phe}. X-ray crystallography has been used to determine a number of transfer RNA structures. The best characterized structure is that of *phenylalanyl transfer RNA* (tRNA^{phe}). A tube-format structure is shown below.

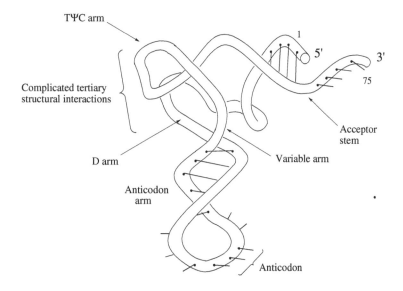

The *anticodon loop* and *arm* are located across the "bay" from the acceptor stem. Each arm of the L-shaped tRNA is double-stranded and forms a short stacked *right-handed helix* that looks much like typical *A-form* DNA.

Aminoacylation of tRNA

The *amino acid acceptor stem* is attached covalently to an *amino acid* by a specific catalytic *Aminoacyl tRNA Synthetase*. These acceptor-specific proteins catalyze the attachment of each specific type of amino acid to its tRNA. These enzymes recognize both the *codon* loop region and amino acid acceptor stem.

The amino acid is linked to the tRNA, via its *carboxylate. Attachment can occur at* either the 2' or 3' hydroxyl group on the ribose of the 3'-terminal adenylate. The primary transcript is processed, by *adding a CCA trinucleotide at the 3' end*, to produce the *matured* tRNA, by a *CCA terminal transferase*. The 5' nucleotides of tRNAs are phosphorylated.

Decoding mRNA The anticodon trinucleotide sequence forms a *duplex structure* with the *codon* of the messenger RNA. The recognition patterns for codons with specific amino acid *isoacceptor tRNA species* are characterized by the *Genetic Code*. More than one tRNA exists for most of the 20 amino acids.

The code is not entirely universal. For example, some codons encode a different tRNA in mitochondria. A second level of information control is called *codon usage bias*, in which one compartment or cell type will favor certain codon(s) to encode a given amino acid, other will favor a different choice of codon(s).

Table 2. Predicted base pairing between the 5' (wobble) position of the anticodon and the 3' position of the codon.

Nucleotide at 5' (wobble) position of anticodon	Nucleotide at 3' position of codon
C	G
A	U
U	A or G
G	U or C
I*	U, A or C

Modified Nucleotides

Transfer RNAs are decorated with a large number of chemical modifications. Specific modified nucleotides are always present in two arms of tRNA: (1) the invariant *ribothymidine* (T or rT) – *pseudouridine* (ψ) – cytidine sequence in the *TψC loop*, at the end of the *TψC arm*. The *D arm* is named after its invariant *dihydrouridine*. Each tRNA has a *variable arm*, which can be anywhere from 3 to 21 nucleotides in length, located between the *anticodon arm* and the TψC arm. Most tRNAs are 73 to 95 nucleotides in length. In one odd case, smaller tRNAs occur in mitochondria that are missing most of their D-arm.

The secondary structure of a tRNA is affected by each covalent modification. Two examples are:

(*1*) The modified nucleotide *dihydrouridine* is a non-planar, non-aromatic ring, which is chemically unstable. Yet, for subtle structural reasons, it is there.

(2) The modified nucleotide *pseudouridine* (Ψ) occurs in the conserved TΨCG loop of tRNAs and is structurally odd. The ribose C1' is attached to a *base ring carbon*, not to nitrogen, as in the conventional uridine-ribose glycosidic bond. The result is that the carbonyl oxygens are "presented" at a very different angle from those in the unmodified nucleotide. This change supports the hydrogen-bond pattern that stabilizes the central domain of the tRNA tertiary structure.

17.4 Messenger RNA (mRNA)

Most students have had a genetics course and are aware of the complicated nature of *protein synthesis* (translation). The ribosome reads the mRNA and the tRNA transfer amino acids, one by one, onto the growing nascent protein chain.

Coupled Transcription-Translation. A prokaryote-specific *coupled transcription-translation* mode of protein synthesis is shown.

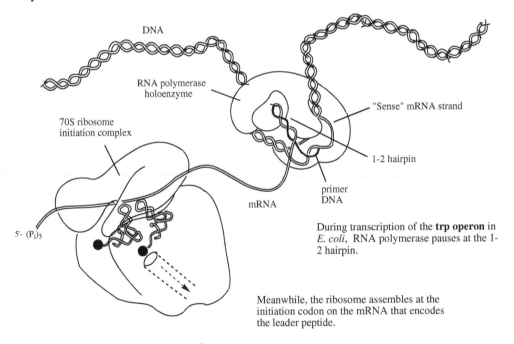

(*Adapted from Moran et. al. Biochemistry* (2nd ed.), 1996, p.30.34; Fig. 30-30.)

Coupled transcription-translation makes the pausing mechanism described below possible. Two things are occurring simultaneously:

(*1*) the mRNA transcript is being produced by the *RNA polymerase* complex, and

(2) the ribosome has actively engaged the mRNA in the *decoding process*.

As the *nascent polypeptide* is synthesized, it emerges from the ribosome through the *exit tunnel*, as indicated by the arrow.

Note that eukaryotic mRNAs are not transcribed while being translated. Eukaryotic mRNAs are transcribed in the nucleus, transferred to the cytoplasm, and extensively processed on the way. They are then translated in the cytoplasm, typically on rough endoplasmic reticulum. The mechanism shown in the figure typically involves the use of a polycistronic mRNA.

Polycistronic mRNA *Prokaryotes* make mRNAs that contain more than one gene sequence. These *polycistronic genes* encode more than one protein or RNA product, sometimes some of each. These proteins or RNAs are typically involved in the steps of one metabolic pathway. As a result, production of all of the enzymes and components required to support and regulate the pathway is controlled by regulating one combined gene, one transcription process and one translation process, if they all occur. It provides a way to coordinate the production schedule of all of the components, and regulate everything *en masse*, through one set of switches.

Polycistronic mRNAs can encode several enzymes, several rRNAs, several tRNAs, and combinations of two or more of these species. Transcription from these genes is regulated *coordinately* (all at once) by protein repressors and activators, as well as by other mechanisms.

The best known example of an operon that is encoded by a polycistronic mRNA is the *Lac operon*, which is involved in uptake and metabolism of lactose by bacteria. Since most students have encountered it at this point, we do not recount the many details that govern the regulation of transcription in that system.

Pausing. The Tryptophan Operon

The *tryptophan operon* encodes several enzymes and *transcriptional repressor* proteins. All of the components are involved in the biosynthesis of tryptophan, and are all co-expressed from the same mRNA. This example of regulation requires coordination between the availability of tryptophanyl tRNAs, the position of the mRNA on the ribosomes, and the structure adopted by the mRNA. As a result, protein synthesis can titrate the *availability of tryptophan*.

The availability of aminoacylated ("charged") transfer RNA, tryptophanyl-tRNA^{trp} depends on whether or not sufficient trp is available for the aminoacyl tRNA synthetase to aminoacylate the tRNA. The mRNA can form two alternative secondary structures. Each leads to a different outcome.

(*1*) When charged tRNA is in short supply, a particular hairpin structure can form in the *leader sequence* called the pause hairpin.

(*2*) When the sufficient tRNA is available, that is, when the amino acid is available, a different hairpin arrangement forms, the *termination hairpin*.

(*3*) When amino acid and charged tRNAs are in short supply, the *pause* hairpin forms and the ribosome pauses. As a result, the *termination* hairpin is blocked from forming, and translation of the polycistronic mRNA produces the encoded proteins. A shortage of the amino acid assures that the enzymes are made, which replenishes amino acid via biosynthesis.

(*4*) When the concentration of amino acid is high and pausing does not occur, the termination hairpin forms, which stops production of the enzymes.

Structural details of the hairpin-forming sequences in the two configurations are shown, for example, in Figure 22.28 on p. 690 of Moran *et al., 2012*.

The Shine-Dalgarno Sequence

This is a conserved sequence in prokaryotic mRNAs that directs their recognition by ribosomes. It operates by forming a duplex between the mRNA and the 3'-end of the 16S ribosomal RNA sequence.

The Literature. Many more details about related subjects such as operon structure, TATA boxes, transcriptional termination sequences, the details of translation, and so on, can be found in *Molecular Biology of the Cell* by Alberts *et al.*, 2002, the pair of books entitled *Molecular Biology* by Weaver 2001 and Cox, Doudna, and O'Donnel 2011, the latest edition of *The RNA World*, the Cold Springs Harbor Press monographs, and the *Current Protocols* series.

17.5 Eukaryotic Messenger RNA

Post-transcriptional Modifications
The following structural elements typically occur in eukaryotic mRNA.

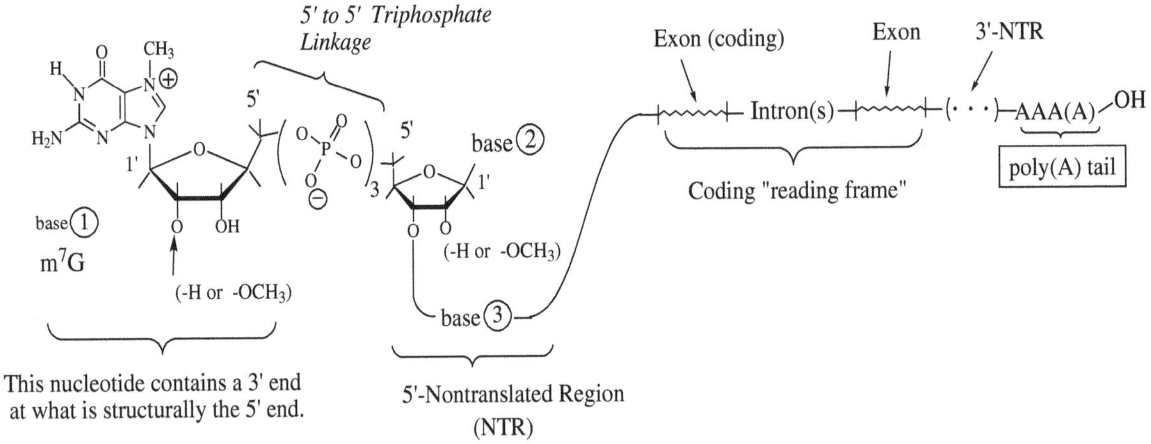

The components are:

(*1*) The *Cap Structure* contains a 5'-terminal m^{7}G that is covalently linked to the second nucleotide through a 5'-to-5' phosphodiester bond. Note that this resembles the case in NADH, except that NADH contains a 5'-to-5' diphosphate linkage, unlike the monophosphate linkage in the mRNA cap structure.

(*2*) *Introns and Exons*. Eukaryotic RNAs are initially transcribed as *genes in parts*. The *spliceosome*, which catalyze intron removal, is composed of a set of proteins and small nuclear RNAs (snRNAs). The remaining mRNA is composed of two or more *exon* sequences.

(*3*) *Poly(A) Tail*. A *poly(A) tail* is added to the 3' end of mRNAs.

 (*i*) The poly(A) sequence forms a cyclic structure with the 5' end of the mRNA. This system regulates the lifetime of the mRNA molecule. Shortening of the tail is analogous to a train ticket being punched by the conductor. When the tail is too short, the mRNA is degraded and turned over.

 (*ii*) The poly(A) tail is not a universal feature of all mRNAs. For example, histone mRNAs contain a Watson-Crick hairpin structure instead.

(*4*) *One mRNA per Gene Sequence*. Eukaryotes typically encode one gene per mRNA, they are *monocistronic*.

Exceptions to the Eukaryotic One-Gene-per-mRNA Scenario
(*1*) *Intron-Based Genes*. The typical *one gene per mRNA* scenario in eukaryotes does not occur when the mRNA contains an *intron* (*intervening sequence*) that encodes an RNA or protein other than the normal exon-derived product. These *intron-derived genes* are sometimes used to make a functional product RNA or protein.

(*2*) *Alternative Splicing*. Another exception occurs when an intron is removed at one of two different *alternative splicing* sites. This produces two different mRNA products. The products are partially the same and have different lengths and nucleotide sequences. The two alternative gene products will have different 3'-ends. The two proteins will differ in that the shorter sequence will produce a truncated (shortened) gene product.

(*3*) *Minus-strand Product Synthesis*. Transcription can occur from the *minus strand* as well, leading to a mechanism that produces two different complementary RNAs from the same gene.

Intron Removal from Precursor mRNA
Introns are removed from mRNA by a process that involves *catalytic participation* by specific functional groups in the *pre-spliced mRNA*. The *spliceosome* is a multiprotein, multi-RNA complex that assembles on the transcript and *removes the intron* in a site-specific manner. The basic reaction is:

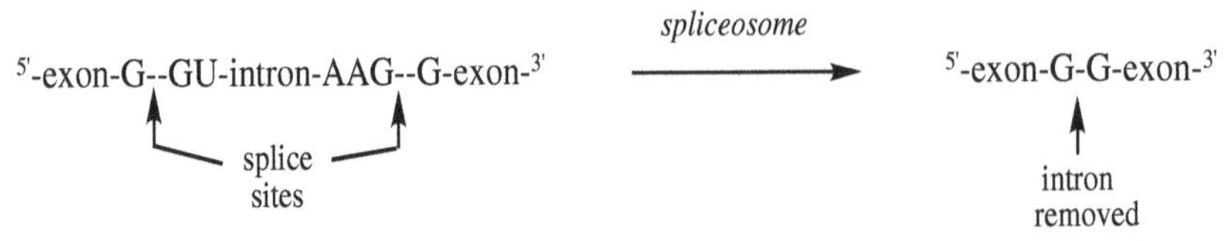

The steps in the mechanism are shown below.

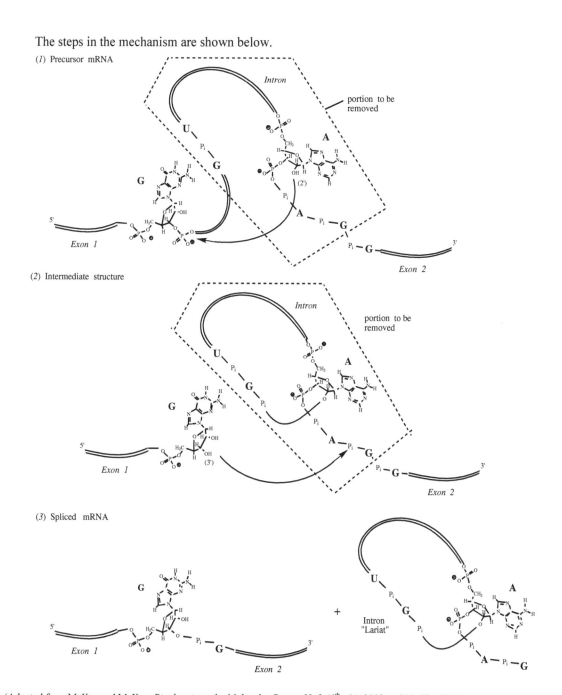

(Adapted from McKee and McKee, *Biochemistry the Molecular Basis of Life* (4th ed.), 2009, p. 716; Fig. 18-42.)

17.5.4 Autoimmune Antibodies. Characterizing Splicing Complexes

The body sometimes (incorrectly) makes antibodies that recognize substances we're supposed to have. This, in turn, wrongly tells the cells, which typically eliminate foreign substances by the immune response, that these *self* components are *not self*. The incorrect immune response against the self antigens of the patient is called *autoimmune disease*. Examples are *Systemic Lupus Erythromatosis* (SLE), arthritis and Crohn's disease.

The splicing apparatus proteins and snRNAs have been extensively characterized because they are common antigens for the autoimmune antibodies produced by SLE patients. Many of the early discoveries regarding the assembly pathways in mRNA splicing were characterized using immunochemical techniques and the unique *autoimmune antibodies* they make. These *reagent antibodies* have been captured as *monoclonal antibodies*, and can be produced in large amounts from *hybridoma cells* in *tissue culture for experimental analyses*.

17.6 Alkaline Hydrolysis of RNA

Hydroxide ions in alkaline solution can remove a proton from the 2'-hydroxyl group of any accessible RNA nucleotide. The nucleotide is converted to an alkoxide, a strong nucleophile, which attacks the phosphorous atom of the 3'-terminal backbone phosphate. This results in two steps: (*1*) release of the 5' oxygen of the next adjacent nucleotide and (*2*) formation of a 2'-, 3'-cyclic nucleoside monophosphate intermediate. The 2'-, 3'-cyclic intermediate is unstable in alkaline solution, so cleavage occurs. Another hydroxide ion from the solution attacks the intermediate and catalyzes hydrolysis, forming either a 2'- or 3'-nucleoside monophosphate.

(*1*) When the polymer reacts, rapid breakdown of the chain occurs because internal phosphodiester bonds are cleaved, leading to many subfragment products.

(*2*) Since the methyl group cannot be extracted from the 2' oxygen, *methylation of the 2' hydroxyl oxygen* renders the phosphodiester linkage *resistant to alkaline hydrolysis*.

See the *DNA Preparation: Phenol-Chloroform Extraction* section in the *Biotechnology* chapter for use of this principle in the preparation of RNA-free DNA in practical lab situations.

17.7 Small Interfering RNA

Small interfering RNAs (siRNA) molecules target an *Argonaut siRNA-dependent ribonuclease* complex to *cut a specific RNA sequence*. These siRNAs are encoded in the genomes of many, if not most organisms (including humans) as *micro RNAs*. They are not translated and interfere with protein expression by either directing the destruction of a transcript, or preventing its use in translation.

(*1*) Bases 2-8 of the guide RNA form an A-form double helix with the target RNA. Prior to target RNA binding, bases 2-6 in the siRNA RISC complex are exposed as an "activated" probe called the "seed region."

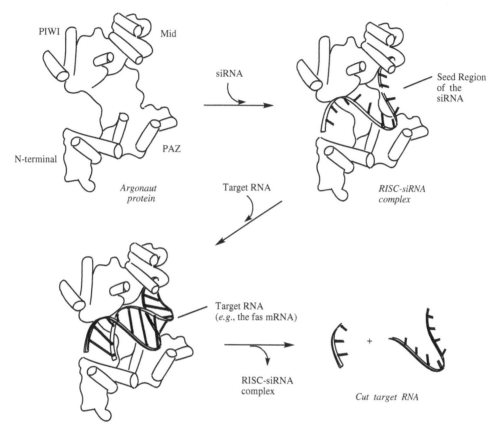

(*2*) The *RNA-induced Silencing Complex* (RISC) can find and cleave its target almost 10-fold faster than the lone guide RNA can anneal to its target. For details, see Pratt, A. and MacRae, I. (2009) The RNA-induced Silencing Complex: A Versatile Gene-Silencing Machine, *J. Biol. Chem.* 284, 17897–17901.

(*3*) This technique was used in a medical application to destroy the mRNA for the *fas* ligand on the *surface* of *T-lymphocyte* cells. This treatment prevents the worst damage to the liver caused when Hepatitis A or C virus infection occurs.

The human RISC contains a siRNA that is complementary to the fas mRNA cut site. When the target RNA binds, the argonaut ribonuclease cuts it, destroying the fas mRNA so the fas ligand protein is not made by the T-lymphocytes.

With no fas ligand made, they cannot bind the fas receptors on the liver, so they do not enter it. If this occurred, it would initiate an apoptosis (programmed cell death) response. The ability to cut the fas mRNA allows one to prevent the most deadly and acute damage to the liver. See Lieberman, J., Master of the Cell in *The Scientist*, Apr. 2010, pp. 43–48.

(*4*) In another application, a siRNA was designed to silence one of the genes of the *respiratory syncytial virus* (RSV), which is the leading cause of infant hospitalization, yet is fairly harmless in healthy adults. The siRNA was administered in a nasal spray and only 44% of those who received the RSV siRNA spray became infected with RSV, compared with 71% in the placebo group. This approach is also being tested as a means to *protect lung transplant patients*. See Holmes, B., Gene Silencing Prevents Human Disease, in *The Scientist*, May 2010, p. 10.

Chapter 18

Biotechnology

Biotechnology involves combining the concepts biochemistry has uncovered with the natural inclination scientists have to tinker with their systems to make new tools.

For example, Wendell Lim and colleagues were investigating the light-sensitive plant protein *Phytochrome* and its binding partner *Phytochrome Interaction Factor* (PIF). *In response to red light*, they found that *the two bind* and then *translocate into the nucleus*. The researchers decided to connect PIF to a *cytoskeletal protein*. They had to do some *genetic engineering* with both phytochrome and PIF first, but with the right *constructs*, when they turned on the red light, the cytoskeletal protein was activated and *reshaped the cell*. They had created a way to turn on *cytoskeletal* protein-protein interactions *using light*. The researchers have made the two plasmid DNAs that encode the mutant protein constructs available on a nonprofit basis. (See Lim *et al., Nature* 461, 997–1001, 2009.)

The route to their discovery involved first characterizing their idea then deciding it would be interesting to try the light-dependent PIF function in a *non-natural task*, reshaping the cell. They had to modify their natural proteins to accomplish their task, but were able to develop an engineered light-sensitive intracellular *switching* system. Biotechnology is the playing field where biology meets nanotechnology.

The essential tools used to accomplish cloning are: (*1*) restriction endonucleases, (*2*) single-nucleotide resolved gel electrophoresis, (*3*) the ability to propagate DNA sequences and test their *in vivo* functions using plasmid DNA, and (*4*) the ability to replicate plasmid DNA after *transforming* it into a bacterial cell. Each is described below.

18.1 Restriction Endonucleases

This class of enzyme catalyzes *DNA restriction* by recognizing the DNA sequence and cleaving both strands to produce two duplex fragments containing *staggered ends*, also called *sticky ends*.

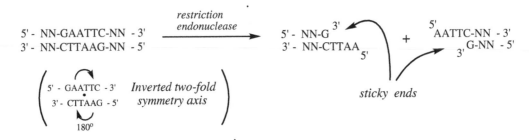

The NN sequences indicate the rest of the strands on either side of the restriction site.

(*1*) In most circumstances, four- to eight-base pair segments operate as recognition sites with a 2-fold axis of symmetry. The paired sequences read the same forward and backward on the complementary strand. These kinds of sequences are known as palindromes. Examples are BIB, DEED, RADAR, MADAM I'M ADAM, and A MAN, A PLAN, A CANAL, PANAMA.

(*2*) When placed at the ends of full-length DNAs from chromosomes, plasmids, and so on, these staggered ends act like Velcro that only binds the specific *complementary* staggered end. The specificity in splicing two pieces of DNA together offered by this technique is one basis of cloning technology.

Restriction Inhibition

The host bacterium can evade its own restriction enzymes by protecting its restriction sites with methylation. The host bacterium DNA is methylated at the restriction enzyme recognition site. Foreign or invader DNAs are cut because they are *not* methylated at the restriction endonuclease cutting sites. Methylation distinguishes the host cell DNA from that of invaders.

The mechanism involves two steps:

(*1*) Following DNA replication, the GAATC site is hemimethylated.

(*2*) A *methyltransferase* catalyzes methylation of the second adenine residue in the recognition site.

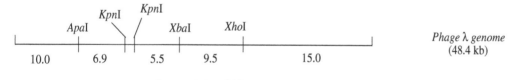

Restriction Mapping

Restriction enzymes can be used to map DNAs. Amino acid sequence-specific proteases were discussed in the Protein section. Different proteases (*e.g.*, trypsin, chymotrypsin) can be used to create overlapped fragments. The same idea can be done using restriction enzymes. Computers can reassemble the fragment information to make sense of it. As a result, one can use broken up preparations of DNA, yet still figure out how they were linked together within the original unbroken DNA.

DNA shears apart each time it is ejected from or drawn into a pipette tip, so we are rarely working with DNAs that are the size of full-length eukaryotic chromosomes. This shearing can be avoided by preparing cells, nuclei, and chromosomes in gels directly then carrying out electrophoresis to analyze them. The *Alternating-Field Gel Electrophoresis* method is used to study multichromosomal preps, such as in *chromosomal typing* analyses.

A *restriction map* of *Bacteriophage Lambda (λ) DNA* is shown below. Five different "restriction enzymes" cleave at the sites shown. Digestion, this enzymatic cleavage, according to the map shows that the λ DNA is cut with the enzyme *Apa*I, forming two fragments that are 10.00 kilobase pairs (kb) and 38.4 kb long.

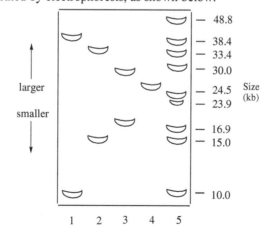

The *λ DNA genome* is a linear, double-stranded DNA that consists of approximately 48,000 bp (48.4 kb). A restriction digest is made by treating this DNA with various restriction enzymes. By measuring the sizes of the resulting fragments, it is possible to develop a map of the cleavage sites. The DNA fragments in a restriction digest are separated by electrophoresis, as shown below.

When λ DNA is subjected to gel electrophoresis, one can distinguish the DNA product fragment by size. The smaller pieces have less molecular weight and therefore travel to the bottom of the agarose gel faster. Lane 1: *Apa*I digestion. Lane 2: *Xho*I digestion. Lane 3: *Kpn*I digestion; we are unable to see the smallest fragment (1.5 kbp). Lane 4: *Xba*I digestion; the two fragments are close in size and not well resolved. Lane 5: Intact λ DNA and an assortment of fragments from the other lanes.

Specificities. The following table lists the *recognition sequences* and *cutting sites* of many common *restriction enzymes*:

Enzyme	Recognition sequence	Source
*Bam*HI	5' G↓GATCC 3' 3' C CTA↑GG 5'	*Bacillus amyloliquefaciens H*
*Bgl*II	5' A↓GATCT 3' 3' TCTA↑GA 5'	*Bacillus globigii*
*Eco*RI	5' G↓AA^{m}TTC 3' 3' CTT↑A^{m}AG 5'	*Escherichia coli* RY13
*Eco*RII	5' ↓CC^{m}TGG 3' 3' GGAC^{m}C↑ 5'	*Escherichia coli* R245
*Hae*III	5' GG↓CC 3' 3' GG↑CC 5'	*Haemophilus aegyptius*
HindII	5' GTPy↓PuA^{m}C 3' 3' CA^{m}Pu↑PyTG 5'	*Haemophilus influenzae* R_d (Note that the sequence can vary.)
*Hin*dIII	5' A^m↓AGCTT 3' 3' TTCGA↑A^m 5'	*Haemophilus influenzae* R_d
*Hpa*II	5' C↓CGG 3' 3' CCG↑G 5'	*Haemophilus parainfluenzae*
*Not*I	5' GC↓GGCCGC 3' 3' CGGCC↑GC 5'	*Nocardia otitidis-caviar*
*Pst*I	5' CTGCA↓G 3' 3' G↑ACGTC 5'	*Providencia stuartii* 164
*Sma*I	5' CCC↓GGG 3' 3' GGG↑CCC 5'	*Serratia marcescens* S_b

18.2 Cloning in a Nutshell

Basic Process. First, the gene you want to clone is cut with *two* restriction endonucleases then spliced into the *linearized target plasmid*. That plasmid must first be prepared by cutting with the same two restriction enzymes. This ensures that the upstream control sequences and the gene body sequences are aligned to produce the mRNA, with all of the appropriate transcription and translation control sequence elements in proper order.

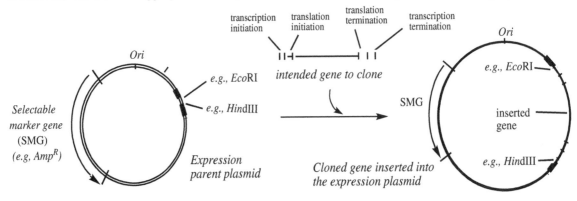

(*1*) A *selectable marker gene* encodes a protein that destroys the antibiotic being used for selection purposes. If the bacterium contains the plasmid, it will be able to live in the presence of the antibiotic. The bacterium is *antibiotic resistant*.

(*2*) The *gene expression cassette* fragments are situated on upstream and downstream from the polycloning site, where the gene to be expressed is inserted.

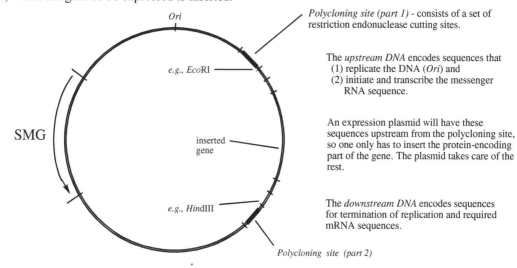

Polycloning site (part 1) - consists of a set of restriction endonuclease cutting sites.

The *upstream DNA* encodes sequences that
  (1) replicate the DNA (*Ori*) and
  (2) initiate and transcribe the messenger RNA sequence.

An expression plasmid will have these sequences upstream from the polycloning site, so one only has to insert the protein-encoding part of the gene. The plasmid takes care of the rest.

The *downstream DNA* encodes sequences for termination of replication and required mRNA sequences.

Polycloning site (part 2)

Transcription and Translation Sequences. The mRNA must contain sequences that initiate, elongate, and terminate (*1*) transcription by the cellular RNA polymerase, and (*2*) translation by the ribosome, thereby producing the encoded protein.

18.3 DNA Preparation: Phenol-Chloroform Extraction

Manipulations involved in actually cloning a gene can be very difficult because, as a coworker once lamented, "everything's invisible!" Well, not always. Sometimes one can see the pellet when one ethanol-precipitates a plasmid prep.

The *phenol-chloroform DNA preparation* method is very practical. One colleague tells me the technique swept through the labs in Germany where he was working when it was first introduced. This *single-step* method uses phenol/chloroform reagent, which is very effective in breaking apart most membrane and cellular components by alkaline hydrolysis and then separating them from the DNA.

(*1*) The base phenol converts to phenolate, an effective nucleophile. Phenolate catalyzes alkaline hydrolysis, which breaks down many types of bonds in membrane components. Proteins are denatured. DNA is not susceptible to base-catalyzed hydrolysis under these somewhat moderate conditions. RNA will only degrade after extensive incubation.

(*2*) Chloroform is polar enough to solubilize more polar materials liberated by breakdown of triacylglycerol components and is effective in undoing the "hydrophobic effect," which is a major stabilizer of membrane structure.

Phospholipids

Proteins

RNAs

For many more details about the practical techniques, ideas, and manipulations used in cloning, expression, purification and detection of DNAs, RNAs and proteins, see Hardin *et al.*, *Cloning, Gene Expression and Protein Purification*, Oxford University Press, 2001.

18.4 Polymerase Chain Reaction (PCR)

Purpose
The goal is of doing PCR is to *synthesize large quantities of a specified DNA sequence with precisely correct ends*.

Contrast with Cloning
Cloning and PCR are used to make large amounts of predetermined DNA sequences. Plasmids are used in *DNA cloning* techniques to carry particular sequences in particular genetic contexts, allowing one to produce mRNAs, proteins and other *gene products*. This can be done in the test tube (*in vitro*) but is most easily accomplished *in vivo*. The unique advantage of a plasmid versus a PCR product is that one can produce a desired product in the cell directly. This allows one to assess the biological effect of that product in a sort-of-normal cell environment.

The difficulty with plasmids in producing large amounts of DNA (versus PCR) is that: (*1*) cells must be grown, (*2*) the DNA must be extracted from the cells, (*3*) the plasmids must be separated and purified, (*4*) the desired fragment in the plasmid must be cut out using restriction endonuclease enzymes, and (*5*) the DNA fragment must be purified (usually on a gel). As a result, PCR has become the method of choice in many, many types of experiments.

On the other hand, cloning is (*1*) relatively easy, (*2*) provides a permanent record of the DNA (when stored as a *glycerol stock* solution), and (*3*) can generally be sequenced more easily than PCR fragments.

Mechanism of PCR
The technique involves *three steps*, which are repeated for a series of 20 to 30 cycles.
(*1*) *Denaturation* of the DNA template strands. This is typically done at 95 °C.
(*2*) *Anneal the two primer DNAs*, one to the 5' terminal point of the sequence to be copied on each template strand. This is done at 55 °C, but is varied to optimize product formation.
(*3*) *Polymerization* to produce the two product strands.
(The cycle is repeated)

DNA Polymerase Mechanism

Reaction temperatures are optimized to:
(*1*) denature the initial template sufficiently,
(*2*) allow the primers to bind, and
(*3*) assure that no competing secondary structures form.

The primer-template DNA-DNA binding temperature is kept high enough to assure that only *stringently selected* primer-template double helices form.

Note. Two terms are used to refer to *double helix formation*: annealing and hybridization.

Reaction Components
(*1*) *Primer DNAs*: They "bracket" the sequence that is to be "amplified" (made in large amounts).
(*2*) *Template DNA*: Can be a plasmid, chromosome, or a biological sample. Chromosomes in blood and saliva swab samples are used in CGI-type applications.
(*3*) *Buffer*: Required to support the DNA polymerase activity.
(*4*) Mg^{2+}: This metal is required to assure, as much as possible, that the product has sufficient *fidelity*. The 3' to 5' exonuclease activity of the DNA polymerase carries out this assurance reaction/process.
(*5*) *Deoxynucleotide Monophosphates* (dNTPs): *i.e.* dATP, dCTP, dGTP and dTTP. Other nucleotides are included in specific applications, such as body labeling the DNA probe molecule with *digoxigenin* (see below).
(*6*) *Taq DNA Polymerase*. This enzyme is produced from the thermophilic bacterium *Thermus aquaticus*. Taq DNA polymerase does not have the *3' to 5' exonuclease activity* and is very *error-prone*. It has been replaced by *Vent* and *Pfu* DNA polymerases.

PCR Reaction Steps

(*1*) First Cycle: Duplex Template.

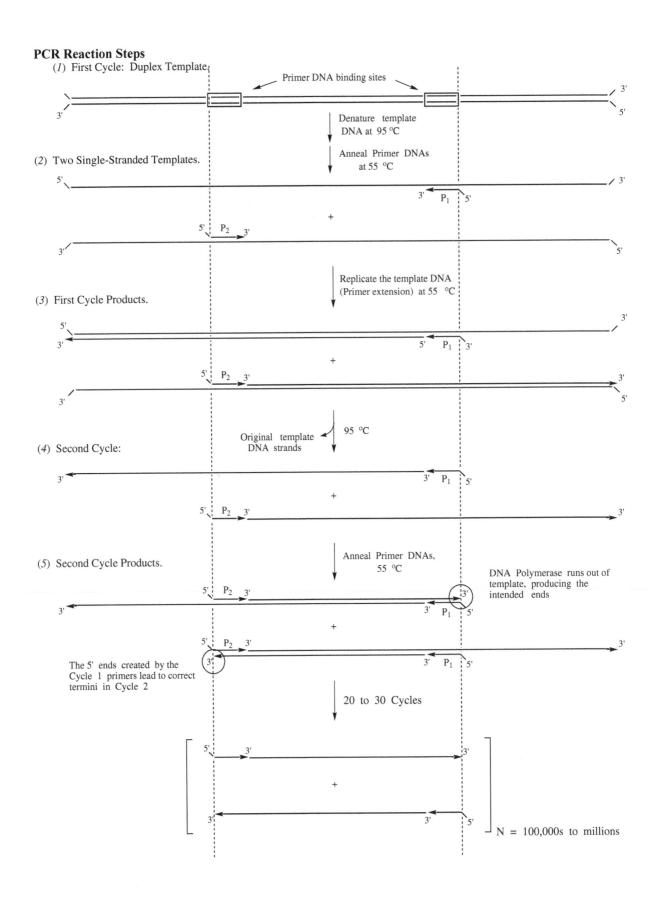

(*2*) Two Single-Stranded Templates.

(*3*) First Cycle Products.

(*4*) Second Cycle:

(*5*) Second Cycle Products.

The 5' ends created by the Cycle 1 primers lead to correct termini in Cycle 2

18.5 Probe DNA

A probe DNA is a *known sequence* one uses to search for the Watson-Crick complementary sequence, in either RNA or DNA, depending on the goal of the experiment.

(*1*) One usually labels the probe with either radioactive [^{32}P]-phosphate, a fluorescent molecule (a fluorophore), the antibody-binding *antigen* digoxigenin, or the avidin-binding molecule biotin (see the *Coenzyme* section).

(*2*) One then *anneals* the probe to the sample, meaning that the probe strands *form duplexes with* strands from denatured duplexes in the sample.

(*3*) The purpose is to detect sequences that are complementary to the probe. If they are present in the sample of interest, the probe binds to them. Since the probe is labeled, one also labels the fragment containing the complementary sequence. Therefore, the third step involves detection using a piece of film, a phosphorimaging apparatus, and so on.

This tool is used in a number of techniques, such as *Northern* and *Southern Blot hybridization analyses*, *Mobility Shift protein-binding assays*, and *Colony Hybridization*. These methods have been used extensively to study *developmental gene expression*, to determine the identity of a DNA in a plasmid, to determine if a cell harbors a specific plasmid, and so forth.

Probe Construction: PCR from a Plasmid Template Using Digoxigenin-dUTP

Creating a probe involves using PCR. *Body labeling* of duplex DNA strands from a plasmid that contains a specified Geminivirus DNA segment. The label digoxigenin (DG) is incorporated in each strand randomly in place of some of the deoxythymidines.

The structure of alkaline-labile *Digoxigenin-11-Deoxyuridine Triphosphate* (DIG-11-dUTP, aka DIG) is:

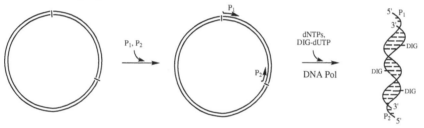

Illustration of Probe DNA Use: Southern Blot Hybridization and Detection Using an Immunoconjugate

Prehybridization and Hybridization. To perform *stringency control* during duplex formation, one does a *prehybridization wash* of the DNA on the blot filter. This is followed by *hybridization* of the probe to the DNA on the blot filter.

Next, one *washes* the probe- DNA complexes on the blot filter using *Standard Saline Citrate* (SSC) buffer containing *sodium dodecylsulfate* (SDS). Next one *blocks* the blot filter (*i.e.* the sites that do not contain probe-viral DNA duplexes) with *Bovine Serum Albumin* (BSA).

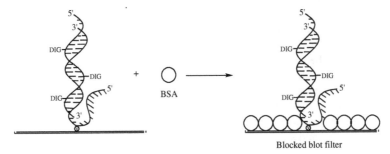

Immunoconjugate Detection of Bound Probe DNA. To detect hybridized DNAs, one then adds the *immuno-conjugate*, a covalent antibody-enzyme complex, to the blot filter.

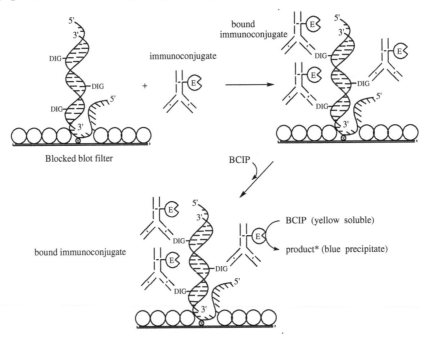

Colorimetric Label Detection is accomplished using the BCIP reaction as shown below. Appearance of a blue precipitate on the blot membrane indicates that the DNA being sought with the probe is present at that position on the gel pattern.

Chapter 19

Metabolism

19.1 Overview

The following map shows the general flow of components, the key intermediate species, the process names, the energy sources produced by the *catabolic processes* and the sources of precursors used to make the four groups of macromolecules. Many other important and sometimes subtle cross-connections are not indicated. For example, the nucleotide bases are made using an intermediate from the urea cycle and an amino acid.

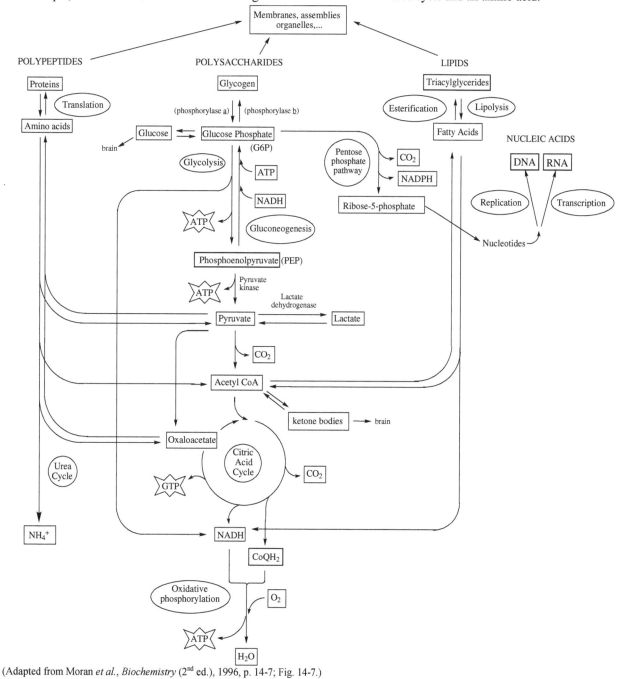

(Adapted from Moran *et al.*, *Biochemistry* (2nd ed.), 1996, p. 14-7; Fig. 14-7.)

19.2 Metabolic Pathway Types

Linear Pathway. An example is the *biosynthesis of proline*, the product of each reaction in the pathway is used as the substrate for the next step. Another key pathway of this type is *glycolysis*, which is described in detail in that chapter.

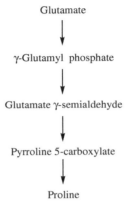

Glutamate

γ-Glutamyl phosphate

Glutamate γ-semialdehyde

Pyrroline 5-carboxylate

Proline

Cyclic Pathway. A series of reactions form a closed loop. Each turn is completed by regenerating the original intermediate. The *Krebs Cycle* metabolizes an acetyl group to form carbon dioxide and other materials then reforms the initial component, oxaloacetate, in a series of ten reactions. Two other examples described in later chapters are the *Malate-Aspartate Shuttle and* the *Urea Cycle.*

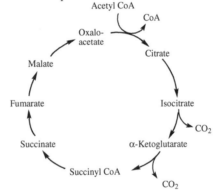

Spiral Pathway. Fatty acid biosynthesis is the classic spiral metabolic pathway. The enzyme complex lengthens the alkane chain by two carbon units with each repeated set of reactions, each turn of the spiral cycle.

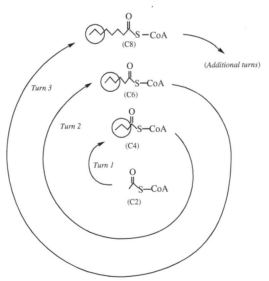

19.3 Energy Conservation

Phosphorylation Energy. In catabolic reactions, conserved energy is stored in the form of nucleoside triphosphates and reduced coenzymes. The relation between use and conservation of phosphorylation-dependent energy in phosphate-containing compounds is:

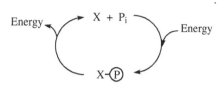

where X is the molecule that accepts the phosphate.

For example, *glucose* is phosphorylated by *hexokinase*, which energizes the compound in preparation for later reactions in glycolysis. In those reactions, the phosphate acts as a *good leaving group*. The energy is provided by, for example, transfer of phosphate from ATP, an excellent leaving group. That is the essence of energy transfer in many biochemical reactions.

Electron-Transfer (Reducing) Energy. In the same way, energy is captured by reduced coenzymes in oxidation reactions.

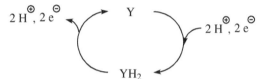

where Y signifies the oxidized component and YH_2 its reduced component. For example, during the reduction step in *fatty acid biosynthesis*, extra reducing power is "packed" into the alkene chain by converting it to an alkane.

Linkage. These two modes of energy transfer and use are directly linked by events that occur in the mitochondrion. ATP is used as an energy carrier. It is made, that is, receives its energy, from the reduced coenzymes that feed electron transport. *Reducing equivalent carriers* NADH and $CoQH_2$ provide the *reducing potential energy* that drives the proton pumps, which, in turn, drive oxidative phosphorylation, the process that makes ATP.

19.4 Key Pathways/Reactions

(1) Glycolysis is the universal pathway for glucose catabolism. A single 6-carbon molecule, glucose, is split into two 3-carbon molecules, namely pyruvate, which can be further oxidized in the citric acid cycle. Glycolysis functions without oxygen to generate ATP. This pathway produces three common products: lactic acid, ethanol, and pyruvate.

(2) The *Citric Acid Cycle* metabolizes the acetyl CoA produced from pyruvate in glycolysis and requires oxygen to run the cycle. Each acetyl CoA that enters the cycle is fully oxidized to carbon dioxide while conserving energy in the form of NADH, $CoQH_2$, and ATP or GTP.

(3) Glucose Metabolism comes in other forms as well including the synthesis and degradation of glycogen, a polymer used to store glucose. The degradation of glycogen is not merely the reverse of its synthesis.

The enzymes involved in *Glycogen Synthesis and Degradation* act in response to metabolic signals in such a way that the rates of these opposing reactions can be directly correspondent with the demands of the entire cell or organism.

(4) Gluconeogenesis is the process responsible for the synthesis of glucose from three-carbon compounds such as pyruvate, lactate, and glycerol.

(5) The Pentose Phosphate Pathway oxidizes glucose producing ribose instead of ATP, in preparation for nucleotide and nucleic acid synthesis, and NADPH, which is needed for biosynthetic reduction reactions. Because this pathway is used for both synthesis and reduction reactions, it is considered both an *anabolic* and *catabolic pathway.*

(6) Electron Transport involves a series of electron-transfer reactions in which input NADH and $CoQH_2$ is eventually used to reduce molecular oxygen, the terminal acceptor of the reducing equivalents. Electron

transport processes drive a set of *proton pumps,* which *produce a proton gradient* across the inner mitochondrial membrane in eukaryotes, or cell membrane in the case of prokaryotes.

(7) Oxidative Phosphorylation. As the protons collect on one side of the membrane, a transmembrane potential energy accumulates, which drives the membrane-spanning F_0F_1 *ATP Synthase* complex to phosphorylate ADP and thereby produce ATP.

$$ADP + P_i \rightarrow ATP + H_2O \tag{1}$$

(8) The *Photosynthesis* process *captures light* energy in the form of reducing equivalents, then uses them to make ATP in a manner similar to the mitochondrial electron transport-oxidative phosphorylation pathway.

Chapter 20

Bioenergetics

20.1 Reaction Equilibria: Standard and Actual Free Energies

Reactions in most cellular niches occur at concentrations far from the *1 M*, which is the case for all reactants and products in the *standard state* value of the *Gibbs free energy change* ($\Delta G^{0\prime}$). Consider the generic reaction:

$$A + B \leftrightarrow C + D \tag{1}$$

The *Gibbs free energy* (ΔG) is calculated by subtracting the *sum of the free energies* of the *products* from the sum of the free energies of the *reactants*.

$$\Delta G = (G_C + G_D) - (G_A + G_B) \tag{2}$$

Adding the *actual* free energy contribution to the *standard (state) free energy*:

$$\Delta G = (G_C^{0\prime} + G_D^{0\prime} - G_A^{0\prime} - G_B^{0\prime}) + RT \ln ([C][D]/[A][B]) \tag{3}$$

Defining $\Delta G^{\circ\prime}$ as $G_C^{0\prime} + G_D^{0\prime} - G_A^{0\prime} + G_B^{0\prime}$, one gets:

$$\Delta G = \Delta G^{0\prime} + RT \ln ([C][D]/[A][B]) \tag{4}$$
$$\text{(actual)} \qquad \text{(standard)}$$

At *equilibrium poise*, the ratio of concentrations in eq. 4 is equal to the *equilibrium constant* K_{eq}.

When the concentrations of products and reactants reach *equilibrium (poise)*, the *rates* and *extents* of *forward* and *reverse reactions* satisfy the *ratio* dictated by K_{eq}. The result is that ΔG must equal zero. The relation between *free energy* and *equilibrium constant* is:

$$\Delta G^{0\prime} = -RT \ln K_{eq} \tag{5}$$

If $\Delta G^{\circ\prime}$ is known, one can use *eq. 5* to calculate K_{eq}. One can also do the reverse calculation, use K_{eq} to calculate $\Delta G^{0\prime}$. A logarithmic relationship exists between $\Delta G^{\circ\prime}$ and K_{eq}, so small change in $\Delta G^{0\prime}$ results in large changes in K_{eq}.

Free energy changes that occur under cellular conditions (ΔG) must be negative in order for an enzymatic reaction to occur. The $\Delta G^{\circ\prime}$ values of many metabolic reactions have *standard $\Delta G^{0\prime}$* values with *positive* signs. The low substrate and product concentrations under cellular conditions produce a significant difference between the ΔG and $\Delta G^{0\prime}$ values of many biochemical reactions.

K_{eq}, is defined as the ratio of substrates to products when ΔG is 0.

$$K_{eq} = ([C][D]/[A][B]); \Delta G = 0 \tag{6}$$

When the reaction components are not under standard state conditions, the ratio of products over substrates *changes*. The change in free energy that corrects for the difference between $\Delta G^{\circ\prime}$ and ΔG is calculated as follows:

$$\Delta G = \Delta G^{0\prime} + RT \ln Q \tag{7}$$
$$\text{(actual)} = \text{(standard)} + \Delta\Delta G$$

Where ΔG is the *steady state Gibbs free energy*, $\Delta G^{0\prime}$ is the value for the same equilibrium under standard-state conditions, Q is the *mass action ratio*, and $\Delta\Delta G$ is the *perturbation* to the equilibrium relative to the standard state poise. This is the true driving force in a steady state situation, where a process is fed and subject to a steady input at some given rate.

$$Q = ([C]^\prime [D]^\prime/[A]^\prime [B]^\prime) \tag{8}$$

The value of *Q* determines the difference in free energy between standard and actual values, indicating the difference between the reaction conditions and equilibrium poise if all of the components were present at 1 M.

One determines the spontaneity of a reaction based on ΔG rather than $\Delta G^{0'}$. The degree of spontaneity determines the predisposition of the reaction direction and potential to undergo concentration-dependent changes in ratio of products to reactants. Remember that if ΔG is > 0, the reaction proceeds toward reactants rather than products.

To *partition* means to shift a distribution, governed by a change in equilibrium poise, to establish a new intended ratio of products over reactants, the *actual* state, rather than the *standard* state. This is significant because biochemically important concentrations can span a wide range, from picomolar to molar, with many in the µM to mM range.

The equations in this section were used to adjust the pattern of standard state free energies for the reactions in glycolysis to the actual values (see below). This correction converts a seemingly meaningless pattern into a very interpretable one, which reveals the principle that drives the flow of components *through* the entire pathway, even through near equilibrium steps.

20.2 Metabolically Irreversible and Near Equilibrium Reactions

Reactions Types
Metabolic reactions can be separated into two types.

(*1*) A *near-equilibrium reaction* has a Q value close to K_{eq}. This leads to the net steady-state ratio of reactant to product concentrations in a living cell. These reactions have small actual free-energy changes (ΔG) and are often strongly reversible.

(*2*) A *metabolically irreversible reaction* is strongly displaced from equilibrium, *i.e.* Q is far from K_{eq}. These differences are commonly up to 100- to 1000-fold. As a result, the ΔGs are large and negative for all metabolically irreversible reactions.

Regulation
The amount of enzyme present in a cell to catalyze a metabolically irreversible reaction is insufficient to drive a near-equilibrium reaction. Most pathways are *regulated* through enzymes that catalyze *metabolically irreversible* reactions. In this sense, they act as a bottleneck to metabolic traffic that slows the *flux of metabolites* through reactions later in the pathway.

Pathways are not well controlled by *near-equilibrium reactions*. Because they are so close to equilibrium, the flux through this step cannot be easily increased. Large changes in substrate and product concentration are the only control mechanism for near-equilibrium reactions. On the other hand, these concentrations have virtually no effect on metabolically irreversible reactions, but are better controlled by effectors that adjust the rate of catalysis.

When a pathway's flux is altered, the concentrations of intracellular metabolites vary in a small range of about 2- or 3-fold. Given the abundance of enzyme present to catalyze most near-equilibrium reaction, the substrate and product concentrations restore readily back to equilibrium poise after perturbation.

Enzyme and Pathway Reversibility
The *near equilibrium reaction enzymes* can catalyze reactions *in both directions*. We will see that this is useful, as illustrated with the *glycolysis* and *gluconeogenesis* pathways, which use the same near-equilibrium reaction enzymes in both directions. As a result, gluconeogenesis only requires enzymes to replace glycolytic enzymes to catalyze the metabolically reversible reaction steps. Four enzymes in gluconeogenesis replace three enzymes use in glycolysis to accomplish this reversal of pathway flux.

Standard Free Energies of Hydrolysis for Common Metabolites
The following table compares the energies of some standard metabolic compounds.

Metabolite	$\Delta G^{0'}_{hydrolysis}$ *(kJ mol^{-1})*
Phosphoenolpyruvate	-62
1,3-*Bis*phosphoglycerate	-49
Acetyl phosphate	-43
Phosphocreatine	-43
Pyrophosphate	-33

ATP to AMP + PP$_i$	-45 *(kJ mol⁻¹)*
ATP to ADP + P$_i$	-30
Glucose 1-phosphate	-21
Glucose 6-phosphate	-14
Glycerol 3-phosphate	-9

(Adapted from Moran *et al.*, *Biochemistry* (2nd Ed.), 1996, p. 14-21, Table 14-3.)

Note that *phosphoenolpyruvate* (PEP) is nearly twice as energetic as ATP. Two steps in gluconeogenesis replace one step in glycolysis. These steps use two ATPs to reverse the step in glycolysis in which PEP transfers its phosphate to make ATP. More insights are gained by looking at these values for the glycolysis reactions, especially after correcting them to be consistent with intracellular conditions.

20.3 Energies and Regulation of Glycolysis

Comparison of ∆G and ∆G°′ for the Reactions of Glycolysis
The free energies of the glycolytic reactions, calculated for the *standard* and *actual states*, are listed in the following table. One reaction, the *interconversion* catalyzed by *triose phosphate isomerase*, is not shown.

Enzyme	∆G$^{0'}$ *(kJ mol⁻¹)*	∆G *(kJ mol⁻¹)*
1. *Hexokinase*	-17	-33
2. Glucose-6-phosphate isomerase	+2	near equilibrium (~0)
3. *Phosphofructokinase-1*	-14	-22
4. Aldolase	+24	~0
5. Triose phosphate isomerase	+8	~0
6. Glyceraldehyde-3-phosphate dehydrogenase	+6	~0
7. Phosphoglycerate kinase	-19	~0
8. Phosphoglycerate mutase	+5	~0
9. Enolase	+2	~0
10. *Pyruvate kinase*	-32	-17

(Adapted from Moran *et al.*, *Biochemistry* (2nd Ed.), 1996, p. 15-21, Table 15-2 and p. 15-20, Fig. 15-23.)

The enzymes shown in italics catalyze the *three metabolically irreversible steps*. Each catalyzes a *kinase* reaction. They drive the flux of components through the pathway.

The information is presented graphically in the following plots, which show ∆G and ∆G$^{0'}$ for each of the ten numbered reactions of glycolysis. The effect of the three kinase reactions is obscured when one plots the *standard state ∆G^{0}'* values. The pattern is revealed when the adjusted *actual free energies* are calculated.

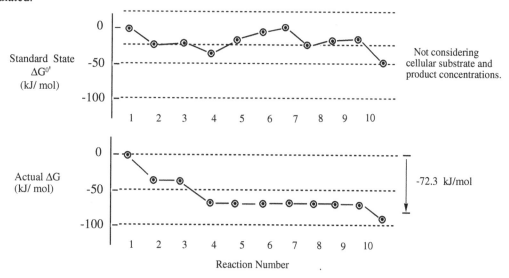

Some of the reactions of glycolysis have positive standard free-energy changes. This means that when the enzyme is *saturated with substrate* the reaction will flow toward reactants, instead of products. Note that the actual ΔGs are generally much smaller than the standard state values.

Regulatory Enzymes

(*1*) Hexokinase, phosphofructokinase-1, and pyruvate kinase all have large actual ΔG values, so they are all subjected to stringent regulation. Regulatory enzymes commonly have *oligomeric* quaternary structures. Of the ten enzymes in glycolysis, only phosphoglycerate kinase is monomeric. The others are either dimers or tetramers. Each has sites for regulatory effector binding near the interface between the subunits. The three regulatory kinases and two enzymes in glycolysis are subject to cooperative substrate binding. Hexokinase is present in both monomer and dimer forms, but is more active as the dimer.

(*2*) Weak binding to a substrate increases the *catalytic efficiency* of an enzyme. To obtain the most effective response to substrate changes and reestablishing equilibrium, the K_M value for an enzyme should be close to its cellular substrate concentration. This is not true for allosterically regulated enzymes. The value for K_M may change allosterically, so the cellular substrate concentration must be shifted away from K_M, in a direction that depends on whether up- or down-regulation is to be deployed.

Chapter 21

Bioelectrochemistry

21.1 Redox Reaction Principles

Reduction potential energies can be measured quantitatively using an electrochemical cells. This is illustrated in the oxidation-reduction reaction where an electron pair is transferred from an atom of *oxidized* zinc (Zn) to a *reduced* copper ion (Cu^{2+}):

$$Zn + Cu^{2+} \leftrightarrow Zn^{2+} + Cu$$

(*1*) This reaction is performed using two solutions, which separate the *anode* from the *cathode*, and divide the total reaction into two *half reactions* (see below). Two electrons are given up by each zinc atom, which is the *reducing agent or reductant*, at the anode. A wire travels through a voltmeter, which connects the zinc in one solution to the copper in the other. The electrons flow through this wire to the *cathode*, where copper, the *oxidizing agent or oxidant*, is reduced to metallic copper.

(*2*) The two solutions are kept electrically neutral by a *salt bridge*, which is made of a tube filled with electrolytes containing a porous partition. The salt bridge allows the nonreactive counterions to flow through an aqueous conduit between the two solutions. This allows accurate *voltmeter* readings by separating the ion flow from the electron flow.

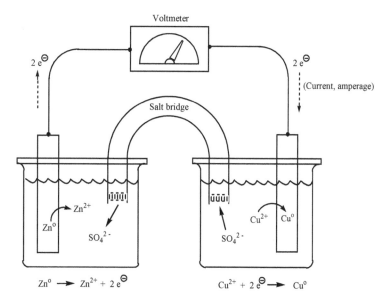

More negative potentials are assigned to reaction systems that have an increasing tendency to donate electrons. Thus, electrons flow spontaneously from *more negative* to *more positive* reduction potentials.

21.2 Redox Energetics: The Nernst Equation

Electromotive Force

The current flows through the circuit indicating that Zn^{2+} is more easily oxidized than Cu^{2+}. The purpose of the voltmeter is to calculate the *potential difference*, the difference between the reduction potentials of the left and right reactions. The *electromotive force* is the potential that is measured.

The *reference half reaction* corresponds to the *oxidation of hydrogen* (H_2). Hydrogen's reduction potential under the standard condition ($E^{0'}$) is subjectively set at 0.0 volts (V).

The standard reduction potential of any given half-reaction can be determined using a reference half-cell and a sample half-cell to create an oxidation-reduction couple. The reference half-cell is composed of a solution of 1 M H^+ and 1 atm of H_2 (g). The sample half-cell includes 1 M each of the oxidized and reduced species. When standard conditions are present, the concentration of the hydrogen ion in the sample half-cell is 10^{-7} M (pH 7). The voltmeter measures the difference in reduction potential between the reference and sample half-reactions. Because the standard reduction potential of the reference half-reaction is 0.0 V, one determines the value of the *sample* half-reaction.

Reaction systems that are *more likely to donate electrons* have more *negative* potentials. Therefore, electrons spontaneously flow towards more positive reduction potentials.

Standard Reduction Potentials and Gibbs Free Energies
The standard reduction potential associated with electron transfer can be used to calculate $\Delta G^{0\prime}$ using the following equation:

$$\Delta G^{0\prime} = -n \, \mathscr{F} \, \Delta E^{0\prime} \qquad (1)$$

where n is the number of electrons transferred, $\mathscr{F}$ is *Faraday's constant* (96.48 kJ V^{-1} mol^{-1}); and $\Delta E^{0\prime}$ is the *difference* between the *standard reduction potentials* of the *oxidized* and *reduced species* (in V).

$$\Delta G^{0\prime} = -RT \ln K_{eq} \qquad (2)$$

$$\Delta E^{0\prime} = - (RT/n \, \mathscr{F}) \ln K_{eq} \qquad (3)$$

Standard reduction potentials for some important biological half-reactions are given in the following table.

Reduction half-reaction	$E^{0\prime}$ (V)
Acetyl CoA + CO_2 + H^+ + 2 e^- → Pyruvate + CoA	- 0.48
Ferredoxin (spinach), Fe^{3+} + e^- → Fe^{2+}	- 0.43
2 H^+ + 2 e^- → H_2 (at pH 7)	- 0.42
α-Ketoglutarate + CO_2 + 2 H^+ + 2 e^- → Isocitrate	- 0.38
Lipoyl dehydrogenase (FAD) + 2 H^+ + 2 e^- → Lipoyl dehydrogenase (FADH$_2$)	- 0.34
$NADP^+$ + 2 H^+ + 2 e^- → NADPH + H^+	- 0.32
NAD^+ + 2 H^+ + 2 e^- → NADH + H^+	- 0.32
Lipoic acid + 2 H^+ + 2 e^- → Dihydrolipoic acid	- 0.29
Glutathione (oxidized) + 2 H^+ + 2 e^- → 2 Glutathione (reduced)	- 0.23
FAD + 2 H^+ + 2 e^- → $FADH_2$	- 0.22
FMN + 2 H^+ + 2 e^- → $FMNH_2$	- 0.22
Acetaldehyde + 2 H^+ + 2e^- → Ethanol	- 0.20
Pyruvate + 2H^+ + 2 e^- → Lactate	- 0.18
Oxaloacetate + 2 H^+ + 2 e^- → Malate	- 0.17
Cytochrome b_5 (microsomal), Fe^{3+} + e^- → Fe^{2+}	0.02
Fumarate + 2 H^+ + 2 e^- → Succinate	0.03
Ubiquinone (CoQ) + 2 H^+ + 2 e^- → $CoQH_2$	0.04
Cytochrome b (mitochondrial), Fe^{3+} + e^- → Fe^{2+}	0.08
Cytochrome c_1, Fe^{3+} + e^- → Fe^{2+}	0.22
Cytochrome c, Fe^{3+} + e^- → Fe^{2+}	0.23
Cytochrome a, Fe^{3+} + e^- → Fe^{2+}	0.29
Cytochrome f, Fe^{3+} + e^- → Fe^{2+}	0.36
NO_3^- + e^- → NO_2^-	0.42
Photosystem P700	0.43
Fe^{3+} + e^- → Fe^{2+}	0.77
½ O_2 + 2 H^+ + 2 e^- → H_2O	0.82
Photosystem II (P680)	1.1

(Adapted from Moran *et al.*, *Biochemistry* (2nd Ed.), 1996, p. 14-27, Table 14-4.)

The Nernst Equation

In most cases, the concentrations of the species under consideration will not be 1 M, so one must adjust to the conditions of interest using the *Nernst Equation*. Similar to the relation between ΔG and $\Delta G^{0\prime}$, the *actual reduction potential* (ΔE) can be calculated from the *change in standard reduction potential* ($\Delta E^{0\prime}$) and the *partition coefficient* (Q) for the redox reaction.

$$\Delta E = \Delta E^{\circ\prime} - (RT/n\,\mathcal{F})\ \ln\ ([A_{ox}][B_{red}]/[A_{red}][B_{ox}]) \tag{4}$$

At 298 K, equation 4 reduces to:

$$\Delta E = \Delta E^{\circ\prime} - (0.026/n)\ \ln\ Q \tag{5}$$

where Q is the ratio of actual concentrations of reduced and oxidized species.

When a reaction is not under standard conditions, one can calculate the electromotive force using the Nernst equation and actual reactant and product concentrations. Remember that a positive ΔE signifies a spontaneous reaction.

Example Calculation

(*1*) Consider the net reaction for electron transport, a redox reaction between NADH and O_2. The oxidized half-reaction will have a *more negative* standard reduction potential. Here, *NADH* will be *oxidized* and *oxygen* will be *reduced*.

$$NAD^+ + 2\ H^+ + 2\ e^- \rightarrow\ NADH + H^+ \qquad E^{\circ\prime} = -0.32\ V$$

$$\tfrac{1}{2}\ O_2 + 2\ H^+ + 2\ e^- \rightarrow\ \ H_2O \qquad E^{\circ\prime} = 0.82\ V$$

The *net reaction* is:

$$NADH + \tfrac{1}{2}\ O_2 + 2\ H^+ \rightarrow NAD^+ + H_2O \qquad E^{\circ\prime} = 1.14\ V$$

Using equation 1:

$$\Delta G^{\circ\prime} = -n\ \mathcal{F}\ \Delta E^{\circ\prime} = -(2)(96.48\ kJ\ V^{-1}\ mol^{-1})(1.14\ V) = -220\ kJ\ mol^{-1}$$

The $\Delta G^{\circ\prime}$ for ATP synthesis from ADP and P_i is 30 kJ mol^{-1}, so the energy released during oxidation of NADH under cellular conditions is sufficient to drive formation of several ATP molecules.

(*2*) A larger set of comparative calculations are presented in the *Review* problems. They compare the redox capabilities of NAD^+ and FAD, and use selected values from this table. The conclusion agrees with the concept presented in the *Coenzyme* chapter, that FAD is used to accomplish *more difficult* redox reactions than NAD^+.

21.3 Electron Transport Chains

A series of *coupled redox reactions* occur in the electron transport chains of both mitochondria and chloroplasts. NADH and $CoQH_2$ are oxidized by the electron transport chain in each case. The energy is captured and converted by trans-membrane proton pumps into a proton gradient. These gradients drive ATP synthesis by $F_0\ F_1$ ATPases.

Oxygen is the terminal electron acceptor of the reducing equivalents that are delivered to electron transport by NADH and $CoQH_2$. Due to the involvement of O_2, they are called *respiratory* electron transport chains.

Changes in ΔG and ΔE produced across a *combined pair* or *series of reactions* can be calculated as the sum of the ΔG or ΔE values for the individual reactions in the chain. Since the values are additive, the total energy for the chain is the sum of each individual sub-reaction.

Two examples of composite plots that illustrate the patterns of potential energies in redox chains are shown in the *Electron Transport* section, for mitochondrial electron transport, and the *Photosynthesis* chapter, for the *Z-Scheme* in chloroplasts.

Chapter 22

Glycolysis

22.1 Reactions 1 Through 10

The following reactions occur. Pyruvate can be used in three different pathways, producing the three indicated products.

Glucose $\rightarrow$ Pyruvate $\rightarrow$ Acetyl-CoA, Lactate or Ethanol

Glycolysis consists of 10 reactions that can be separated into two phases called the *hexose* and *triose stages*. The hexose stage makes up the first four steps. At step 4, a C-C bond of fructose 1, 6-bisphosphate is cleaved and all subsequent pathway intermediates are triose phosphates. The two triose phosphates formed from fructose 1, 6-bisphosphate in step 4 undergo interconversion to produce the single triose phosphate, glyceraldehyde-3-phosphate, which continues through the remainder of the pathway.

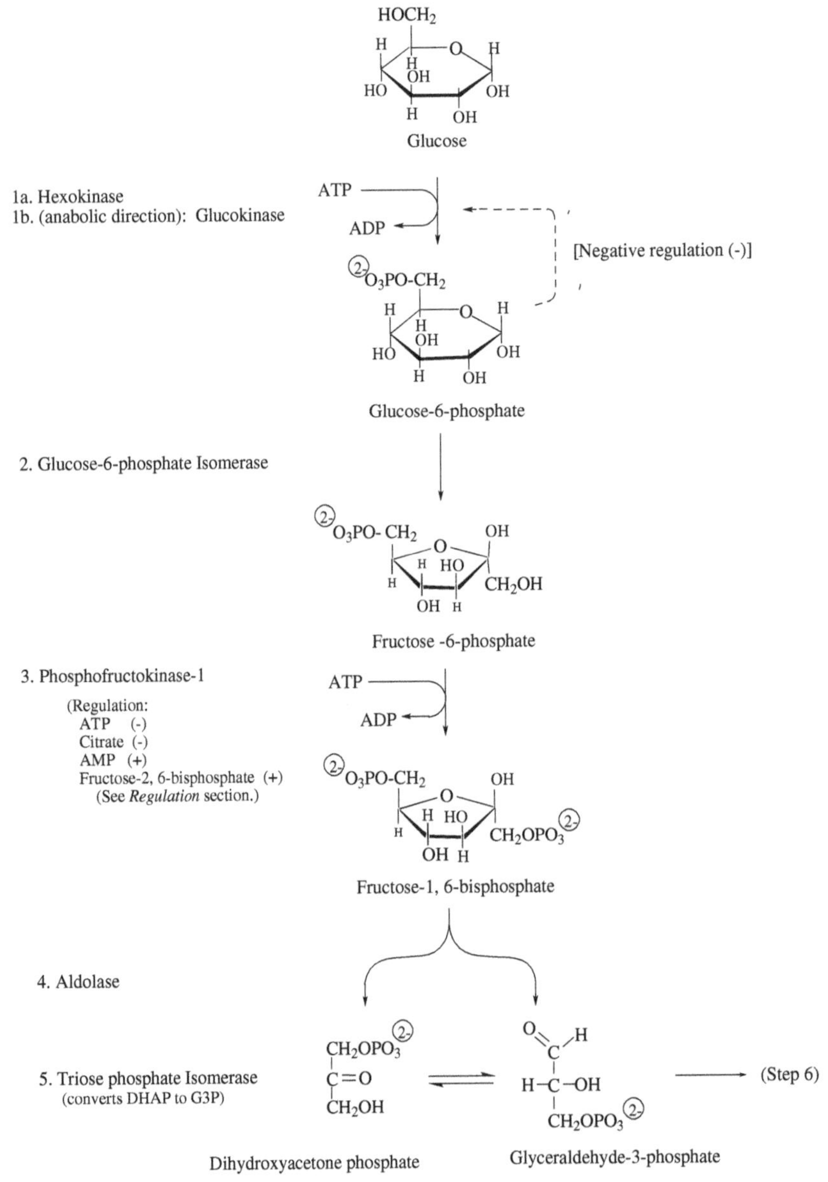

Glucose

1a. Hexokinase
1b. (anabolic direction): Glucokinase

ATP
ADP

[Negative regulation (-)]

Glucose-6-phosphate

2. Glucose-6-phosphate Isomerase

Fructose -6-phosphate

3. Phosphofructokinase-1

(Regulation:
ATP   (-)
Citrate (-)
AMP (+)
Fructose-2, 6-bisphosphate (+)
(See *Regulation* section.)

ATP
ADP

Fructose-1, 6-bisphosphate

4. Aldolase

5. Triose phosphate Isomerase
(converts DHAP to G3P)

(Step 6)

Dihydroxyacetone phosphate

Glyceraldehyde-3-phosphate

(continued)

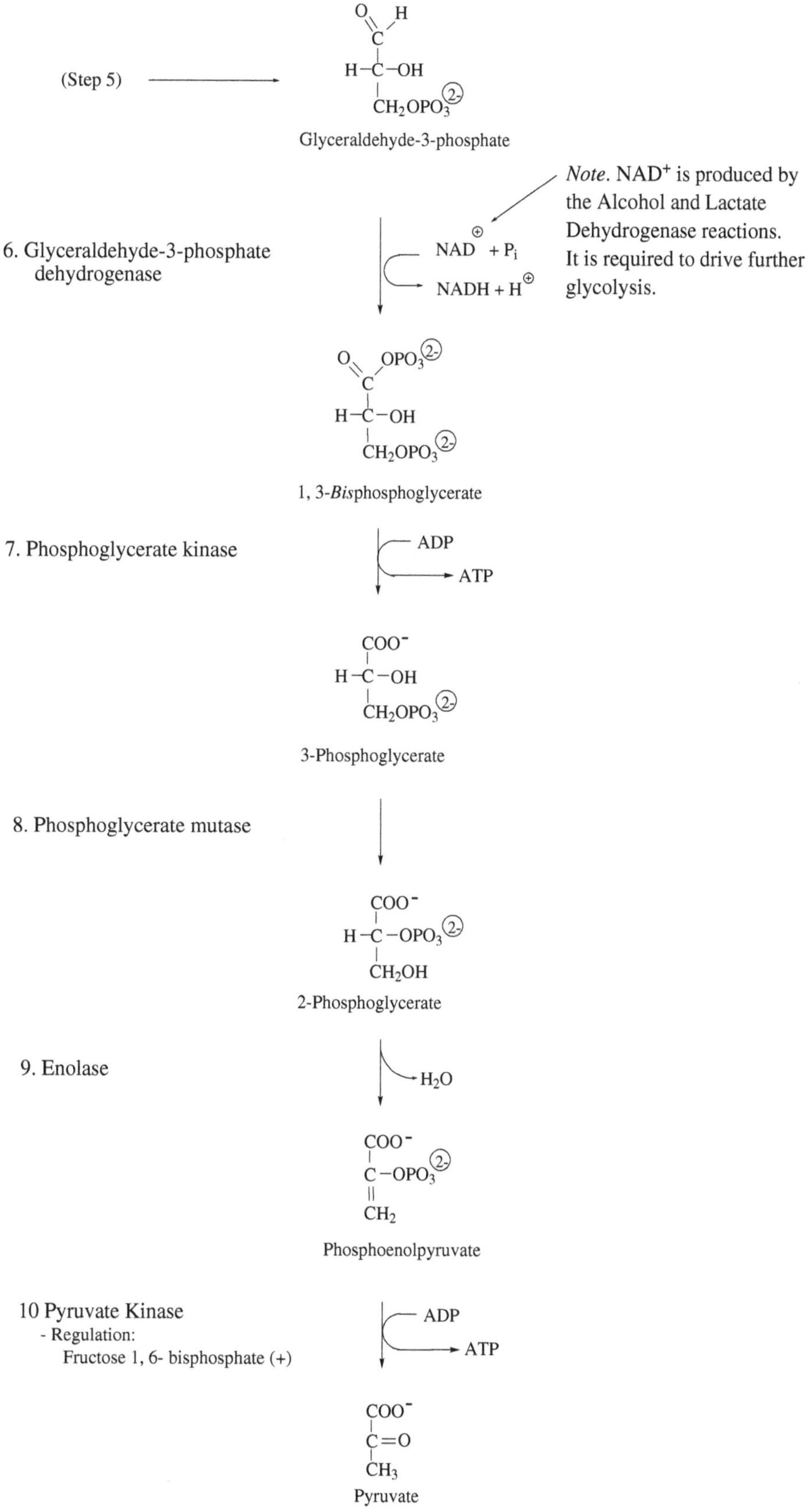

$$\text{(Step 5)} \longrightarrow$$

```
    O   H
     \ //
      C
      |
  H—C—OH
      |
  CH₂OPO₃²⁻
```

Glyceraldehyde-3-phosphate

6. Glyceraldehyde-3-phosphate
   dehydrogenase

NAD⊕ + Pᵢ
NADH + H⊕

Note. NAD⁺ is produced by the Alcohol and Lactate Dehydrogenase reactions. It is required to drive further glycolysis.

```
   O    OPO₃²⁻
    \\  /
      C
      |
  H—C—OH
      |
  CH₂OPO₃²⁻
```

1, 3-*Bis*phosphoglycerate

7. Phosphoglycerate kinase

ADP
ATP

```
   COO⁻
    |
  H—C—OH
    |
   CH₂OPO₃²⁻
```

3-Phosphoglycerate

8. Phosphoglycerate mutase

```
   COO⁻
    |
  H—C—OPO₃²⁻
    |
   CH₂OH
```

2-Phosphoglycerate

9. Enolase

H₂O

```
   COO⁻
    |
    C—OPO₃²⁻
    ‖
   CH₂
```

Phosphoenolpyruvate

10 Pyruvate Kinase
   - Regulation:
     Fructose 1, 6- bisphosphate (+)

ADP
ATP

```
   COO⁻
    |
    C=O
    |
   CH₃
```

Pyruvate

The reactions and enzymes that catalyze them are:

Reaction	Enzyme
(1) Glucose + ATP → Glucose 6-phosphate + ADP + H$^+$	Hexokinase, glucokinase
(2) Glucose 6-phosphate ↔ Fructose 6-phosphate	Glucose 6-phosphate isomerase
(3) Fructose 6-phosphate + ATP → Fructose 1, 6-*bis*phosphate + ADP + H$^+$	Phosphofructokinase-1
(4) Fructose 1, 6-*bis*phosphate ↔ Dihydroxyacetone phosphate + Glyceraldehyde 3-phosphate	Aldolase
(5) Dihydroxyacetone phosphate ↔ Glyceraldehyde 3-phosphate	Triose phosphate isomerase
(6) Glyceraldehyde 3-phosphate + NAD$^+$ + P ↔ 1, 3-*Bis*phophoglycerate + NADH + H$^+$	Glyceraldehyde 3-phosphate dehydrogenase
(7) 1, 3-*Bis*phophoglycerate + ADP ↔ 3-Phosphoglycerate + ATP	Phosphoglycerate kinase
(8) 3-Phosphoglycerate ↔ 2-Phosphoglycerate	Phosphoglycerate mutase
(9) 2-Phosphoglycerate ↔ Phosphoenolpyruvate + H$_2$O	Enolase
(10) Phosphoenolpyruvate + ADP + H$^+$ → Pyruvate + ATP	Pyruvate kinase

ATP is consumed then produced during glycolysis. These gains and losses are:

ATP consumed per glucose: - 2 *hexose* stage
ATP produced per glucose: +4 *triose* stage
Net ATP produced per glucose: **+2**

The steps in the triose stage of glycolysis occur twice per molecule of glucose metabolized, so all yields must be multiplied by two.

22.2 Regulation: Activation and Inhibition

Allosterism.
The following *allosteric regulators* affect the indicated sites: (+) is activation; (−) is inhibition.

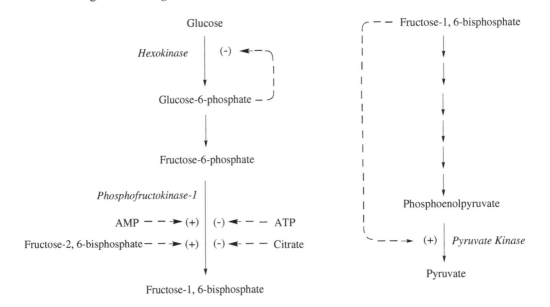

Glucagon

This protein hormone offsets the action of *Insulin*. Instead of causing glucose usage, Glucagon limits it and promotes the storage of glucose as polymeric *Glycogen* in the liver. *Glucagon action* involves the following steps.

(*1*) When the blood glucose concentration is low, *glucagon release* triggers an increase in the activity of *Protein Kinase A*, as described in the *Signal Transduction* chapter.

(*2*) *Protein Kinase A* catalyzes *phosphorylation* of *Phosphofructokinase 2 (PFK-2)*, which inactivates the enzyme and slows Glycolysis.

(*3*) The resulting *decreased concentration of fructose-2, 6-bisphosphate*, a potent activator of PFK-**1**, decreases the activity of PFK-1. The result is a net *activation* of *Gluconeogenesis*.

Fructose-1, 6-bisphosphate   Fructose-2, 6-bisphosphate

22.3 Four Fates of Pyruvate

Lactate

In mammals, the production of lactate in muscle cells is naturally followed by its subsequent *reconversion into pyruvate*. During exercise, lactate builds up in the muscle cells and is transported out, carried through the bloodstream, then into the liver where the enzyme *lactate dehydrogenase* converts lactate into pyruvate. *Oxygen* is required for further *metabolism of pyruvate*. When the oxygen supply in the tissues is low, glycolysis occurs under *anaerobic conditions* to produce lactate. Lactate can accumulate in the blood to sufficiently high levels to produce a condition called *lactic acidosis*, which can cause a dangerous drop in the pH of the blood.

Pyruvate   *Lactate Dehydrogenase*   Lactate

The net reaction for the breakdown of glucose to lactate is:

$$\text{Glucose} + 2\ P_i + 2\ ADP \rightarrow 2\ \text{Lactate} + 2\ ATP + 2\ H_2O$$

As with NAD^+, glycolysis requires that ATP be regenerated to continue flux through the pathway. Adding the reaction for ATP consumption to the reaction above, one obtains:

$$2\ ATP + 2\ H_2O \rightarrow 2\ ADP + 2\ P_i + 2\ H^+$$

Note that lactic acid, not lactate, is the product.

$$\text{Glucose} \rightarrow 2\ \text{Lactate} + 2\ H^+ \rightarrow 2\ \text{Lactic Acid}$$

Metabolic and Food Applications

Lactic acid is the cause of the *Ache in Muscles* during and after *Exercise*.

During *Fermentation* of milk sugars to lactic acid by bacteria, the acid denatures the proteins. As a result, proteins such as β-lactoglobulin, the main component of whey protein, form curds. *Curdling* is required in the production of cheese and yogurt.

Ethanol

Anaerobic Metabolism

Pyruvate can be anaerobically metabolized to produce ethanol in yeast. The net reaction for conversion of pyruvate to ethanol is:

$$Glucose + 2\ P_i + 2\ ADP + 2\ H^+ \rightarrow 2\ Ethanol + 2\ CO_2 + 2\ ATP + 2\ H_2O$$

Under anaerobic conditions, pyruvate is converted to ethanol and CO_2 by yeast cells during oxidation of NADH. Two reactions take place. First, the enzyme *pyruvate decarboxylase* functions to decarboxylate pyruvate into acetaldehyde. Secondly, acetaldehyde is reduced to ethanol by the enzyme *alcohol dehydrogenase* and the cofactor NADH.

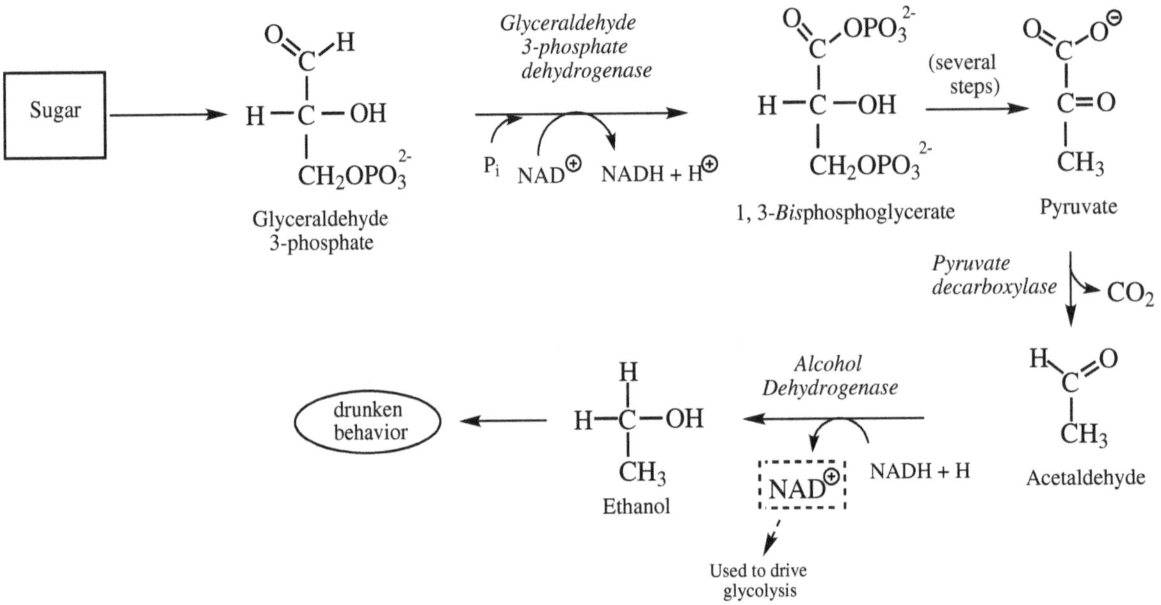

Metabolic Purpose

Reduction of NAD^+ to NADH occurs, so NAD^+ is required as a reactant for glycolysis to proceed continuously. The metabolic purpose of making ethanol is to *regenerate more NAD$^+$*, so more glycolysis can occur under anaerobic conditions. When oxygen is present, NADH oxidation occurs by oxidative phosphorylation, an aerobic process. In the absence of oxygen, the production of ethanol or lactate consumes NADH, regenerating the essential NAD^+ to drive more glycolysis. This makes metabolic sense because:

(*1*) The conditions are anaerobic, so, the yeast cell sacrifices the NADH. It cannot cash it in anyway. Because no O_2 is available, and it is the *terminal acceptor of reducing equivalents in electron transport*, that process cannot occur. Since electron transport drives formation of the proton gradient that fuels oxidative phosphorylation, the latter is also shut down.

(*2*) The anaerobic alternative is to squeeze more metabolic energy out by making more NAD^+ and driving more glycolysis.

Commercial Applications

These reactions are practical in their commercial role of *beer* and *bread manufacturing.* In the brewery, the conversion of pyruvate to ethanol produces *carbon dioxide.* This CO_2 is captured and used to *carbonate* the alcoholic brew, producing the foamy head and effervescent tingle. In the bakery, carbon dioxide causes the bread dough to rise.

Acetyl CoA

Pyruvate Translocase. Getting Acetyl CoA Into the Mitochondrion. Glycolysis and the Krebs cycle are linked through the *bridge reaction,* in which pyruvate is converted to acetyl CoA. Pyruvate from glycolysis is transported from the cytosol to the mitochondria, where the Krebs cycle occurs, by a *symport* called *pyruvate translocase*, which also requires H⁺.

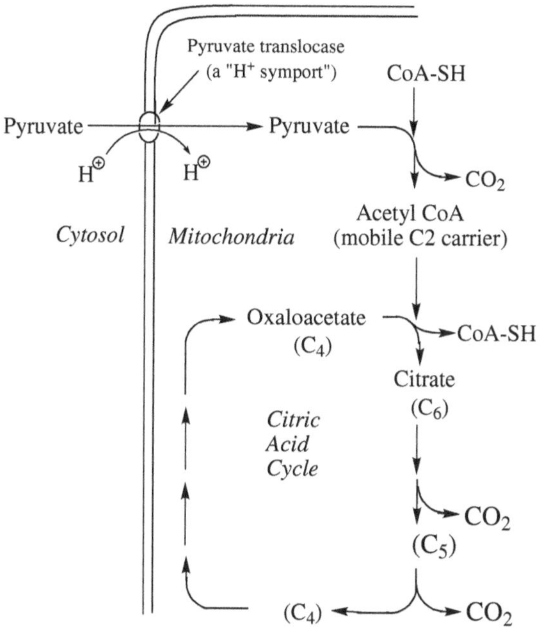

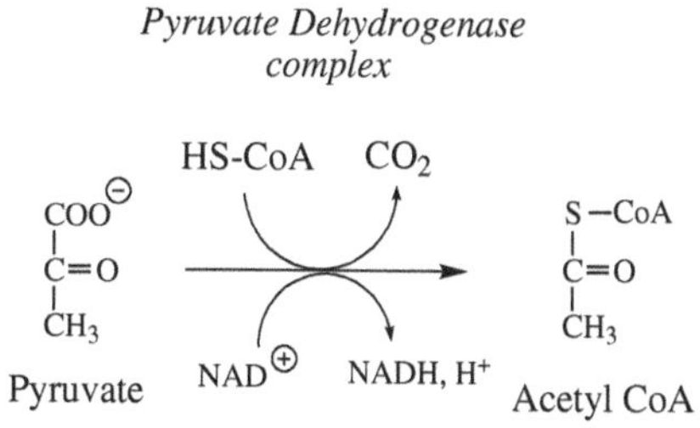

Pyruvate Dehydrogenase: Coenzyme Paradise

The *pyruvate dehydrogenase complex* converts pyruvate to acetyl CoA. Five coenzymes are involved. *Exercise*: Identify them and review the details of each individual redox reaction. What is oxidized and what is reduced? What are the products?]

The pyruvate dehydrogenase complex is a *heterotrimer*, with each successive enzyme catalyzing the next reaction in the cyclic sequence. The product of the first reaction is immediately used as substrate by the next. Metabolites are directed through the system by covalent binding to a flexible prosthetic group bound to the enzyme E_2.

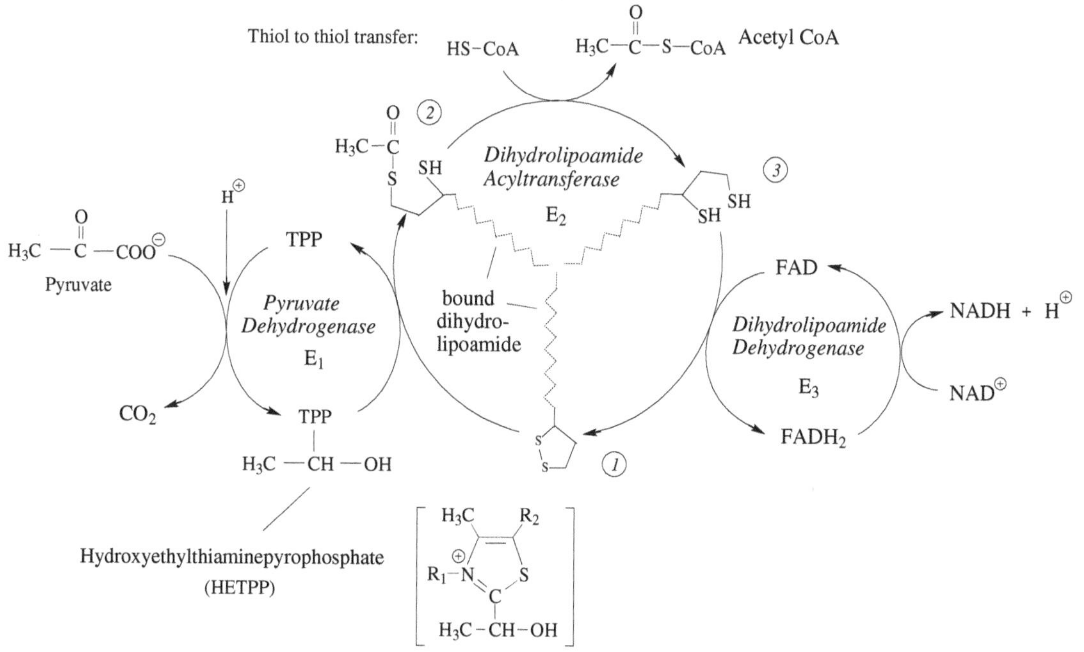

Oxaloacetate. This compound is produced in the first step of *gluconeogenesis* by *Pyruvate Carboxylase*. See the *Biotin* section of the *Coenzymes* chapter for details.

Chapter 23

The Krebs Cycle

23.1 Pathway

This pathway, which is also called the *Citric Acid Cycle* and *Tricarboxylic Acid (TCA) Cycle*, involves the eight consecutive reactions shown below. Each turn of this cyclic pathway begins with the linkage of the two-carbon acetyl group from acetyl CoA with oxaloacetate. The overall process releases two molecules of CO_2, produces three NADHs and one $CoQH_2$, transfers a phosphoryl group to GDP or ADP, and regenerates the original catalyst *oxaloacetate*.

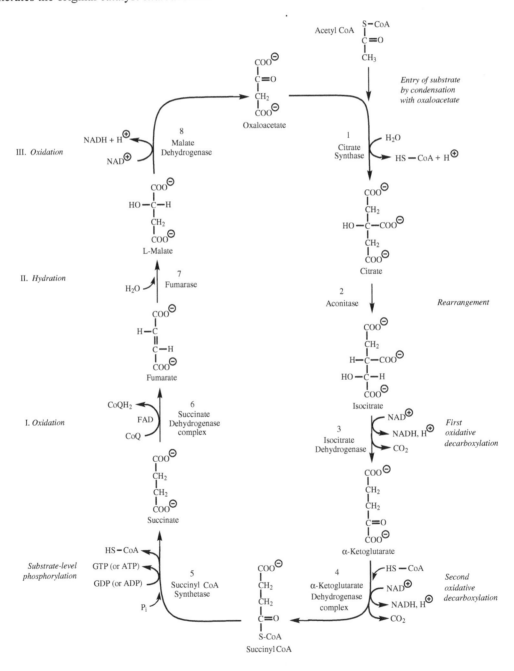

Source Remark

The conceptual logic, graphics and reactions in the metabolism sections generally follow the treatment of Horton *et al*, *Principles of Biochemistry* (4th ed.), 2006. The original material was transcribed from the facts presented by Moran *et al*, *Biochemistry* (2nd ed.), 1994.

Stepwise Reactions

Reaction	Enzyme
(*1*) Acetyl CoA + Oxaloacetate + H_2O → Citrate + CoA-SH + H^+	Citrate synthase
(*2*) Citrate ↔ Isocitrate	Aconitase (Aconitate hydratase)
(*3*) Isocitrate + NAD^+ → α-Ketoglutarate + NADH + CO_2	Isocitrate dehydrogenase
(*4*) α-Ketoglutarate + CoASH + NAD^+ → Succinyl CoA + NADH + CO_2	α-Ketoglutarate dehydrogenase complex
(*5*) Succinyl CoA + GDP (or ADP) + P_i ↔ Succinate + GTP (or ATP) + CoA-SH	Succinyl-CoA synthetase
(*6*) Succinate + CoQ ↔ Fumarate + $CoQH_2$	Succinate dehydrogenase complex
(*7*) Fumarate + H_2O ↔ L-Malate	Fumarase (Fumarate hydratase)
(*8*) L-Malate + NAD^+ ↔ Oxaloacetate + NADH + H^+	Malate dehydrogenase

23.2 Reactions

Net Reaction

Acetyl CoA + 3 NAD^+ + CoQ + GDP (or ADP) + P_i + 2 H_2O →

CoA-SH + 3 NADH + $CoQH_2$ + GTP (or ATP) + 2 CO_2 + 2 H^+

Unique and Common Characteristics

(*1*) The net reaction shows that each molecule of acetyl CoA produces three molecules of NADH, one molecule of $CoQH_2$, and one molecule of GTP or ATP. The citric acid cycle is referred to as a *multistep catalyst* because during each turn, in which one acetyl CoA is oxidized to CO_2, *oxaloacetate is regenerated* at the end of the cycle.

(*2*) *Succinyl CoA Synthetase* catalyzes the *substrate-level phosphorylation* reaction (step 5). To accomplish the reaction, a histidine in the active site is first phosphorylated then transfers the phosphate to either GDP or ADP, forming the trinucleotide.

(*3*) The series of three linearly connected reactions 6, 7, and 8 accomplish the same task as three analogous reactions that occur during the breakdown of fatty acids in the *β-oxidation pathway*.

23.3 Yields

Krebs Cycle Yields

Both NADH and $CoQH_2$ are oxidized by electron transport. ATP is produced concurrently via oxidative phosphorylation.

Krebs Cycle Reactions:

Reaction	Energy-Yielding Product	ATP Equivalents
Isocitrate Dehydrogenase	NADH	2.5
α-Ketoglutarate Dehydrogenase complex	NADH	2.5
Succinyl-CoA Synthetase	GTP or ATP	1.0
Succinate Dehydrogenase complex	$CoQH_2$	1.5
Malate Dehydrogenase	NADH	2.5
Total:		**10**

(*1*) Each molecule of *NADH* oxidized to NAD^+ produces *2.5 molecules of ATP* by oxidative phosphorylation. Each molecule of *$CoQH_2$* oxidized to CoQ produces *1.5 molecules of ATP*.

(*2*) Given these figures, oxidation of one *acetyl CoA* by the *Krebs Cycle* produces approximately *10 molecules of ATP*.

Total ATP Produced Per Glucose

Each of the three central metabolic pathways yield products that contribute to the total number of ATP produced per glucose. Glycolysis produces a net gain of 2 ATPs, 2 NADHs and 2 pyruvates. The bridge reaction produces 2 NADH. Combining these ATPs with those generated by two rounds of the Citric Acid cycle, the total yield is about **32** molecules of **ATP** per glucose.

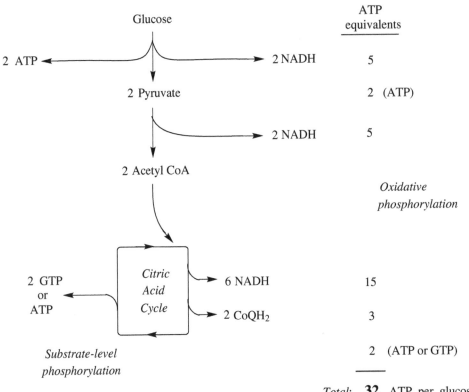

Reaction	ATP Equivalents
Glycolysis	2 + 5 (from 2 NADH)
Pyruvate Dehydrogenase complex (Pyruvate x 2)	5
Krebs Cycle (Acetyl CoA x 2)	20
Total:	**27 + 5 = 32**

23.4 Cellular Redox Potential

The calculations above focus on ATP formation. The cell is sensitive to the *ATP charge* it maintains.
Disulfide Oxidation

Cells must also maintain the proper *redox state*. The complicated set of redox reactions in the pyruvate dehydrogenase mechanism requires that the sulfhydryl components can form the proper oxidized and reduced species, and that appropriately sized pools of them accumulate. If the cellular redox potential is incorrect, these structures may not form as easily as is required for efficient metabolism. The result is a *redox imbalance*, analogous to a pH imbalance during *lactic acidosis*.

Regulating Systems in Prokaryotes and Eukaryotes

The *redox potential* is regulated in prokaryotes and eukaryotes using different redox state maintenance systems. The bacterial system involves a small protein called *Thioredoxin* and the accessory protein *Thioredoxin Reductase*. In eukaryotes, the small tripeptide compound *glutathione* is adjusted by accessory *reductase* and *oxidase* proteins to maintain the proper redox state. The redox state in *E. coli* encourages *disulfide formation*; most eukaryotic cells favor a state in which most *sulfhydryls* are in the *free* state.

23.5 Regulation

Overview

Regulation occurs at: (*1*) the step that feeds the cycle, the bridge reaction, and (*2, 3*) the two *oxidative decarboxylation* steps. As expected, the energy-producing steps are regulated.

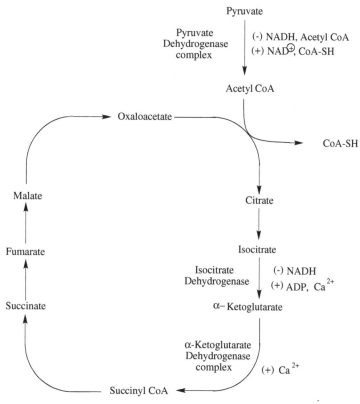

Isocitrate Dehydrogenase

Mammalian *isocitrate dehydrogenase* is allosterically activated by Ca^{2+} and ADP. The enzyme is inhibited by NADH. Mammalian isocitrate dehydrogenase is not covalently modified. In contrast, *E. coli* isocitrate dehydrogenase is *inhibited* by *phosphorylation* at a particular serine.

The protein containing the kinase that *inhibits* isocitrate dehydrogenase can also *reactivate* the enzyme using its *phosphatase activity*. The latter activity is located on a separate domain that catalyzes *hydrolysis of phosphoserine, reactivating* isocitrate dehydrogenase. The kinase and phosphatase activities are regulated reciprocally, *i.e.* an inhibitor of the kinase activates the phosphatase, and *vice versa*.

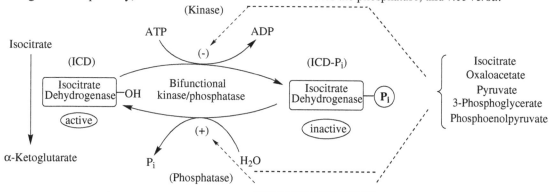

See the *Phosphorylation-Dephosphorylation* section of the *Enzyme* chapter for the mechanistic consequences of phosphorylation.

α-Ketoglutarate Dehydrogenase is also activated by Ca^{2+}.

Chapter 24

Gluconeogenesis

24.1 Reactions

The two offsetting pathways are shown below. *Four reactions in gluconeogenesis* (catalyzed by the enzymes shown on the right) replace the offsetting *three reactions in glycolysis* (catalyzed by the enzymes listed on the left).

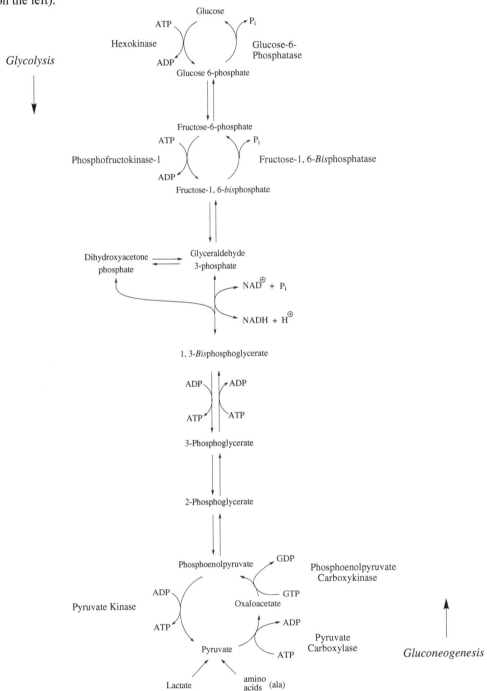

24.2 Regulation

Fructose-2, 6-Bisphosphate

The hormone *glucagon* regulates the fructose-6-phosphate / fructose-1, 6-*bis*phosphate futile cycle as follows. Under catabolic condition, fructose-2, 6-bisphosphate is produced and acts as a potent *activator* of *Phosphofructokinase-1. As a result, glycolysis predominates and gluconeogenesis is inhibited.*

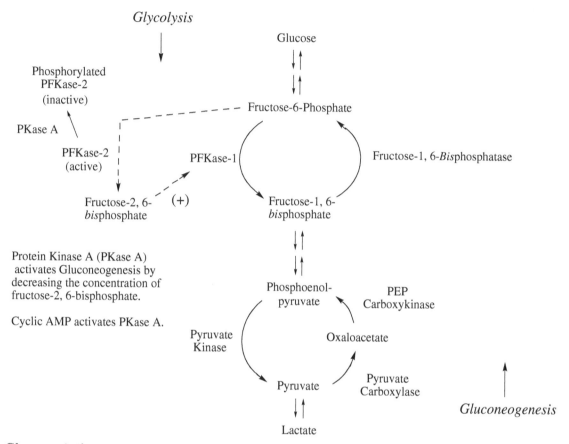

Glucagon Action

In contrast, when glucagon binds its *receptor, cyclic AMP* (cAMP) is produced. This *second messenger* then activates *Protein Kinase A*, which catalyzes *phosphorylation* of *Phosphofructokinase-2.* The resulting decrease in *fructose-2, 6-bisphosphate* concentration removes this potent activator of Phosphofructokinase-1 and *relieves the inhibition of fructose-1, 6-bisphosphatase.* The net result is that *cAMP,* which is produced by the extracellular binding of circulating glucagon, *activates gluconeogenesis.* This example shows how an *intracellular metabolic pathway* can be controlled by an *extracellular hormonal stimulus.*

24.3 Sources Used to Produce Glucose

The *principal substrates* that contribute to gluconeogenesis pathway in mammals are *amino acids*, most commonly *alanine* and *lactate*. These amino acids are produced by breakdown of muscle protein. Under particular metabolic conditions, the oxaloacetate produced by the Krebs Cycle is redirected into the gluconeogenic pathway. Because gluconeogenesis is not saturated by the concentrations of amino acids in the blood, an excess of free amino acids build up, causing an increased conversion of amino acids to glucose. In the same way, the concentration of lactate in the blood does not saturate gluconeogenesis.

Pyruvate is produced from alanine by an *aminotransferase* reaction. The lactate that fuels gluconeogenesis is produced in bulk from the muscle, explaining the large mass and high rate of glycolytic activity of *muscle* tissue. As a result, *other tissues* in the body can strongly affect the level of gluconeogenesis in the *liver*.

Chapter 25

Electron Transport and Oxidative Phosphorylation

25.1 Mitochondria in Red and White Muscle

A cell's overall energy requirement depends upon the number of mitochondria present. White muscle tissue contains a *small* number of mitochondria and acquires energy via *anaerobic glycolysis*. For example, an alligator's jaw is composed largely of *white muscle* tissue. It is able to snap open and shut swiftly but is quickly exhausted and unable to sustain this motion for more than a few seconds. Oxygen-enriched red muscle contains many more mitochondria. Red meat is red because it is suffuse with hemoglobins transporting O_2 into the muscle.

25.2 Overall Process

Oxidative phosphorylation is driven by a *proton gradient*, which is produced by the active transport of protons from the mitochondrial matrix to the inner membrane space, through the mitochondrial inner membrane.

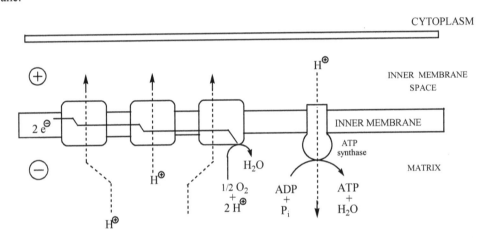

This proton movement creates a gradient, operating as an *aqueous circuit* that connects the electron transport chain to *ATP Synthase*. This is similar to the action of a wire in an electrochemical reaction in that the electrons are passed from the reducing agent, NADH or $CoQH_2$, through the electron transport chain, and finally to the terminal oxidizing agent O_2. This series of *coupled redox reactions* pass the electrons, which drives the proton pumps. The transformed *free energy* is captured by phosphorylation of ADP to produce ATP.

   Two interlinking processes work together to drive oxidative phosphorylation:
(*1*) *NADH and $CoQH_2$* are *oxidized* by the membrane-embedded mitochondrial *electron-transport chain*. This sequence of electron carriers passes the electrons from the reduced coenzymes to O_2, the *terminal electron acceptor* in aerobic metabolism. Coupled to some of the redox reactions in the chain, protons are pumped across the mitochondrial membrane into the inner membrane space. The energy from the oxidation reaction drives the buildup of protons and creating a proton concentration gradient. The *inner membrane space* becomes *more positively charged* than the *matrix*.
(*2*) This accumulation of protons creates a potential energy gradient, called the *protonmotive force*, which supplies the energy that drives the protons back across the inner membrane. The protons flow through the integral membrane enzyme complex called *ATP Synthase* ($F_0 F_1 ATPase$), which phosphorylates ADP to produce ATP.

$$ADP + P_i \rightarrow ATP + H_2O$$

25.3 Chemical Potential Energies That Drive Proton Transport

Source Remark. This section is based on the treatment in Chapter 18 of Moran *et al. Biochemistry* (2nd ed.), 1996.

Free Energy of Proton Transport

The proton motive force is analogous to the electromotive force in electrochemistry. Similar to an electrical circuit, the protons flow from the matrix, through the electron transport-driven proton pumps to the inner membrane space, then back out through the ATP synthase.

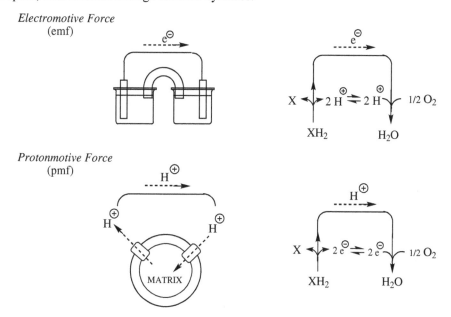

The following reaction shows the reduction of O_2 by a reducing agent XH_2 in an electrochemical cell.

$$XH_2 + \tfrac{1}{2} O_2 \leftrightarrow X + H_2O$$

the reducing agent XH_2 passes electrons along a wire that connects the two electrodes where the redox half-reactions occur.

Electrons pass from the anode, where XH_2 is oxidized,

$$XH_2 \leftrightarrow X + 2 H^+ + 2 e^-$$

to the cathode, where O_2 is reduced.

$$\tfrac{1}{2} O_2 + 2 H^+ + 2 e^- \leftrightarrow H_2O$$

A *salt bridge* connects the two reaction cells, allowing electrons to flow freely between them. The *electromotive force* is the potential energy difference, measured in volts, between the electrodes. The specific *reduction potentials* of XH_2 and O_2 determine the difference in free energy between them, which in turn determines the direction of electron flow and extent of reduction of each oxidizing agent.

Both *chemical* and *electrical potential energy* are generated by the movement of protons across the membrane. *The free energy change* due to *chemical* potential energy is characterized by the difference in proton concentration on each side of the membrane.

$$\Delta G_{chem} = n\,R\,T\,\ln\left([H^+]_{in}/[H^+]_{out}\right) \qquad (1)$$

where n is the number of protons, R is the universal gas constant (8.315 J K^{-1} mol^{-1}) and T is the absolute temperature. Equation 1 can be rewritten as:

$$\Delta G_{chem} = 2.303\,n\,R\,T\,(pH_{in} - pH_{out}) \qquad (2)$$

The *free energy change* due to the *electric potential energy* is characterized by the *change in membrane potential* ($\Delta\psi$).

$$\Delta G_{elec} = z \, \mathcal{F} \, \Delta\psi \qquad\qquad (3)$$

where z represents the *charge* of the transported substance, and $\mathcal{F}$ is Faraday's constant (96.48 kJ V^{-1} mol^{-1}). Because the charge for each proton translocated is 1.0 (n = z), the *total free-energy change for proton transport* is:

$$\Delta G = n \, \mathcal{F} \, \Delta\psi + 2.303 \, n \, R \, T \, \Delta pH \qquad\qquad (4)$$

Protonmotive Force
An expression for the potential that occurs between the two sides of the mitochondrial membrane is obtained by dividing equation 4 by n $\mathcal{F}$. The term $\Delta G / n \mathcal{F}$ is the *proton motive force* (Δp).

$$\Delta p = \Delta G / n \, \mathcal{F} = \Delta\psi + (2.303 \, R \, T \, \Delta pH / \mathcal{F}) \qquad\qquad (5)$$

Note that Δp is not ΔpH. At 25°C, 2.303 RT/$\mathcal{F}$ = 0.059 V and

$$\Delta p = \Delta\psi + (0.059 \, V) \, \Delta pH \qquad\qquad (6)$$

The *proton concentration gradient* is a pH gradient. The other source of free energy in the proton motive force is the *charge gradient*.

25.4 Mitochondrial Electron Transport

The electron transport is composed of *four* complexes which each contain protein subunits and cofactors. Each complex undergoes *cyclic reduction* and oxidation via a *mobile electron carrier, ubiquinone* (CoQ) and *Cytochrome c*, respectively, which provide a link between the subunits. The energies are calculated as described below.

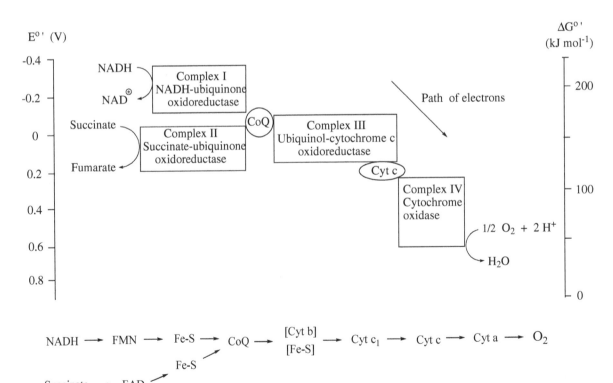

(Adapted from Moran *et al., Principles of Biochemistry* (5th ed.), 2012, p. 424; Fig. 14.6.)

The *standard reduction potential* (V) is directly related to the standard free energy change (kJ mol-1) by the formula:

$$\Delta G^{\circ\prime} = - n \mathscr{F} \Delta E^{\circ\prime} \qquad (7)$$

The energy stored during the electron transfer process occurs in discrete steps, releasing a significant amount of energy, and is stored as a proton concentration gradient.

The following table provides standard reduction potentials of the mitochondrial redox components.

Substrate or complex	$E^{\circ\prime}$ (V)	Substrate or complex	$E^{\circ\prime}$ (V)
NADH	- 0.32	**Complex III**	
Complex 1		Fe-S clusters	+ 0.28
FMN	- 0.30	Cytochrome b_{560}	- 0.10
Fe-S clusters	- 0.25 to - 0.05	Cytochrome b_{566}	+ 0.05
Succinate	+ 0.03	Cytochrome c_1	+ 0.22
Complex II		Cytochrome c	+ 0.23
FAD	0.0	**Complex IV**	
Fe-S clusters	- 0.26 to 0.00	Cytochrome a	+ 0.21
$CoQH_2/CoQ$	+ 0.04	Cu_A	+ 0.24
CoQ^-/CoQ	- 0.16	Cytochrome a_3	+ 0.39
$CoQH_2/CoQ^-$	+ 0.28	Cu_B	+ 0.34
		O_2	- 0.82

(Adapted from Moran *et al.*, *Biochemistry* (2nd ed.), 1996, p. 18-10; Table 18.2.)

The values are only applicable under standard conditions. The *actual reduction potentials* (E) can be calculated from *standard* values ($E^{\circ\prime}$) in a manner analogous to calculating actual free energy changes from standard values.

$$E = E^{\circ\prime} + R\ T\ \ln\ [red]/[ox] \qquad (8)$$

The difference between E and $E^{\circ\prime}$ varies in proportion to the logarithm of the *[reduced] to [oxidized] ratio*. The following table shows standard free energies released in the oxidation reaction catalyzed by each reaction center.

Complex	$E^{\circ\prime}_{reductant}$ (V)	$E^{\circ\prime}_{oxidant}$ (V)	$\Delta E^{\circ\prime}$ (V)	$\Delta G^{\circ\prime}$ (kJ mol^{-1})
I (NADH/CoQ)	- 0.32	+ 0.04	+ 0.36	- 70
II (Succinate/CoQ)	+ 0.03	+ 0.04	+ 0.01	- 2
III (CoQH$_2$/Cytochrome *c*)	+ 0.04	+ 0.23	+ 0.19	- 37
IV (Cytochrome *c*/O$_2$)	+ 0.23	+ 0.82	+ 0.59	- 110

(Adapted from Moran *et al.*, *Biochemistry* (2nd ed.), 1996, p. 18-10; Table 18.3.)

25.5 Electron Transfer and Proton Flow in Complexes I through IV

Complex I

This center contains FMN and a series of Fe-S centers transfer electrons from NADH to CoQ. Two protons are needed for the reduction of CoQ to CoQH$_2$. Approximately *four protons are translocated per pair of electrons transferred in electron transport.*

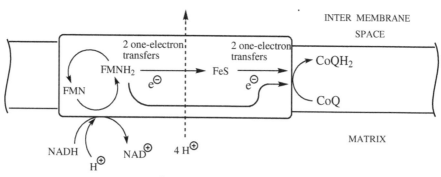

(Adapted from Moran *et al.*, *Biochemistry* (2nd ed.), 1996, pp. 18-13 to -17; Figs. 18-11, -12, -13, and -15.)

Complex II

Electrons enter this complex at the $CoQH_2$ energy level. The proton concentration gradient is not affected by this center.

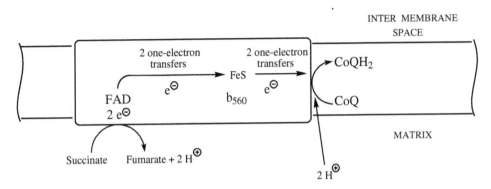

Recall that *Succinate Dehydrogenase* is also a component of the Krebs Cycle. This is that enzyme. Reducing equivalents are transferred from $FADH_2$ to ubiquinone (CoQ), forming the mobile electron carrier ubiquinol ($CoQH_2$).

Complex III

Electrons are transferred from $CoQH_2$ to Cytochrome *c*. When the electrons are passed, two protons move across the membrane from the matrix and two protons are captured by reduction of $CoQH_2$, creating a net gain of *four protons* in the inner membrane space.

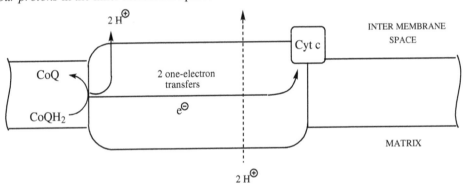

Complex IV

Coupled electron transfer reactions occur between the iron in Cytochrome *c* and the copper in the first of two forms of *Cytochrome a* (A), which passes them to the second copper-containing form (B). Finally, they are passed to oxygen. Half of one molecule and two protons are reduced to form water.

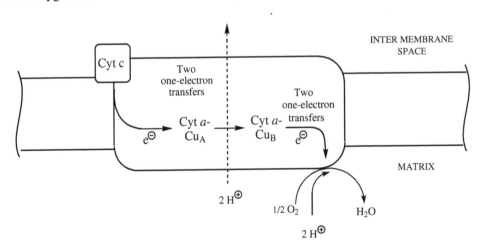

The proton concentration gradient is affected by *Complex IV* in two ways: *(1)* As electrons flow between the a cytochromes, protons are pumped from the matrix to the inner membrane space, and *(2)* Protons are required to form water within the matrix. The net production of proton gradient by this center is nullified.

25.6 Oxidative Phosphorylation

E. coli ATP Synthase

The F_0F_1 ATPase in *E. coli* has a "knob-and-stalk" structure. In *E. coli*, the F_1 component lies on the inside of the plasma membrane, but in mitochondria, F_1 is found on the inside of the inner mitochondrial membrane. The F_0 component forms a proton channel and spans the length of the membrane.

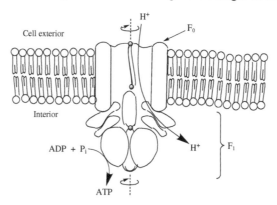

F_1 is composed of three α and three β subunits, which together form the "knoblike" six-sided structure. The "stalk" is composed of one γ, one δ, and one ε subunit, connecting the F_1 and the F_0 base. The F_0 base piece is composed of a complex structure of subunits and structural details that are not easily distinguishable. ATP is produced as the complex rotates around a central axis, powered by the pumping of protons from the inner membrane space through to the mitochondrial or cell interior. In this way, the F_0F_1 ATPase acts as a molecular nanomachine.

Chloroplast ATP Synthase

The ATP synthase is made up of a four-subunit (I, II, III, IV) integral membrane protein complex (CF_0) with two parts: a proton pore which is made of 12-14 copies of subunit III and an external protein complex (CF_1) that generates ATP and is composed of five subunits (α, β, γ, δ and ε). Subunit IV (not shown) binds to both CF_0 and CF_1.

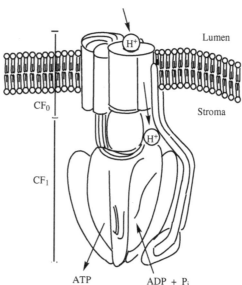

(Adapted from McKee and McKee, *Biochemistry the Molecular Basis of Life* (4th ed.), 2009, p. 475; Fig. 13.6.)

Chemiosmosis

The *Chemiosmotic Theory* proposed by Nobel Laureate Peter Mitchell holds that electron transport in mitochondria produces a proton gradient, which drives ATP biosynthesis (*i.e.,* the mechanism of oxidative phosphorylation).

In 1974, Ephraim Racker and colleagues demonstrated a proof of this principle. Their proton gradient was produced by the protein *bacteriorhodopsin*, which is found in the membranes of the the the archaebacterium *Halobacterium salinarium*. The seven membrane-spanning helices form in this protein (shown below) form a channel that allows *active transport of protons*. A lysine binds one retinal, which can absorb light and undergo the same all-trans to 13-cis photoisomerization reaction that occurs in the opsin-bound coenzyme during visual stimulation. This isomerization releases a proton, which travels up the channel, releasing it on the outside of the membrane.

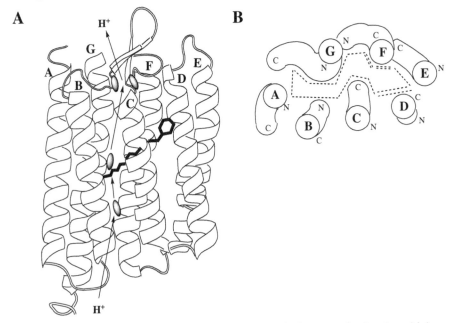

The intent of Racker *et al.* was to couple the proton pump in *bacteriorhodopsin*, which can produce a proton gradient, with a proton gradient-dependent *ATP synthase*, all embedded within a lipid vesicle.

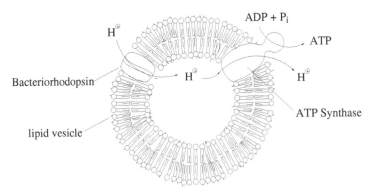

(Adapted from Horton *et al.* Principles of Biochemistry (4th ed.), 2006, p. 461; Box 15.1.)

The result was that bacteriorhodopsin pumped protons into the vesicle, which formed a proton gradient between the inside and outside of the vesicle. This gradient drove the exit of protons through the ATP synthase, catalyzing ATP formation. The essential idea of chemiosmosis was confirmed.

In this system, light-driven isomerization of retinal drives proton gradient formation. In the mitochondrial case, it is driven by electron transport. As described in the *Photosynthesis* chapter, both modes occur in chloroplasts, electron-driven and photoelectron-driven.

Chapter 26

The Malate-Aspartate Shuttle and Proteomics

26.1 Getting NADH into the Mitochondrion: Isozymes

The purpose of this shuttle is to transfer the electron of NADH, which cannot cross the mitochondrial membrane, into the matrix. In the first step, cytosolic NADH is used to reduce oxaloacetate to produce malate, catalyzed by *cytosolic Malate Dehydrogenase*. Next, malate passes through the mitochondrial membrane, in exchange for α-ketoglutarate, catalyzed by the *antiport* called *Dicarboxylate Translocase, and carrying the reducing equivalent it received from NADH.* Within the mitochondrion, malate is oxidized by the *mitochondrial Malate Dehydrogenase* to form oxaloacetate and regenerate NADH, which is used to fuel electron-transport.

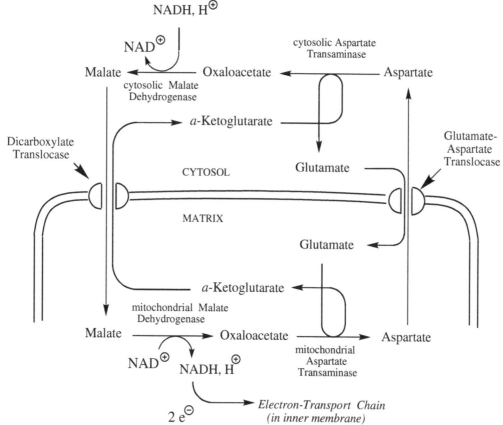

In order to complete the shuttle cycle, oxaloacetate in the matrix must be transported back to the cytosol. Since it cannot move through the mitochondrial membrane, it is converted into aspartate by *mitochondrial Aspartate Transaminase.* Aspartate can pass through the membrane via the antiport *Glutamate-Aspartate Translocase.* Oxaloacetate and glutamate react with mitochondrial aspartate transaminase to produce aspartate and α-ketoglutarate. Glutamate enters the matrix through *Glutamate-Aspartate Translocase.* The α-ketoglutarate exits the mitochondrial matrix via *Dicarboxylate translocase,* allowing entry of more malate. In the last step of the cycle, *cytosolic Aspartate Transaminase,* converts aspartate into oxaloacetate. Note that the cytosolic and mitochondrial versions of the two enzymes Aspartate Transaminase and Malate Dehydrogenase operate in opposite directions.

NADH Yields: Cytosolic Versus Mitochondrial
Reduction of oxaloacetate in the cytosol uses a proton. When malate is reduced in the mitochondrial matrix, the proton is released. Since electron transport oxidizes cytosolic NADH indirectly, it donates one

less proton to the proton concentration gradient than mitochondrial NADH. As a result, the total ATP yield for oxidation of two cytoplasmic NADHs is about 4.5, rather than 5.0 for two mitochondrial NADHs.

Regulation: Oxidation Phosphorylation Is Controlled by Cellular Energy Demand

Oxidative phosphorylation is not regulated simply by feedback inhibition or allosterism. The rate is controlled predominantly by substrate availability and cellular energy demand. Concentrations of intramitochondrial NADH, O_2, P_i, and ADP determine the rate of respiration.

26.2 Isozymes and Proteomics

Isozymes

The Malate-Aspartate Shuttle makes use of two sets of *isozymes*, cytosolic and mitochondrial versions of Malate Dehydrogenase and Aspartate Transaminase. The cytosolic and mitochondrial Aspartate Transaminase and Malate Dehydrogenase operate in opposite directions. They are tuned to catalyze the *opposite reaction* in the two compartments.

Proteins are often made in several isozyme forms in cells. For example, we encountered the adult and fetal versions of hemoglobin. Both are hemoglobin, but each is endowed with specific capabilities that allow the cell to adapt to particular circumstances, e.g., adult and fetal O_2 availability.

In 1984, the front lobby of the developmental biology building at Argonne National Labs contained a 20' by 12' picture of a beautifully resolved summary version of a two-dimensional gel of human serum proteins. It was a fascinating set of *groups of isozymes* in different patterns, with many examples of three or more different resolved isozymes! These isozymes are discussed below.

Motivation for Proteomics

The Malate-Aspartate Shuttle demonstrates the essential *coupling* that occurs *between different cellular compartments*. These couplings are essential. They can be inhibited, enhanced, or disconnected by cellular malfunctions, and so forth.

A typical life science problem might address the following sorts of questions:

(1) How does a particular toxin induce cancer?

(2) How does a plastic additive affect the level of a particular hormone?

To answer such inquiries, one must look at how *complex protein mixtures* change as the toxin is administered during the time frame when it induces its effect(s). In such scenarios, researchers find that a *set of proteins* typically changes, sometimes dozens are affected.

26.3 Characterization by Two-dimensional Gel Electrophoresis

Two-dimensional (2D) Gel Electrophoresis is used to analyze complex mixtures of proteins (Google "O'Farrell gels").

(*1*) Proteins are separated in the first dimension by *isoelectric focusing*, in which the proteins migrate to the position of their respective isoelectric points (pI) on a pre-constructed stabilized pH gradient material.

(*2*) The second dimension involves separating the gels *based in different molecular weights* using *sodium dodecyl sulfate polyacrylamide gel electrophoresis* (SDS-PAGE).

The advantage of this 2D approach over one-dimensional electrophoresis gel is that one can separate and characterize proteins even when they are nearly the same molecular weight, and cannot be resolved on the SDS-PAGE gel.

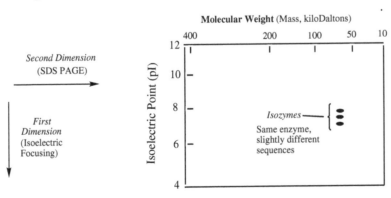

This method allows one to separate large mixtures of proteins and thereby analyze the gene products of that particular genome. 2D Gels have been used extensively to characterize human plasma protein populations. Many of the proteins really consist of two or several isozymes. Such techniques are used in many fields, *e.g.*, medicine, criminology, genealogy, archaeology, migration and habitat mapping, and many others.

26.4 Mass Spectrometry and Proteomics

Proteomics is the study of protein populations and comparisons of these populations after isolation from one or more sources, for example, healthy and specifically infected tissues.

Mass Spectrometry (MS) can be used to determine the individual masses of a set of proteins then separate them into single species, which can then be analyzed for amino acid content, and types and locations of modifications. A single gel spot, corresponding to a selected protein, is taken from the *2D gel*, chemically modified, as required for the MS experiment, then injected into the machine. *Capillary Electrophoresis* is also used as the first separation step, replacing the 2D gel.

Chapter 27

Degradation and Synthesis of Lipids

27.1 Beta Oxidation of Saturated Fatty Acids

The β Oxidation process forms a *spiral metabolic pathway*, with each rotation consisting of *four* enzyme-catalyzed reactions. The products of each round are a single molecule of $CoQH_2$, NADH, acetyl CoA, and a fatty acyl CoA molecule that is two carbon atoms shorter than the molecule that entered that cycle.

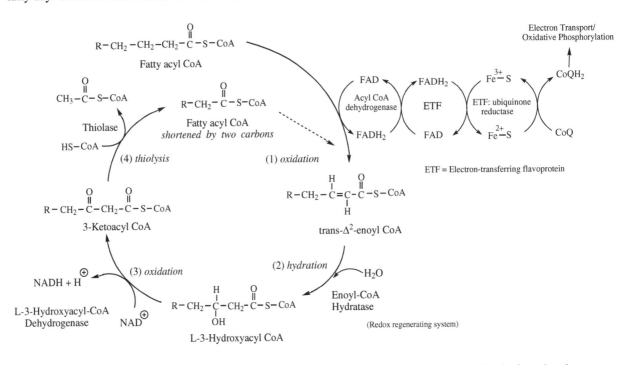

(*1*) In the first step of β oxidation, *enoyl CoA* combines with water to form the L isomer of *3-hydroxylacyl CoA*.

(*2*) Using NAD^+, L-3-hydroxyacyl CoA is then oxidized to *3-ketoacyl CoA*.

(*3*) Next, the nucleophilic sulfhydryl group of Coenzyme A attacks carbonyl C3 of the 3-ketoacyl CoA group, catalyzed by *Thiolase*. This reaction cleaves the methylene-carbonyl C3 bond, releasing acetyl CoA, and producing a fatty acyl CoA molecule *two carbons shorter* than the initial substrate.

(*4*) This shortened acyl CoA molecule becomes the substrate for the next round of β oxidation, continuing the cycle until the fatty acyl CoA molecule is reduced to acetyl CoA.

As the fatty acyl chain shortens, the reactions are catalyzed by particular *isozymes* of acyl-CoA dehydrogenase that are specific for short, medium, long or very long chains.

Note that steps 1 through 3 of β oxidation are the same chemical reactions as in the last three steps of the Krebs Cycle, which convert succinate to oxaloacetate. In addition, these three steps are reversed to accomplish the reverse effect during fatty acid biosynthesis.

27.2 Biosynthesis of Fatty Acids.

Mechanism

The biosynthesis of fatty acids in *E. coli* begins with acetyl CoA and *malonyl CoA*. (*1*) The *loading stage* consists of the esterification of both acetyl CoA and malonyl CoA to ACP. (See the *Coenzyme A* section for the definition of *ACP*.) (*2*) The *condensation stage* is characterized by action of *Ketoacyl-ACP Synthase*, also known as the "condensing enzyme," to accept an acetyl group from acetyl-ACP, thereby releasing ACP-SH. An acetyl group is then transferred to malonyl-ACP by ketoacyl-ACP synthase, forming acetoacetyl-ACP and CO_2. Loss of CO_2, a good leaving group, drives the reaction to completion. The odd point is that each added two-carbon unit in fatty acid biosynthesis is contributed by a *three-carbon* compound.

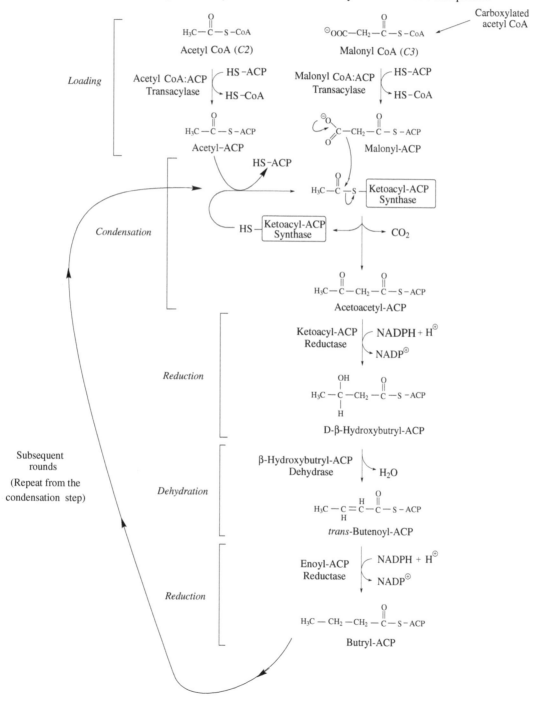

In the *first reduction*, NADPH-dependent Ketoacyl-ACP Reductase catalyzes the conversion of acetoacetyl-ACP to D- β-hydroxybutyrl-ACP. *Dehydration* removes a water molecule from D-β-hydroxybutanoyl-ACP, forming a double bond and producing trans-butenoyl-ACP. In the *second reduction*, trans-butenoyl-ACP is reduced to butyryl-ACP.

The last four stages of this reaction are repeated in the next turn except that *butanoyl-ACP* undergoes the condensation stage with another malonyl-ACP instead of acetyl-ACP. No *loading* is necessary. The third turn uses hexanyl-ACP, the fourth uses octanyl-ACP, and so forth. Trimming the mechanism down to the essential details:

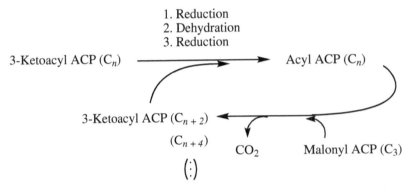

(*1*) Initiation involves condensation of the acetyl and malonyl groups to produce the four-carbon butyryl skeleton.

(*2*) In the elongation stage, three reactions modify the carbonyl-containing precursor, *two reduction* reactions and a *dehydration* reaction.

(*3*) These steps produce *acyl ACP*, which becomes the substrate for the *new condensation reaction* with malonyl ACP, beginning the next turn of the spiral.

Biological Occurrence. Fatty acid synthesis always occurs in the cytosol. In humans, fatty acids are mostly produced in *liver cells, lactating mammary glands*, and to a lesser degree, in *adipose tissue*. The sequence of reactions is catalyzed by a multifunctional enzyme in animals.

Synthesis of Malonyl ACP and Acetyl ACP

Malonyl ACP is the primary substrate for chain elongation in fatty acid biosynthesis. It contributes the two-carbon units in two steps.

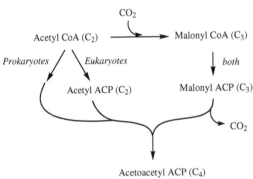

(Adapted from Moran *et al., Principles of Biochemistry* (5th ed.), 2012, p. 476; Fig. 16.1.)

In the first, *acetyl CoA is carboxylated to form malonyl CoA, catalyzed by Acetyl-CoA Carboxylase.* This carboxylation reaction is similar to the mechanism catalyzed by pyruvate carboxylase. The bicarbonate anion (HCO_3^-) is activated by ATP, then transferred to biotin to form *carboxybiotin*. This activated CO_2 is then transferred to acetyl CoA to form malonyl CoA.

Bacterial acetyl-CoA carboxylase is a heterotrimeric complex composed of *Biotin Carboxylase* and a heterodimeric *transcarboxylase*. In eukaryotes, acetyl CoA is converted to acetyl ACP by *Acetyl CoA-ACP Transacylase*. In both types of cells, *Malonyl CoA-ACP Transacylase* catalyzes transfer of the malonyl fragment from CoA to ACP.

Regulation. Acetyl-CoA Carboxylase is metabolically irreversible and the focus of metabolic regulation.

27.3 Length Determination of Fatty Acids

The enzymes that catalyze fatty acid biosynthesis are specialized with respect to preferred chain length. Different isozymes prefer the following size classes: short ($< C_6$), medium (C_6 to C_{12}), long ($> C_{12}$), or very long ($> C_{16}$).

An *elongase reaction* occurs in the synthesis of *arachidonate*, the fatty acid precursor to the *eicosanoids* (step 2; see Sect 12.8). Long-chain fatty acids of length C_{20} and C_{22} are common; longer chains such as C_{24} and C_{26} are less so.

(Adapted from Moran *et al.*, *Principles of Biochemistry* (5th ed.), 2012, p.480; Fig. 16.8.)

27.4 Synthesis of Acidic Phospholipids.

Carbohydrate-carbohydrate joining reactions, such as lactose biosynthesis, require the nucleotide UDP as a cofactor. The UDP is a good leaving group, similar to the loss of pyrophosphate in nucleoside biosynthesis using phosphoribosylpyrophosphate (PRPP). See the *Uridylyl Carbohydrates* and *ATP* sections of the *Coenzymes* chapter for details.

CTP donates a *cytidylyl group* to phosphatidate to create *CDP-diacylglycerol*. In the next reaction, CMP can be displaced by the hydroxyl oxygen of either (*1*) serine to form *phosphatidylserine*, or (*2*) inositol to form *phosphatidylinositol*. The other modified phosphatidyl compounds described in the *Lipids* chapter are made by analogous reactions.

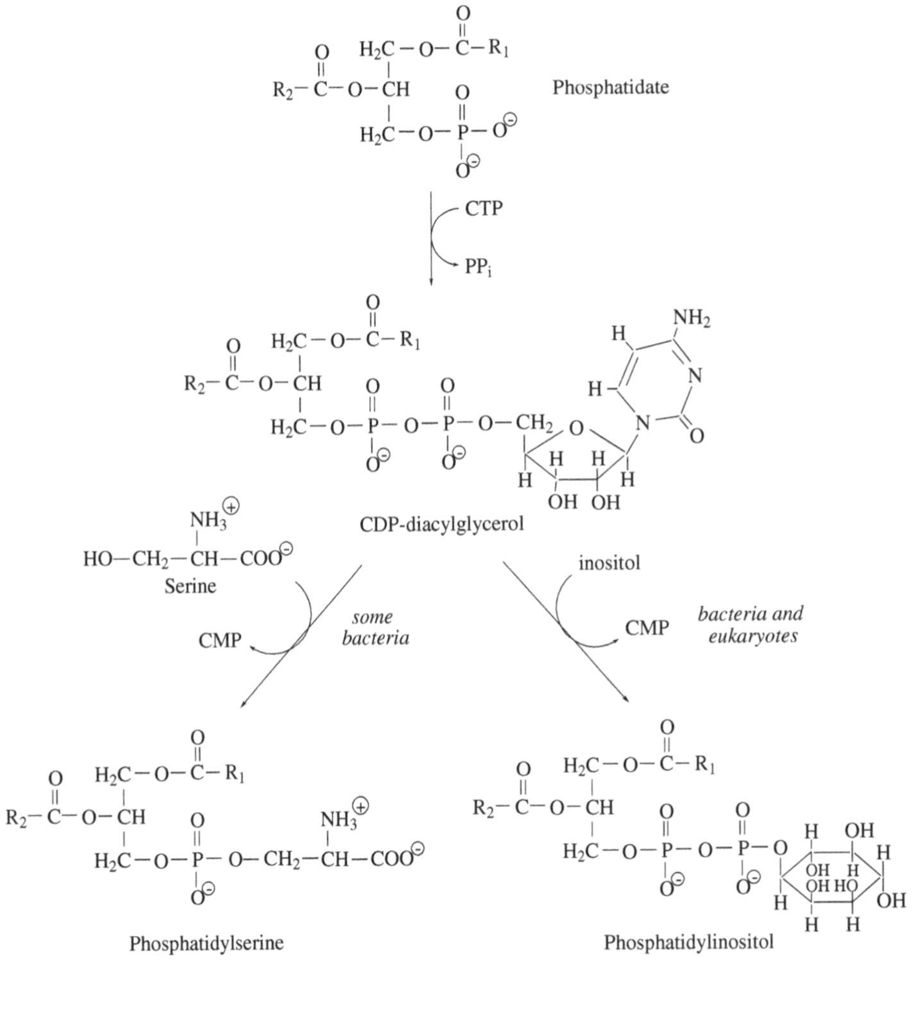

(Adapted from Moran *et al., Principles of Biochemistry* (5th ed.), 2012, p.482; Fig. 16.11.)

27.5 Cholesterol Biosynthesis

Stage 1

The *first stage* in cholesterol synthesis is *making isopentenyl diphosphate*. The steps are:

(*1*) In the first process, three acetyl CoA molecules undergo condensation to produce *HMG-CoA*.

(*2*) Next HMG-CoA is *reduced* to form *mevalonate*. The enzyme that catalyzes this reaction, HMG-CoA Reductase, is the primary control point in cholesterol biosynthesis. The use of *statins* to control cholesterol metabolism through this reaction in humans is described below.

(*3*) Next, mevalonate is phosphorylated, in two sequential reactions, using ATP.

(*4*) Finally, an ATP-dependent decarboxylation reaction produces isopentenyl diphosphate.

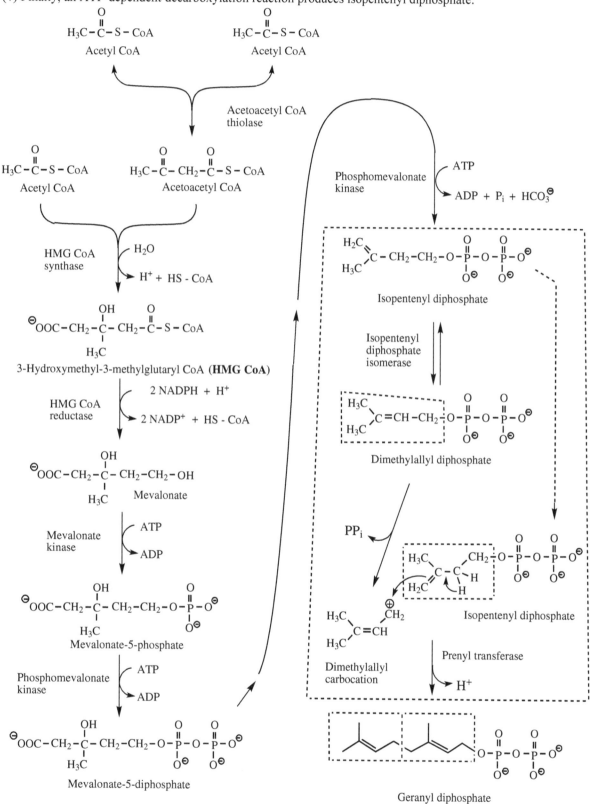

(Adapted from Moran *et al., Principles of Biochemistry* (5th ed.). 2012, p. 490; Fig. 16.17.)

Stage 2

The *second stage* of cholesterol synthesis is marked by isoprene-to-isoprene condensation reactions.

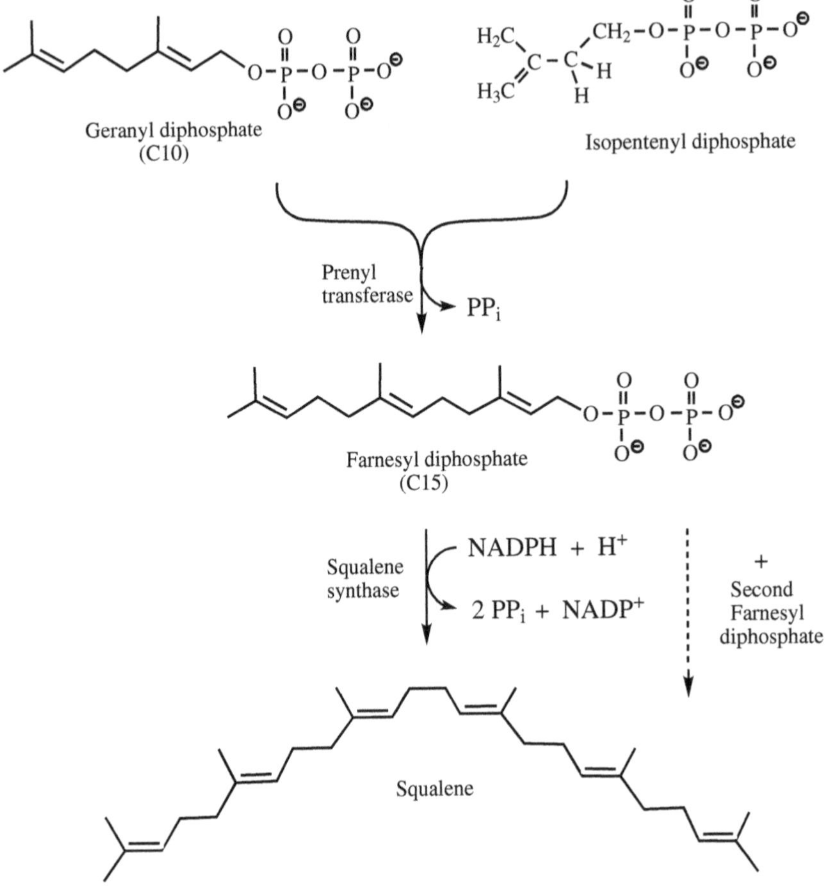

Geranyl diphosphate
(C10)

Isopentenyl diphosphate

Prenyl
transferase → PP$_i$

Farnesyl diphosphate
(C15)

Squalene
synthase

− NADPH + H$^+$

→ 2 PP$_i$ + NADP$^+$

+
Second
Farnesyl
diphosphate

Squalene

(Adapted from Moran *et al. Principles of Biochemistry* (5th ed.), 2012, p. 491; Fig. 16.18.)

Stage 3

The concluding stage of cholesterol synthesis converts squalene to cholesterol. Converting lanosterol to cholesterol is an intensive process, requiring up to 20 steps.

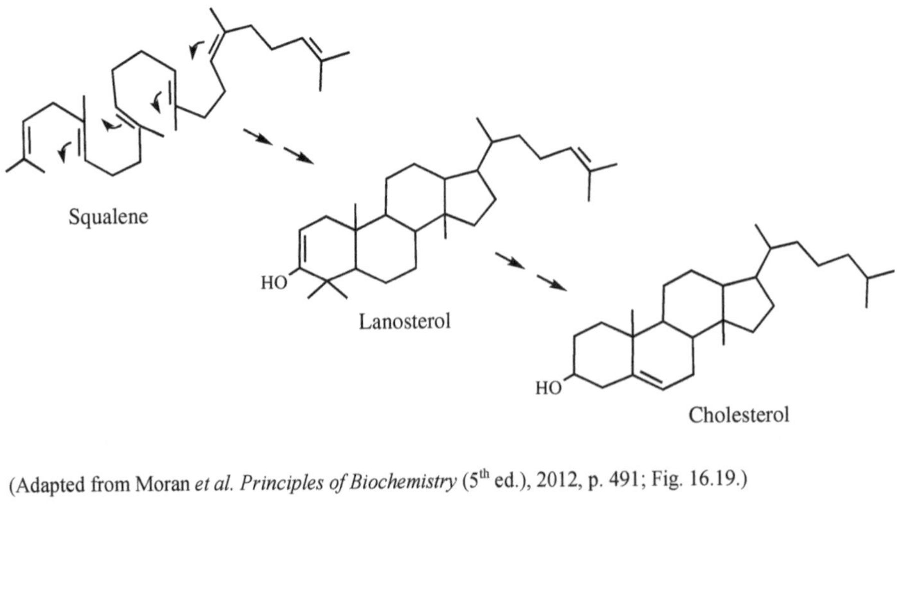

Squalene

HO

Lanosterol

HO

Cholesterol

(Adapted from Moran *et al. Principles of Biochemistry* (5th ed.), 2012, p. 491; Fig. 16.19.)

27.6 Regulating Cholesterol Levels

HMG-CoA Reductase is one of the most highly regulated enzymes known because it *catalyzes the rate-limiting step in cholesterol biosynthesis*. It is controlled by three primary regulatory mechanisms: covalent modification, repression of transcription and control of degradation.

Cholesterol levels in cells are highly regulated because it can suppress transcription. Both intracellular cholesterol in plasma lipoproteins and cholesterol in dietary chylomicrons contribute to transcriptional activation.

By decreasing the serum cholesterol levels in the body, one can decrease their risk of coronary heart disease. A group of drugs known as *statins* are often used to treat *hypercholesterolemia*. They work by competitively inhibiting *HMG-CoA reductase*. The structures of HMG-CoA and two common statins are shown below.

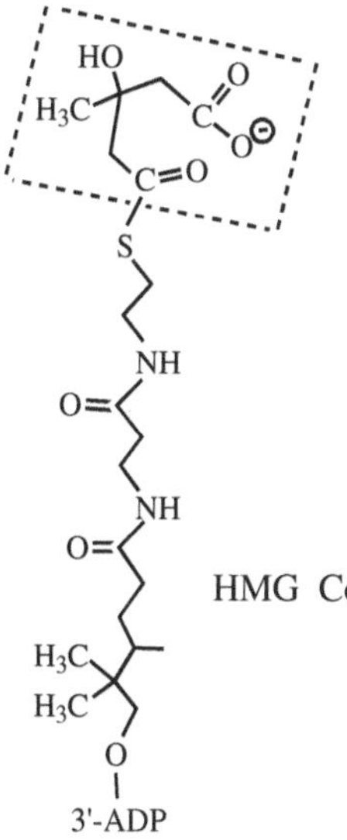

HMG CoA

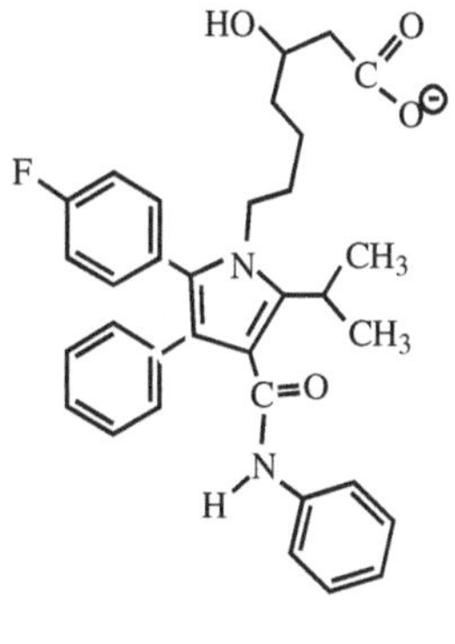

Atorvastin
(Lipitor®)

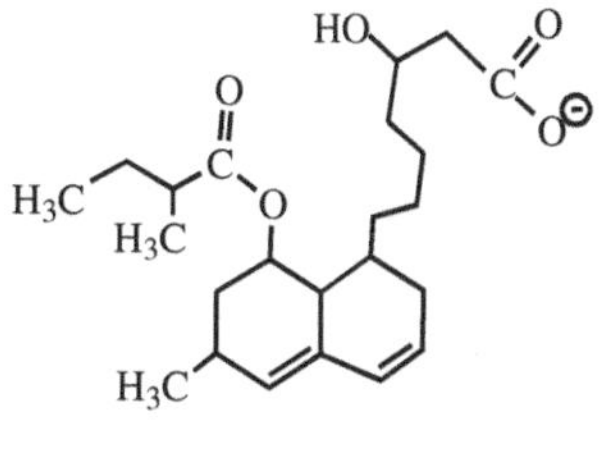

Lovastin
(Mevacor®)

(Adapted from Moran *et al. Principles of Biochemistry* (5th ed.), 2012, p. 492; Box 16.3.)

However, using statins to control cholesterol through inhibition of HMG-CoA reductase can potentially affect other processes. For example it also *inhibits mevalonate* synthesis, which is necessary for make a number of important biomolecules, such as *ubiquinone*.

Chapter 28

Photosynthesis

28.1 Light and Dark Reactions

From childhood, we associate photosynthesis with the green color of plants. Plants use the green molecule *Chlorophyll*, as well as a number of other "accessory pigments," as antennae to capture the energy of light. This energy is converted into NADPH, which is used in the *Ribulose Bisphosphate Carboxylase* (RUBISCO) reaction to "fix carbon," in the form CO_2, to produce the metabolite *3-phosphoglycerate*.

28.2 Photo-Gathering Pigments

Molecules. *Chlorophylls* (Chl) are the primary membrane-embedded light receiving molecules.

Chlorophyll species	R_1	R_2	R_3
Chl *a*	$-CH=CH_2$	CH_3	$-CH_2-CH_3$
Chl *b*	$-CH=CH_2$	$\overset{O}{\overset{\|}{-C}}-H$	$-CH_2-CH_3$
BChl *a*	$\overset{O}{\overset{\|}{-C}}-CH_3$	$-CH_3$	$-CH_2-CH_3$
BChl *b*	$\overset{O}{\overset{\|}{-C}}-CH_3$	$-CH_3$	$-CH=CH_2$

(*1*) β-Carotene - orange and red pigments in carrots and tomatoes

(*2*) Neoxanthin - contains cyclohexane and cyclohexane epoxide ring systems

(*3*) Phycoerythrin - linearized heme ring, pyrrazoles

Absorption Spectra

The following set of spectra show that the photosynthetic antenna pigments span the visible region, assuring that all wavelengths are captured.

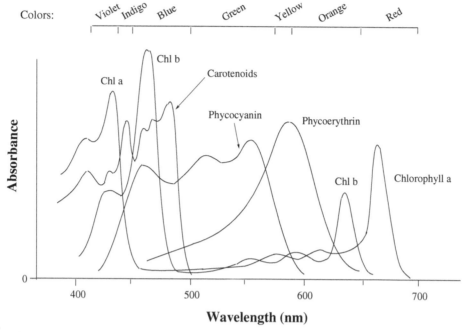

Note. The plant pigments appear as to us visually as the colors that are not absorbed. For example, the *chlorophyll a* and *b* molecules absorb in the blue and red but have a lack of absorbance in the green, so they appear to be green to our eyes.

28.3 Photosynthetic Electron Transport Pathway (Z scheme)

Electrons are transferred between membrane-associated photosynthetic electron carriers depending upon the difference in reduction potential.

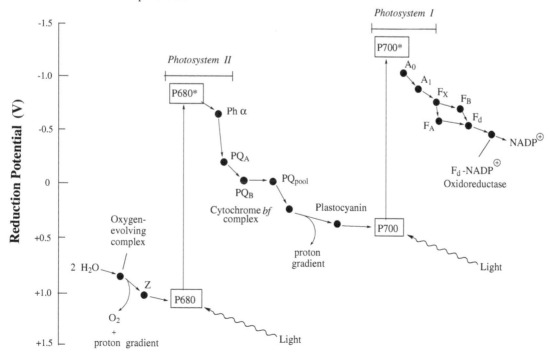

As in mitochondrial electron transport, *electrons flow* naturally from *lower reduction potential* carriers (stronger reductants) *to higher reduction potential* carriers (stronger oxidants).

When *light energy is absorbed* in photosynthesis, the *reduction potential* of *the primary donor* is *forced* to *decrease* dramatically. The photo-excited complexes in the Z-scheme called P680* and P700* have significantly lower reduction potentials than their ground-state equivalents P680 and P700. This indicates that they are stronger reductants (electron donors). The electron is activated to a higher energy state. The two connected consequences of photosynthetic electron transport are:

(*1*) Light drives the *production* of a *proton gradient*.

(*2*) Excess protons are transported from the *granal lumen* to the *stroma*, which drives *ATP production* via oxidative phosphorylation.

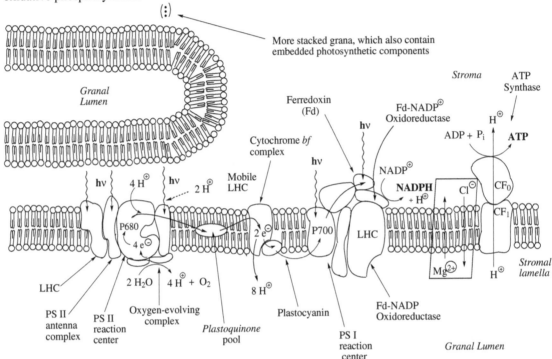

The Chlorophyll Special Pair. The light energy is received by chlorophyll antenna pigments and focused into the pathway shown by the dashed lines. When it reaches the "special pair" of chlorophylls, an electron is contributed to the electron transfer pathway. All of these chlorophylls are held tightly by membrane-bound proteins.

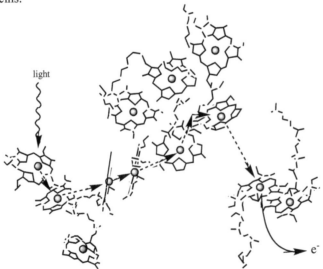

(Adapted from Moran *et al., Principles of Biochemistry* (5th ed.), 2012, p. 446, Fig. 15.3.)

Photosystem I

Photosystem I electron transfer is initiated with the special pair of chlorophylls (P700). The energy is directed through the branch to phylloquinone, then on to the Fe-S clusters and finally to ferredoxin. P700* is reduced by either cytochrome c or plastocyanin.

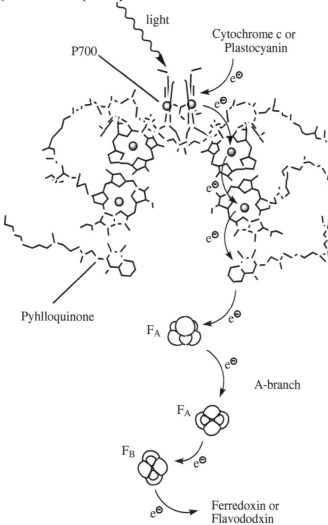

(Adapted from Moran *et al., Principles of Biochemistry* (5th ed.), 2012, p. 451, Fig. 15.9.)

Chapter 29

Carbon Fixation: The Calvin Cycle and C4/CAM Pathways

This pathway was worked out using the unusual radioisotope Carbon-11. The 'usual' tracer isotope used now is C-14. The key effect is capture of gaseous carbon dioxide (CO_2) to form the common metabolic intermediate 3-phosphoglycerate.

29.1 The Dark Reactions: Carbon Fixation

The enzyme complex *Ribulose Bisphosphate Carboxylase* (RuBisCO) catalyzes 'carbon fixation.' Ribulose 1,5-bisphosphate (C5) and three CO_2 molecules form six *3-phosphoglycerates*.

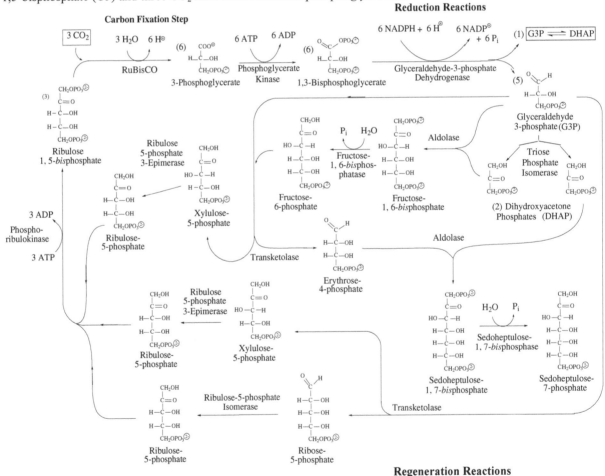

29.2 Biosynthesis of Ribose-5-phosphate

Two *Transketolase*s catalyze carbon skeletal rearrangement steps in the Calvin cycle. Transketolases use Vitamin B_{12} (Cobalamin) to rearrange the skeleton (see the *Coenzyme* section). These rearrangements provide a pool of differently sized skeleton fragments for a variety of biosynthetic reactions.

In one reaction, a Transketolase converts glyceraldehyde-3-phosphate (C3) to xylulose-5-phosphate (C6) and *ribose-5-phosphate* (C5), the sugar subcomponent in RNA, DNA, NADH, and so on.

In another reaction, fructose-6-phosphate (C6) is converted to xylose-5-phosphate (C5) and erythrose-4-phosphate (C4).

The Calvin Cycle can be summarized as follows:

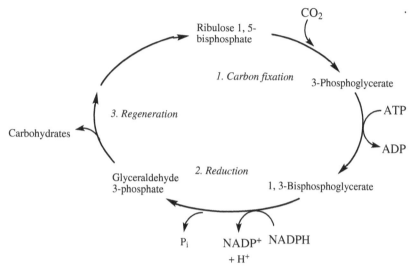

(Adapted from *Moran et al., Principles of Biochemistry* (5th ed.), 2012, p. 462; Fig. 15.20.)

29.3 RuBisCO Mechanism

RuBisCO in plants, algae and cyanobacteria are composed of 8 small and 8 large subunits. The "carbon fixation" mechanism consists of five steps:
(*1*) *Enolization*: keto-enol tautomerism
(*2*) *Carboxylation* (*carbon fixation*): the "ene" electrons attack the carbon of CO_2.
(*3*) *Hydration*: OH^- from H_2O attacks the central carbonyl carbon; the carbonyl oxygen picks up the H^+.
(*4*) *Cleavage*: the "3-gem diol" intermediate cleaves to produce 3-phosphoglycerate (3PGA).
(*5*) *Protonation*: the remaining carbanion abstracts H^+ from an active site base to form a second 3PGA.

(*1*) Carbon is fixed to produce 3-phosphoglycerate, which can interconvert with dihydroxyacetone phosphate, catalyzed by Triose Phosphate Isomerase.

(*2*) Under suitable conditions, these compounds fuel Gluconeogenesis. In woody plants, this provides glucose for the biosynthesis of cellulose, fruit, leaves, wood, and other essential tissues. Plants are our ultimate source of many, if not most, of our essential nutrients and metabolites. Approximately 70% of the atmospheric O_2 is produced by photosynthesis in land plants, especially in tropical forests. Nearly all of our food comes directly or indirectly from plants, energized by sunlight.

(*3*) Reductive conversion of CO_2 to carbohydrate compounds is driven by ATP and NADPH—as a result of the "light reactions." Carbon fixation and carbohydrate synthesis occur in in the cytoplasm of cyanobacteria and the stroma of the chloroplast in algae and plants.

Melvin Calvin: Parking at Berkeley and the Nobel Prize:

As a postdoc at Berkeley, I recall walking past Melvin Calvin's unique plant-green Saab parked in *his exclusive spot*—right next to the Laboratory of Chemical Biodynamics building. Parking at Berkeley was a hassle. If I drove my car, it usually meant a five-block walk, which is no big deal except in the miserable late November rain. I even bought a motorcycle because I could park it under the building I worked in.

One October day, it was announced that Professor Yuan Lee had won the Nobel prize—as had Professor Calvin before him for his Cycle. The student newspaper ran a two-inch headline that read: "Yuan Lee Gets Parking Space!"... At Berkeley, one receives a *lifelong parking space* upon winning the Nobel prize!

29.4 The C₄ and CAM Pathways

Plants that grow in high light intensities and temperatures, *e.g.*, corn, sorghum and maize, have evolved a way to avoid wasteful *photorespiration* (light-dependent water loss) by using a second carbon fixation method called the *C₄ pathway*. The net effect is to increase the ratio of CO_2 to O_2, thereby increasing the activity of RuBisCo and facilitating delivery of CO_2 to the interior of the leaf, where the Calvin cycle is active.

A similar pathway called *Crassulacean Acid Metabolism* (CAM) occurs in succulent plants such as cactus, which grow in an arid climate and must carefully avoid water loss. The solution is that they assimilate CO_2 at night, thereby limiting the water loss that would occur if CO_2 was absorbed during the day. The key is that the plant is covered with a waxy coating and CO_2 exchange occurs through bicellular port structure called *stomata*.

Details of both pathways are described in most standard biochemistry textbooks.

Chapter 30

The Urea Cycle

30.1 Purpose and Reactions

The purpose of the Urea Cycle is to convert ammonium (NH_4^+) to the less toxic *urea*. The overall reaction is:

$$CO_2 + NH_4^+ + \text{aspartate} + 3\,ATP + 2\,H_2O \longrightarrow$$

$$\text{urea} + \text{fumarate} + 2\,ADP + 2\,P_i + AMP + PP_i + 5\,H^+$$

The enzymes that catalyze the individual transformations only occur in liver cells and are distributed in both the cytoplasm and mitochondria, analogous to the reactions in the Malate-Aspartate Shuttle.

(1) The initial steps occur in the mitochondrial matrix and involve the synthesis of *carbamoyl phosphate* from *carbonate* and *ammonium*, the sources of the *carbonyl group* and *one of the nitrogen atoms* in the final urea product. These reactions, which require two ATPs, are shown below.

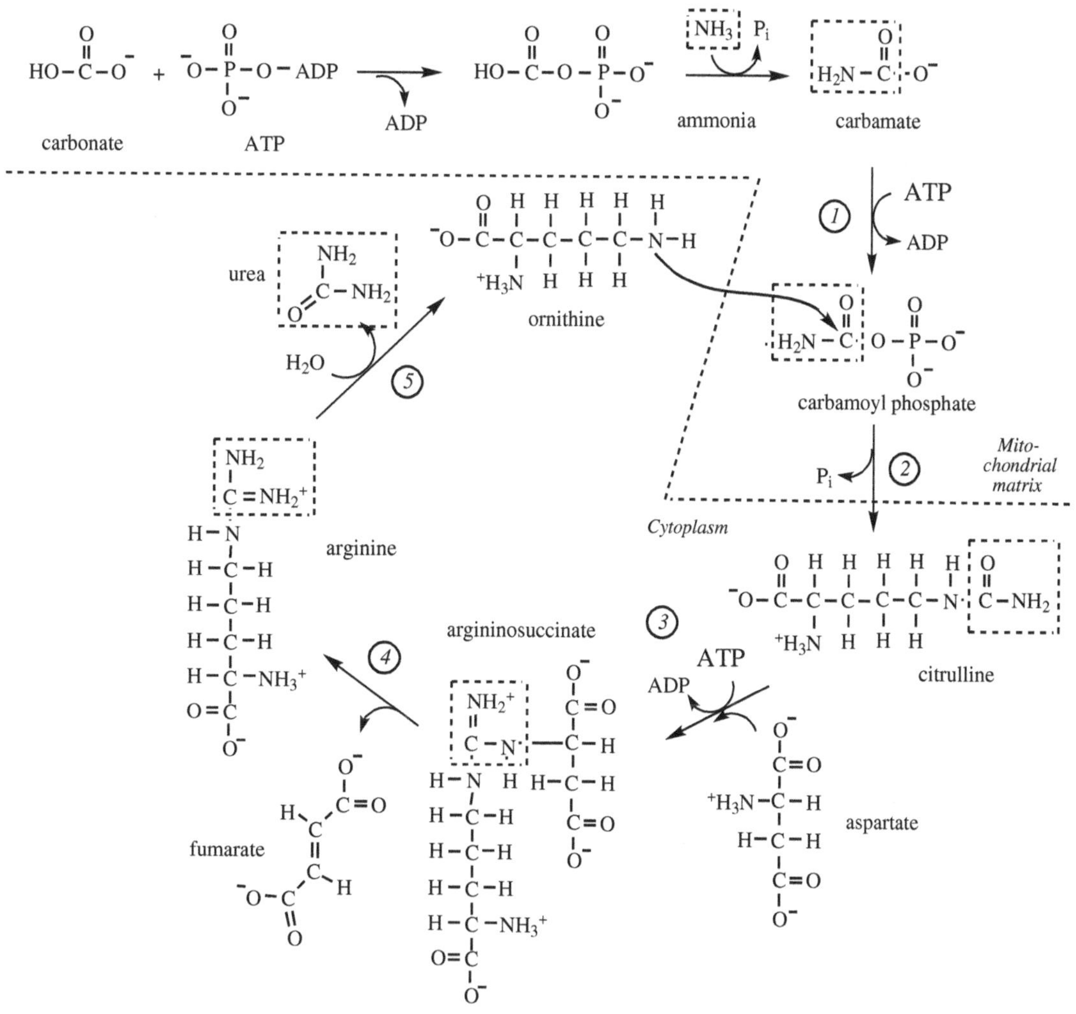

(Adapted from McKee and McKee, *Biochemistry the Molecular Basis of Life* (5th ed.), 2012, p. 546; Fig. 15.3.)

(*2*) Next, carbamoyl phosphate reacts with ornithine to produce citrulline, which is transported out of the mitochondria in exchange for ornithine. This step is driven to completion by loss of phosphate from carbamoyl phosphate. Note that ornithine is regenerated by the cycle, a role analogous to that of oxaloacetate in the Krebs Cycle.

(*3*) In the third reaction, aspartate, which is produced by transamination of oxaloacetate, contributes the second nitrogen atom that ends up in urea.

(*4*) Reaction 4 produces fumarate and arginine, which is cleaved in reaction 5 to yield urea and regenerate ornithine.

The enzymes, which are indicated by the circled numbers, are summarized in the following table.

Number	Enzyme	Reaction
1	Carbamoyl Phosphate Synthase	$HCO_3^- + NH_4^+ + 2\ ATP \rightarrow$ carbamoyl phosphate $+ 2\ ADP + 3\ H^+$
2	Ornithine Transcarbamoylase	carbamoyl phosphate + ornithine $\rightarrow$ citrulline + P_i
3	Argininosuccinate Synthase	citrulline + aspartate + ATP $\rightarrow$ argininosuccinate + $ADP + PP_i$
4	Argininosuccinate Lyase	Argininosuccinate $\rightarrow$ fumarate + arginine
5	Arginase	Arginine + $H_2O \rightarrow$ ornithine + urea

Ornithine is transported in exchange for either H^+ or citrulline. The fumarate is is transported back into the mitochondrial matrix, where it is converted to malate (see the *Malate-Aspartate Shuttle* chapter).

30.2 Regulation

The Urea Cycle is regulated by several interconnected mechanisms.

(*1*) Dietary protein consumption changes the levels of all five of the Urea Cycle enzymes.

(*2*) Glucagon and the glucocorticoid hormones activate their transcription, while insulin represses their production.

(*3*) *Carbamoyl Phosphate Synthetase I* is allosterically activated by *N-acetyl glutamate*, which is made from glutamate and acetyl CoA. *Glutamate* is a major source of excreted NH_4^+. This reaction, which is catalyzed by *N-Acetylglutamate Synthase*, is a good indicator of the cellular glutamate nitrogen load.

(*4*) *N-Acetylglutamate Synthase* is allosterically activated by arginine, so higher arginine levels activate the Urea Cycle at the first committed step.

30.3 Comparative Nitrogen Excretion

Nitrogen is enriched in the nucleic acid bases, especially the purines adenine and guanine. Adenine is converted to hypoxanthine by deamination. The following reactions show how the purines are degraded prior to excretion. Note that different end products exist depending on the type of organism. Marine organisms produce ammonia on surfaces such as gills, where it is washed away before accumulating to toxic levels.

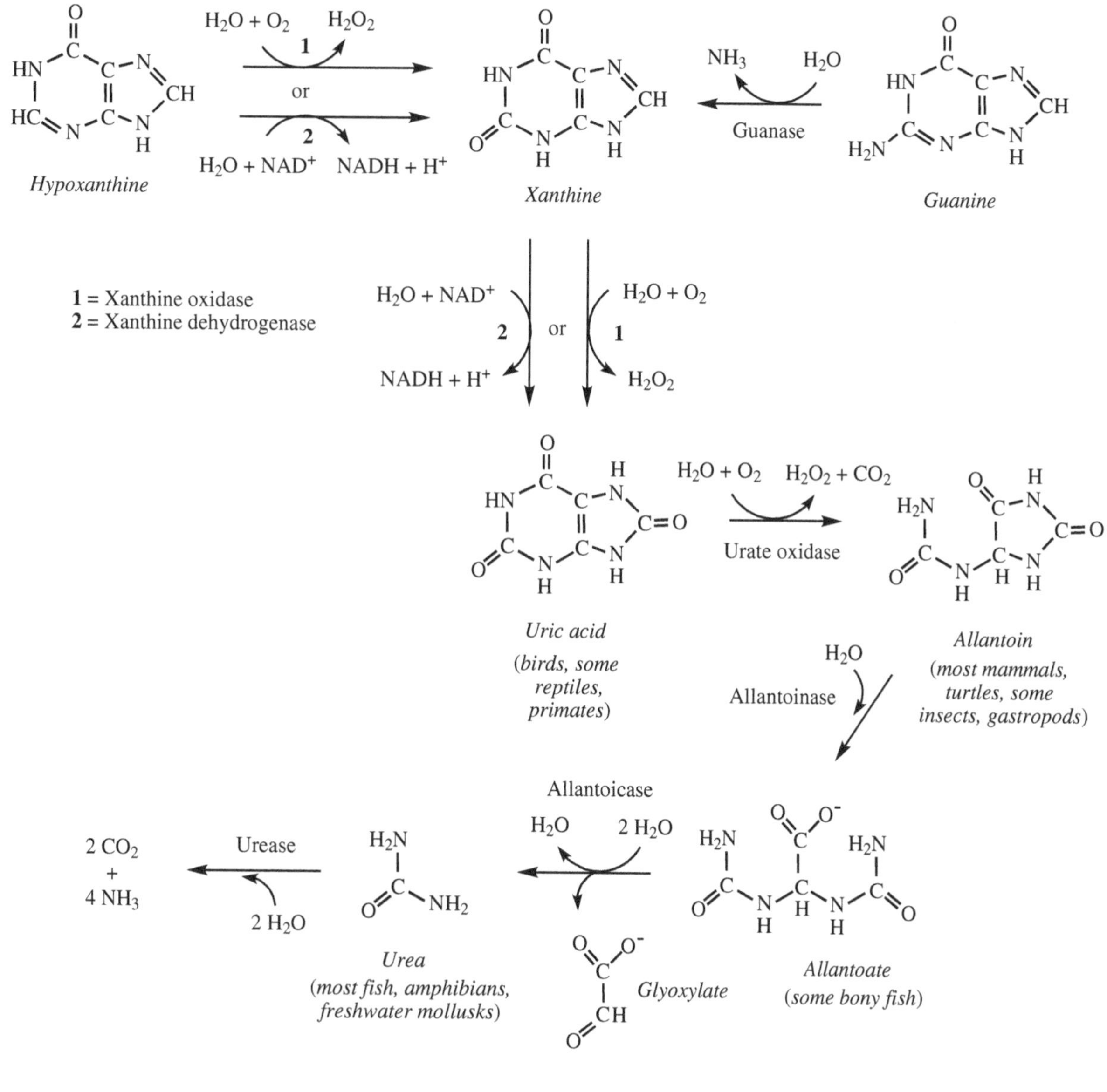
Hypoxanthine
Xanthine
Guanine
Guanase
1 = Xanthine oxidase
2 = Xanthine dehydrogenase
Uric acid
(birds, some reptiles, primates)
Urate oxidase
Allantoin
(most mammals, turtles, some insects, gastropods)
Allantoinase
Allantoicase
Allantoate
(some bony fish)
Glyoxylate
Urea
(most fish, amphibians, freshwater mollusks)
Urease
2 CO2 + 4 NH3

30.4 Protein Degradation and Programmed Cell Death

Eukaryotic cells contain a set of enzymes in both the cytosol and nucleus that link a protein called *ubiquitin*, which is connected via the C-terminal end to the lysine of the target protein. As a result the protein is degraded by a large multi-protein complex called the *proteosome.* ATP is required for the proteosome to degrade the ubiquinated protein.

Programmed cell death is called *apoptosis.* The affected cell decreases in volume, the chromatin becomes fragmented, the plasma membrane becomes damaged and the mitochondria swell. The enzymes that are responsible for cell death are called *caspases*, which generally contain cysteine and react with aspartate residues on the substrate target proteins. Cells can die due to normal developmental turnover, in order to control antibody production and as a result of disease, for example, neurodegenerative disease.

It is interesting to note that another targeting protein analogous to ubiquitin exists. The *SUMO*-mediated pathway targets SUMO-linked proteins for import into the nucleus. The SUMO-protein cross-linkage reaction is called *sumoylation*.

STUDY GUIDE

Review Session for Chapters 1—3

Organelle Functions

Review the organelles and concepts in the *Cell Biology Review.* Example first quiz questions, in our typical format, are shown below. The quiz will be a combination consisting of one of the first three questions, or some variant that addresses the properties and functions of organelles. We typically give some variation of the fourth question, which requires drawing a relatively simple organic molecular structure and describing some typical inherent property.

1. (a) Name and explain the functional process that occurs on the rough endoplasmic reticulum (RER).
(b) What is the name of the "machines" that are present on the RER surface and what do they make?

2. (a) Explain what a chloroplast does.
(b) How are chloroplasts special in a genetic sense?

3. (a) Explain what a mitochondrion does.
(b) How are mitochondria special in a genetic sense?

4. Draw the ethanol molecule. Draw every atom and bond clearly. Circle then label the two parts that differ with respect to solution properties and provide a suitable label for each. (*Hint.* Think about charges and solubility/insolubility.)

Organic Chemical Functional Groups and Their pK$_a$s.

Look up the following items in the textbook, Wikipedia, *etc*. Draw an example of each. Differentiate the *class* of the molecule from a specific *functional group*. What would the pK$_a$ values of each be? Why?

alcohols (pK$_a$ = 15–16 for aliphatic); aromatic (pK$_a$ = 10)

carbonyls, aldehydes, ketones

carboxylic acids (pK$_a$ = 5)

amines (pK$_a$ = 9)

thiols, sulfhydryls (pK$_a$ = 8)

esters

amides

ethers

anhydrides

hemiacetals (acetals)

phosphoryl (pK$_a$ = 2, 6, & 12)

2-mercaptoethanol

urea

guanidinium (and guanidine)

pyrophosphate (pK$_a$ values: 1.52; 2.3; 6.60; 9.25) How much is ionized at pH 7?

acetylphosphate

Chemical Principles

Review the following concepts. Write a carefully worded definition, using proper grammar, punctuation and usage. Include any useful illustration material that will enhance the clarity of your explanation, for example, graph, equation, chemical detail, and/or mechanism.

Oxidation, Reduction

Review the concept associated with the acronym "oil rig." Oxidation is loss of one or more electron(s). Reduction is a gain of electrons. These reactions always involve electron exchange.

Reaction Equilibrium and Thermodynamics

The relation between the equilibrium constant of a reaction and its Gibbs free energy is:

$$\Delta G = - R\, T\, \ln K \qquad\qquad (1)$$

Two contributions lead to the net ΔG are*:*

$$\Delta G = \Delta H - T\, \Delta S \qquad\qquad \text{- the \emph{Gibbs Equation}-} \qquad (2)$$

where ΔH is the *enthalpy change,* and $-T\Delta S$ is the negative product of *temperature* times the *entropy change,* that is, the entropic contribution to ΔG.

Prediction of Reaction Spontaneity.

When $\Delta H < 0$ and $\Delta S > 0$, the reaction is *always* spontaneous.

When $\Delta H > 0$ and $\Delta S < 0$, the reaction is *never* spontaneous.

When $\Delta H < 0$ and $\Delta S < 0$, the reaction is spontaneous at *low temperature.*

When $\Delta H > 0$ and $\Delta S > 0$, the reaction is spontaneous at *high temperature.*

Physical Chemistry Concepts

(*1*) Bond polarity (*i.e.,* the carbonyl C=O bond has a $\delta+$ / $\delta-$ charge distribution)

(*2*) Hydride ($H^{\bullet}$): A hydride is a hydrogen atom with an attached electron. It is *not* H^{+}. This is an electron-free hydrogen nucleus, that is, a proton.

Water-Solute Interactions

(*1*) Hydrophobic effect

(*2*) Hydrogen bonds

(*3*) Van der Waals interactions (London dispersion forces, induced dipole:induced dipole); dipole:induced dipole; dipole:dipole; dipole:monopole (ion)

(*4*) Ionic bonds (salt bridges); (monopole:monopole)

(*5*) Hydrogen bonding: Plot the relation between interaction energy and distance between the two hydrogen bonded atoms.

pH and pK$_a$s

(*1*) A *functional group* is in the *basic* form when the pH is greater than the pK$_a$ of the group. The group is in the *acidic* form when the pH < pK$_a$. To accomplish this transformation, H^{+} dissociates from its conjugate base.

(*2*) Practice drawing the pH titration curves for each amino acid.

Review Session for Chapter 4

The following examples of Quiz 2 questions constitute a review of the structures of the amino acids. The name and score information are included as part of our typical quiz format.

Quiz 2A

1. Draw the structures of the following amino acids at pH 7. Show all bonds and atoms.

(a) L-methionine   (b) L-serine

2. Name (below) *all* of the functional groups in the two molecules drawn in #1, including those on the R group. Be precise with your nomenclature.

(a) L-methionine   (b) L-serine

Quiz 2B

1. Draw the structures of the following amino acids at pH 7. Show all bonds and atoms.

(a) L-histidine   (b) L-glutamate

2. Name (below) *all* of the functional groups in the two molecules drawn in #1, including those on the R group. Be precise with your nomenclature.

(a) L-histidine   (b) L-glutamate

Quiz 2C

1. Draw the structures of the following amino acids at pH 7. Show all bonds and atoms.

(a) L-leucine   (b) L-asparagine

2. Name (below) *all* of the functional groups in the two molecules drawn in #1, including those on the R group. Be precise with your nomenclature.

(a) L-leucine   (b) L-asparagine

Quiz 2D

1. Draw the structures of the following amino acids at pH 7. Show all bonds and atoms.

(a) L-proline   (b) L-aspartate

2. Name (below) *all* of the functional groups in the two molecules drawn in #1, including those on the R group. Be precise with your nomenclature.

(a) L-proline   (b) L-aspartate

Quiz 2E

1. Draw the structures of the following amino acids at pH 7. Show all bonds and atoms.

(a) L-lysine

(b) L-valine

2. Name (below) *all* of the functional groups in the two molecules drawn in #1, including those on the R group. Be precise with your nomenclature.

(a) L-lysine

(b) L-valine

Quiz 2F

1. Draw the structures of the following amino acids at pH 7. Show all bonds and atoms.

(a) L-isoleucine

(b) L-cysteine

2. Name (below) *all* of the functional groups in the two molecules drawn in #1, including those on the R group. Be precise with your nomenclature.

(a) L-isoleucine

(b) L-cysteine

Review Session for Chapters 5 and 6

Look up and define each of the following concepts. These topics are incorporated into example quiz format on the following pages. In subsequent sections, the quizzes will not be shown explicitly.

(*1*) Van der Waals interactions.

(*2*) Hydrogen bonds.

(*3*) Edman degradation. Describe the two different purposes.

(*4*) Disulfide bonds.

(*5*) α-helix. Provide an illustration of the hydrogen bonding pattern as part of your definition.

(*6*) β-sheet. Also provide an illustration of the hydrogen bonding pattern.

(*7*) Hydrophobic effect.

(*8*) Ramachandran plot.

Quiz 3A

1. Define a Van der Waals interaction. Include a description of the relation between interatomic distances and free energies.

2. Define the following terms as completely as you can:
(a) Protein primary structure:

(b) Protein quaternary structure:

Quiz 3B

1. Define a hydrogen bond interaction. Include a description of the relation between interatomic distances and free energies.

2. Define a protein β-sheet; include a drawing of the key hydrogen bonds.

Quiz 3C

1. Define the following terms as completely as you can:
(a) Protein secondary structure:

(b) Protein tertiary structure:

2. Describe Edman degradation. Draw the structure of the key reagent.

Quiz 3D

1. Define a chaotropic agent. Give two specific examples used with proteins.

2. Define a protein α-helix; include a drawing of the key hydrogen bonds.

Quiz 3E

1. Define a disulfide bond. Draw one between two fully drawn amino acids.

2. Explain the purpose of a Ramachandran plot. What are ϕ and Ψ?

Quiz 3F

1. Define a zwitterion. Draw a generic amino acid zwitterion.

2. Define and describe the energetic origins of the hydrophobic effect with respect to protein stability.

Review Session for Chapters 7—8.6

Note. From here on, the quiz questions are presented as lists. Any combination of two or three questions might appear on a given quiz. In order to provide practice at the process of composing answers, none are provided for Review Sessions 4 through 7. To help accelerate the pace in the latter sessions, briefly sketched versions of answers are provided.

1. Define an "initial rate" in the context of enzyme kinetics. Give the relevant equation and name the terms in it.

2. Does an act of catalysis change an enzyme? Explain and give the expression for the generic enzymatic reaction.

3. Define the term "maximum velocity" in the context of enzymatic catalysis. Give the relevant equation and name the terms.

4. Write out the Michaelis-Menten equation and name each term.

5. (a) What are the two definitions of K_M?

(b) Define the Michaelis complex.

6. Define the parameter k_{cat} and provide the key defining equation.

7. Write out the linked chemical equilibrium reactions involving E, S, and other terms that describe simple Michaelis-Menten catalysis.

8. Show how the reactions involving E, S and the inhibitor I differ for competitive, noncompetitive, and uncompetitive inhibition of enzyme catalysis.

9. List the three requirements for enzyme catalysis.

10. How do the *initial rate* and *steady state* ideas coincide in the description of Michaelis-Menten kinetics? Why must they coincide for Michaelis-Menten analysis to work.

Review Sessions for Chapters 8.7—8.8

1. Which external factors limit the rate/velocity of an enzyme-catalyzed reaction? Give three examples and explain how each modulates the catalytic rate.

2. (a) Draw acetylcholine. See part *b* before you start.

   (b) Which atom, on which amino acid in the esterase site of Acetylcholine Esterase, reacts with the substrate? Draw a simple depiction of the enzyme surface beneath your acetylcholine drawing in part (*a*). Show how the recognition and catalytic entities of the enzyme interact with the substrate. Include an arrow to show how the electrons of the key enzymatic atom attack the substrate.

3. (a) Explain in general how the nerve gas antidote pyridine aldoximine methiodide (PAM) reactivates acetylcholine esterase.

   (b) What is the key functional group on the antidote?

   (c) What kind of reaction produces the reactivated enzyme?

4. Draw the generic reaction for the bisubstrate-enzyme "ping-pong" reaction. Define the letters used.

5. What is meant by an enzyme cascade. Include a drawing and indicate how amplification of the initial signal is achieved.

6. Blood clotting is initiated by very few enzymes. Explain how a clot composed of many, many proteins is produced. Include a reaction or reactions and emphasize how such a large number of products is achieved. Name this phenomenon.

7. Define the term zymogen and give one specific example important for food digestion.

8. Describe feedback inhibition. Why is it a logical way to regulate a metabolic pathway?

9. Define allosterism. Describe how it is initiated using the appropriate name.

10. Describe one example of kinase and phosphatase reactions that regulate either glycolysis or the Krebs Cycle. Name the regulated enzyme and explain which form is active and which is inactive.

11. (a) Which global cellular process does Cyclin Kinase regulate? Which two amino acids can be phosphorylated?

   (b) Describe the composition of the Mitosis Promoting Factor complex phosphorylation/dephosphorylation state of the activated form.

Review Session for Chapter 9

1. Give two examples of *reversible* factors that control the catalytic capability of an enzyme. Explain each briefly.

2. Give two examples of *irreversible* factors that control the catalytic capability of an enzyme. Explain each briefly.

3. List the two chemical modes of catalysis. Define each briefly.

4. List the two binding modes of catalysis. Define each briefly.

5. Explain how uncatalyzed and catalyzed reaction coordinate diagrams differ. Explain each change you show.

6. Define a nucleophile and explain what happens in a nucleophilic substitution (S_N2) reaction. Provide a mechanism diagram.

7. (a) Define an electrophile. (Think about how it reacts in contrast to a nucleophile.)

   (b) Explain how to create an electrophile via acid-base catalysis from histidine. Explain why it is an electrophile.

8. (a) Why is histidine used so often by enzymes to carry out acid-base catalysis?

(b) Explain why enzymes typically have an optimal pH.

9. Draw and label the participants in the *catalytic triad* of amino acids typically present in the active site of Serine Proteases.

   (b) Explain how the three amino acids collaborate to accomplish acid-base catalysis.

10. (a) What parameters appear in the Arrhenius equation? Which variable changes and what causes it to change? What are the constants and/or energies?

   (b) Is it a thermodynamic equation or a kinetic equation? Explain. (Hint. The changing variable allows one to distinguish between an equilibrium- or rate-dependent equation.

11. Draw the structure of ATP. Include all atoms and bonds.

12. Why is Mg^{2+} typically required to achieve optimal activity with ATP cosubstrate in enzyme-catalyzed reactions?

Review Session for Chapters 10—11.4

1. Define the term coenzyme. Give one example and describe how the coenzyme is used in a typical biochemical application.

2. Briefly describe one example in which a heavy (transition) metal is used to facilitate a biochemical reaction.

3. What is ATP used for in most biochemical applications? Give two different examples that involve different parts of the molecule.

4. (a) Draw NAD^+ (NADH) in both the oxidized and reduced forms. Use the contraction "R" to signify the common parts of the structure in the latter drawing (i.e., only show the part that changes, connected to "R").

   (b) Describe how this coenzyme is used in a typical biochemical application.

5. Draw $FAD/FADH_2$ in both oxidized and reduced forms.

(b) Describe how this coenzyme is used in a typical biochemical application.

6. (a) Draw pyridoxal phosphate prior to and after forming a Schiff base with the ε-amino group of a lysine residue from an enzyme E.

   (b) Describe how this coenzyme is used in a typical biochemical application.

7. (a) What chemical group does coenzyme A typically carry in the course of its biochemical function? Draw the covalent complex in an abbreviated manner that emphasizes the ligand-carrying functional group of the coenzyme.

   (b) Why is this function such a crucial link in intermediary metabolism?

8. Explain how the noncovalent complex between biotin and avidin is used in biotechnology to capture ligand-binding entities, such as DNA-binding proteins. Provide a step-by-step procedure and diagram that illustrates your explanation.

9. (a) What is the crucial function of N^5, N^{10}-methylenetetrahydrofolate in the production of DNA?

   (b) Explain how our understanding of this function can be used in a strategy for anticancer chemotherapy.

10. (a) How does a ketose differ from an aldose? Give one example of each.

    (b) How does a pyranose differs from a furanose? Give one example of each.

11. (a) What is UDP-galactose and how is it used to make lactose?

    (b) Draw the structure of lactose.

12. How does cis-retinal function in transducing the signal of a photon of light into a chemically recognizable form?

Review Session for Chapters 10—11.4: Key

Note Regarding the Key Material. The *Key* information provided in the remaining sections of the Review material contain rudimentary answers, which should not be misconstrued as sufficiently detailed for use on formal quizzes and tests. Fully developed explanations should be written (and revised) as a means to practice fleshing out informative definitions at the level expected on a formal quiz or test. Students ask me regularly "Do I need to include (some given piece of an explanation or definition)?" My answer is "Will it make it more complete and clear? If so, include it." Our goal is to get you to write excellent answers, which will in turn earn an excellent grade. More importantly, since you wrote it out, you'll be better prepared to remember and use the information in the future.

Structure	*Functional Part*	*Purpose*
1. coenzyme	*e.g.,* NAD^+, FAD, PLP, CoA, ATP (cosubstrate)	Organic cofactor (vitamin) that aids in the catalysis of an enzymatic reaction.
2. heavy metal	*e.g.,* Molybdenum—Xanthine Oxidase (purine catabolism) Zinc—Carbonic Anhydrase; Zinc Finger proteins Cobalt—vitamin B_{12} (one carbon transfers) Iron—Fe_xS_y clusters – redox reactions The following metals are not transition metals: Mg^{2+}, Ca^{2+}, K^+, Na^+.	Metallic atom (transition metal – 3rd period) used as the focus of an enzymatic reaction
3. ATP	cosubstrate; activated monophosphate , pyrophosphate glucose-6-phosphate phosphoribosyl pyrophosphate (PRPP)	ATP + glucose → ADP + glucose-6-P_i ATP + ribose-5-P_i → AMP + phosphoribosyl pyrophosphate
4. NAD^+	 $S_{reduced}$ + $NAD^+_{oxidized}$ ⇌ $P_{oxidized}$ + $NADH_{reduced}$ + H^+	acceptor/donor of reducing equivalent (redox partner) (glycolysis, Krebs) → electron transfer
5. $FAD/FADH_2$		acceptor/donor of reducing equivalent (redox partner) (glycolysis, Krebs) → electron transfer

6. PLP	 (*e.g.*, Acetoacetate Decarboxylase)	R_1, R_2 from substrate removed; weakens bonds α – to the Schiff base with PLP
7. pyruvate	 (*e.g.*, the Pyruvate Dehydrogenase "Bridge Reaction")	takes acetyl group derived from glycolysis and uses it as "fuel" for the Krebs Cycle
8. Biotin		1. Capture a ligand binding protein (usually) and 2. Capture the complex and purify it from everything else by avidin-biotin affinity chromatography
9. Tetra-hydrofolate (THF)	(a) Contributes a methyl to group form dTMP in biosynthesis from dUMP. The THF analog methotrexate inhibits Dihydrofolate Reductase, which makes the N^5,N^{10}-THF for Thymidylate Synthetase. (b) This allows preferential killing of cells that are rapidly incorporating dT into DNA. Cancer grows more rapidly than most "normal" cells, results in preferred cancer cell destruction.	

10. (a) ketose versus aldose (b) pyranose (5 C's) versus furanose (4 C's)	(a) *ketose* *aldose* CH₂OH versus D-glyceraldehyde dihydroxyacetone (b) *pyranose* 5 C's & 1- O in ring *furanose* 4 C's & 1-O in ring (α -D-glucose) (β-D-ribose)	
11. UDP-galactose & lactose	cosubstrate; activates galactose for di- or oligosaccharide synthesis *UDP-galactose* *lactose*	
12. Retinal	*cis*-retinal → absorbance, photoisomerization → *trans*-retinal → chemical signal → receptor linked to optic nerve → visual sensation	

Review Session for Chapters 11.5—12.5

1. (a) Draw the chair and boat conformations of the β-D-glucopyranose structure.

   (b) Explain why a polysaccharide chain would turn directions much more in one case than in the other and which conformational isomer would be expected to produce the most radical change in chain direction.

2. (a) Why is the hexa-atomic ring of inositol more stable than that of galactose?

   (b) Given what you know about the biosynthesis of lactose from UDP-galactose and glucose, why would you expect inositol to be less likely to polymerize than glucose or galactose? (Hint: What are their relative capabilities in serving as nucleophilic centers?)

3. (a) What does NAG-α (1→6)-gal-α (1→4)-glc-β (1→4)-gal mean?

   (b) Draw the molecule with emphasis on the meaning of α and β.

4. (a) How does glycogen differ from the amylopectin in starch?

   (b) How would this facilitate the function of glycogen?

5. How does penicillin work selectively on bacteria but not harm us significantly? (What does it do?)

6. (a) How do extra-cellular surface carbohydrates regulate osmotic pressure around cells?

   (b) What advantage does this impart to cells?

7. Explain how the terms "export" and "clearance" apply to the functional status of carbohydrate-bearing proteins.

8. (a) How does Phospholipase C produce two different second messengers in the corresponding signal transduction pathway?

   (b) Draw an appropriate reactant.

9. What function does chondroitin sulfate serve in cartilage and skeletal joints?

10. (a) Assuming equal length, do saturated or unsaturated fatty acids have a lower melting temperature (T_m)?

    (b) What is the nature of the reaction/equilibrium one measures? What molecular property causes the differences in T_m values?

11. How do lipid bilayers and micelles differ? Include a simple drawing.

12. (a) Draw the general structure of a phosphatidyl choline molecule.

    (b) Name the functional groups.

Review Session for Chapters 11.5—12.5: Key

1. (a) Sugar pucker conformations:

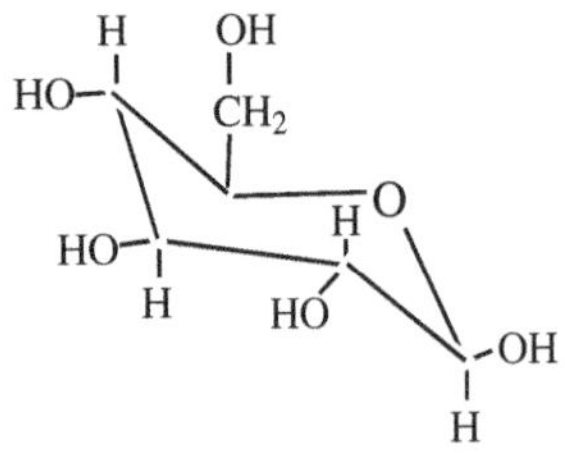
H
OH
HO
CH2
O
HO
H
H
HO
OH
H

chair
conformation

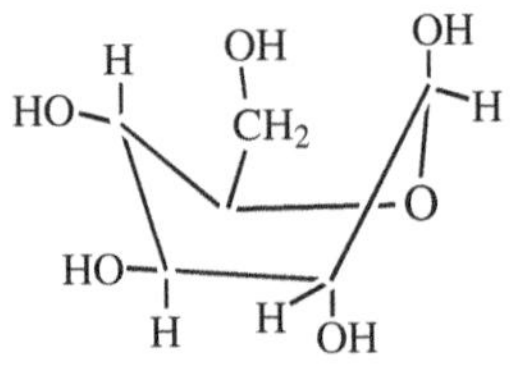
H
OH
OH
HO
CH2
H
O
HO
H
H
OH

boat
conformation

(b) The boat C1 and C4 hydroxyls produce a kink in the polysaccharide that is more pronounced than in the chair polymer.

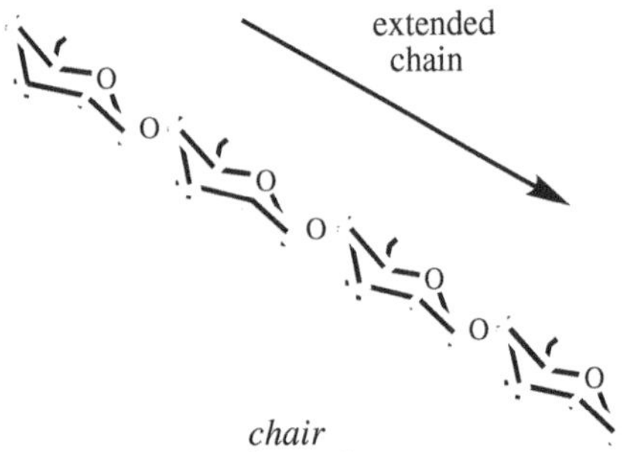
extended
chain

chair
conformation

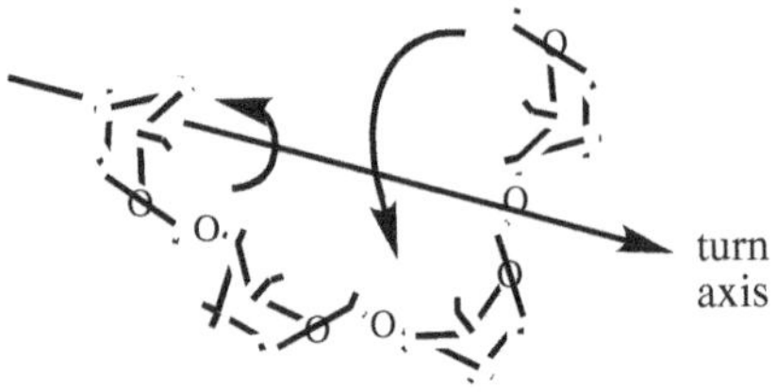
turn
axis

boat
conformation

2 (a) Galactose forms the pyranose via acetal – H^+ catalysis; a reversible reaction. As a result the ring is easily hydrolyzed to form the linear form. In contrast, the cyclohexane ring of inositol must be oxidized by a more powerful oxidant than 10^{-7}M OH^-/H^+ solution. The six OHs on inositol provide plenty of routes to ring opening, but it's harder to do than with galactose.

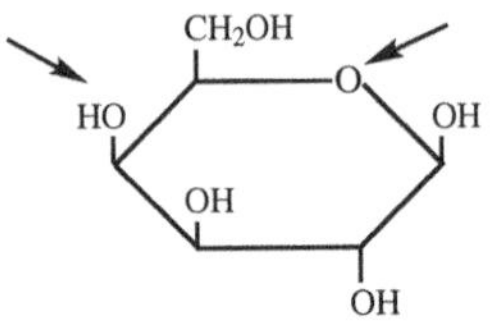
CH2OH
O
HO
OH
OH
OH

β-D-galactose

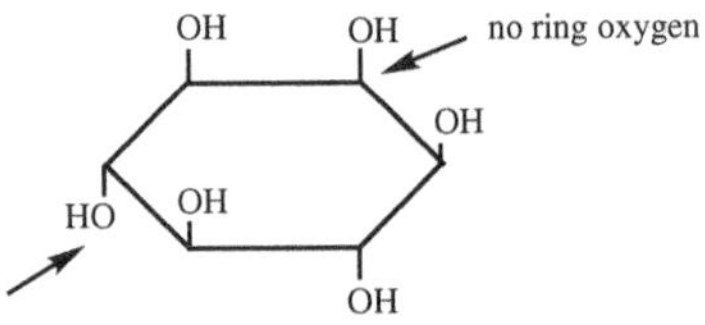
OH
OH
no ring oxygen
OH
HO
OH
OH

inositol

(b) Eliminating UDP by nucleophilic attack at C1 is more likely for galactose than with inositol because there is more electron withdrawl by the ring oxygen, which is present in galactose but not in inositol.

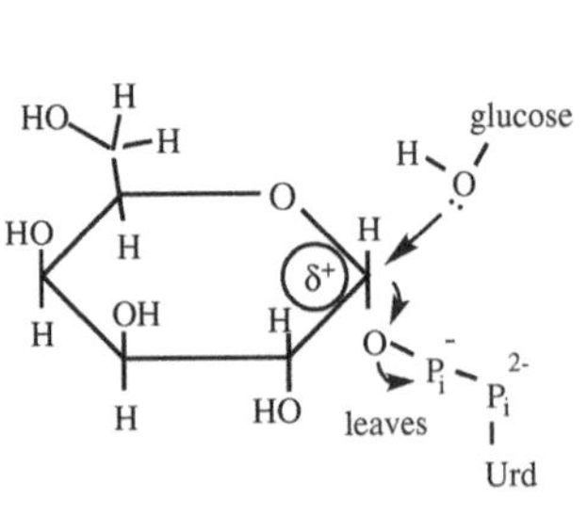
glucose
leaves
Urd

galactose

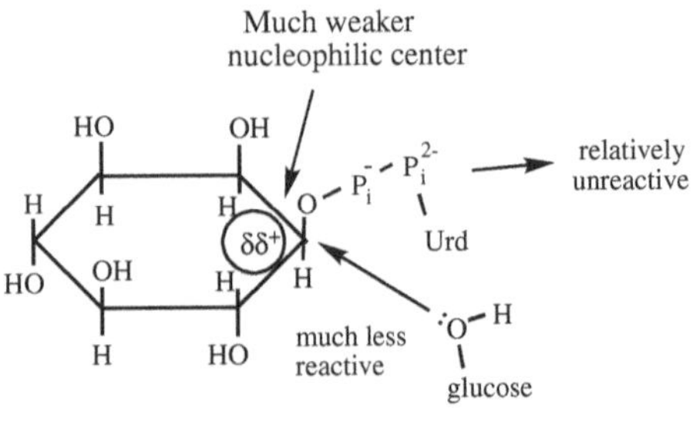
Much weaker
nucleophilic center
relatively
unreactive
Urd
much less
reactive
glucose

inositol
(protected phosphate)

3 (a) α (1→6) refers to the linkage between C1 of the first residue and C6 of the second residue, with the linked O pointing down from the 1st to second residue. β (1→4) refers to pointing up at C1.

(b) The molecule

. NAG - α(1➤ 6) - gal - α(1➤ 4) - glc - β (1➤ 4) - gal is:

Key:

NAG   *N-acetylglucosamine*
gal   *galactose*
glc   *glucose*

The α and β linkages are shown.

4 (a) Glycogen contains more α (1→6) linkages and on a percentage basis fewer α (1→4) linkages than amylopectin. It is more branched.

(b) Glycogen functions to store glucose units in the liver that can be mobilized during glycolysis as fuel when needed. A higher branching number results in a higher storage density, resulting in more efficient storage, allowing simultaneous liberation of terminal glucoses from more branches.

5. Penicillin inhibits linkage of the second peptide in the peptidyl-glycan cell coat cross-linkage of the bacterial cell wall. Human cell membranes are constructed of cytoskeletal elements that differ significantly from those of bacteria. As a result, the bacteria are killed while human cells remain relatively unaffected.

6 (a) The osmotic pressure of the concentrated poly-glucose polymers drives them to absorb water from the bulk bathing extra-cellular fluids. Water tries to dilute the carbohydrates, which cannot be diluted because they are interconnected and connected to the extracellular surface. These carbohydrates thereby both wet the cell and inflate it. This helps cell-cell adhesion and fluid behavior, making them communal and wanderers with the same forces: hydrogen bonding and osmotic pressure.

(b) Protection, polyfunctionality, recognition of "self" and "non-self" entities, sequestration, and so on.

7. CHOs function to tag and chaperone proteins in cells, between cells and in circulating cells. This amounts to export. For example, antibodies are CHO tagged. The CHO also determines the "age" of a protein. When the CHO is shortened (but still present), it targets certain proteins for destruction and clearance.

8. (a) Phospholipase C (PLC) functions in the phospholipids – IP_3 signal transduction pathway by hydrolyzing the IP_3-DAG connection. Each component is a second messenger, IP_3 activates Ca^{2+} influx (indirectly activating Ca^{2+} sensitive protein kinase DAG directly activates PKC).

8 (b)

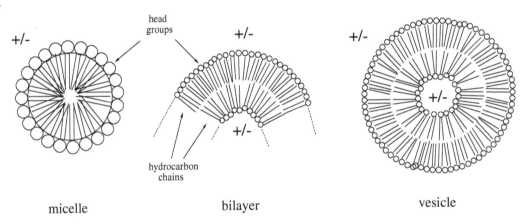

phospatidyl inositol triphosphate
(reactant)

9. Chondroitin sulfate is one of several glycosaminoglycans that forms a core of the cushion material in cartilage, bone and cornea. These materials absorb water and buffer against shocks, bumps, and so on.

10 (a) Unsaturated fatty acids; double bonds disorder the interaction stacking/aggregation tendencies and thereby reduce their stabilities in liquid crystal formation.

(b) One follows "melting" from the liquid crystal to the disordered (dissolved) emulsion (solution). At lower temperatures, chains form ordered lattices; at higher temperatures, the hydrophobic effect is overcome (water is released from the surrounding solved) and these lattices melt.

11.

head groups

+/-   +/-   +/-

+/-

+/-

hydrocarbon chains

micelle   bilayer   vesicle

A bilayer has two layers, back to back, of hydrocarbon chains in the center of a sandwich with outer layers composed of the polar headgroups of the phosphatidyl-X lipid molecules.

The micelle is a single droplet of phospholipid molecules with the solvent-exposed surface composed of the polar headgroups.

12 (a) and (b)

fatty acids

glycerol   phosphate   choline

Review Session for Chapters 13—14.7

1. Explain five details that describe the fluid mosaic membrane model.

2. (a) Draw and label the structures of the four nucleic acid bases in DNA.

   (b) Label the bases drawn in part (a).

3. (a) Draw and label a RNA trinucleotide using three different bases.

   (b) Label the bases and other structural subunits drawn in part a.

4. Draw the structures of the two Watson-Crick base pairs. Scientists typically use the symbol ••• to designate hydrogen bonds.

5. Define a glycosidic bond in a nucleoside.

Review Session for Chapters 13—14.7: Key

1. The "fluid mosaic" model includes the following details:

(a) It is supported by and assembled within/around the lipid bilayer. The mixed composition "liquid crystal" is interrupted by integral membrane proteins, cholesterol, lipids, phospholipids, terpenes, glycosphingolipids, and so forth.

(b) Peripheral membrane proteins are not covalently bound, they are often "anchored" to phospholipid anchor fragments.

(c) Integral membrane proteins – embedded in the lipid bilayer with one or both faces exposed (2 different exposures, inside & out). for example, 7-helix transmembrane motif.

(d) The lipid-anchored proteins are connected by covalent "tails" of
   (i) fatty acids (myristoyl, etc.),
   (ii) polyprenes (farnesyl, squalyl), or
   (iii) glycosphingolipids.

(e) Oligosaccharide tree-like structures are connected to proteins, glycosphingolipids, proteoglycans, and so forth.

2. The Watson-Crick base structures are:

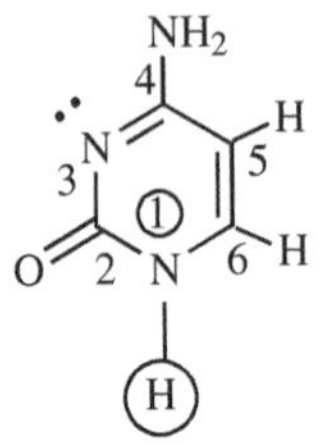

Cytosine (Cyt)

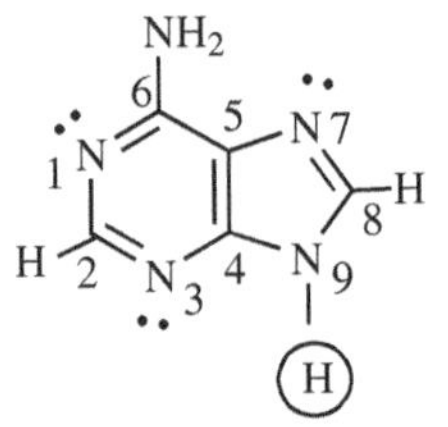

Adenine (Ade)

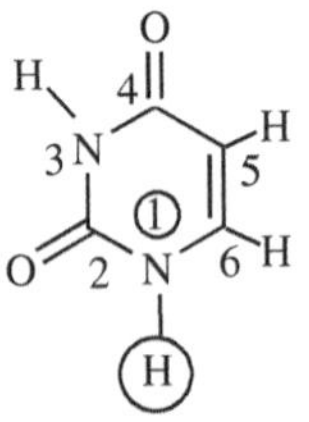

Uracil (in RNA) (Ura)

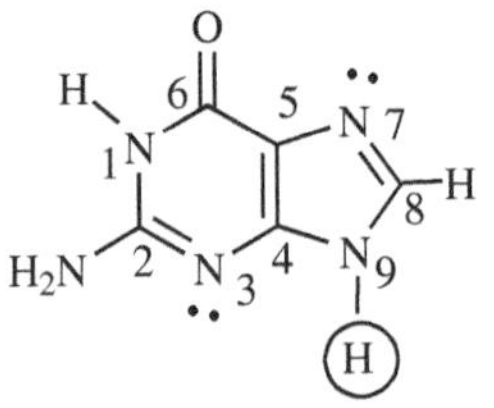

Guanine (Gua)

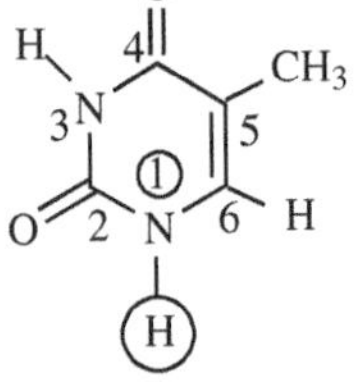

Thymine (in DNA) (Thy)

Adenine (Ade) base pairs with Thymine (Thy) in DNA and Uracil (Ura) in RNA.

Cytosine (Cyt) base pairs with Guanine (Gua).

3. The following figure shows the details of an RNA trinucleotide:

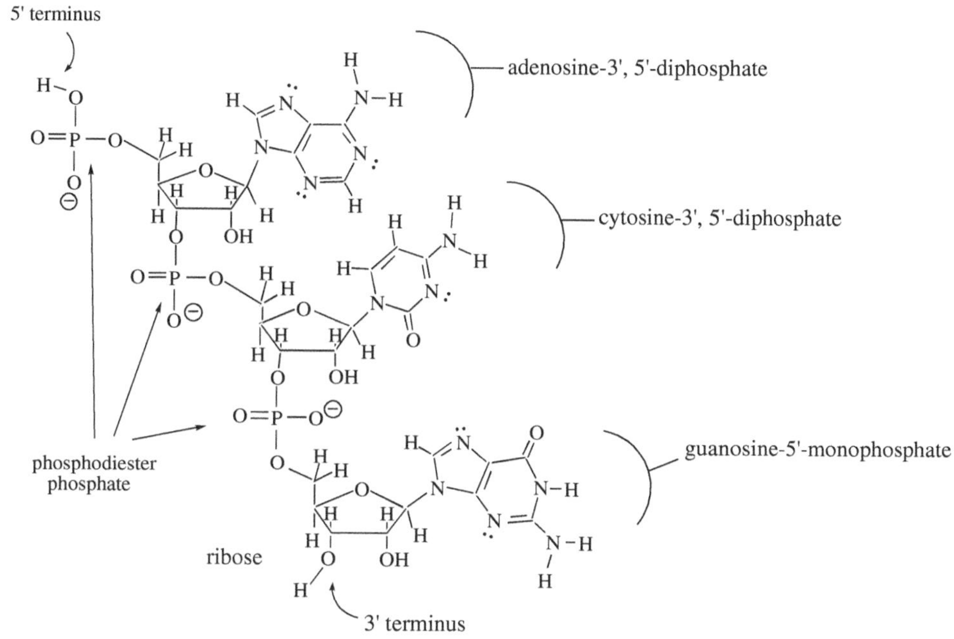

4. The Watson-Crick base pairs are:

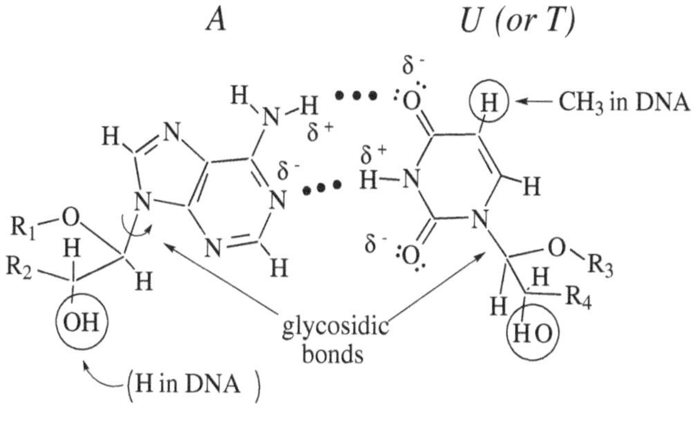

(a) A·U or dA·dT

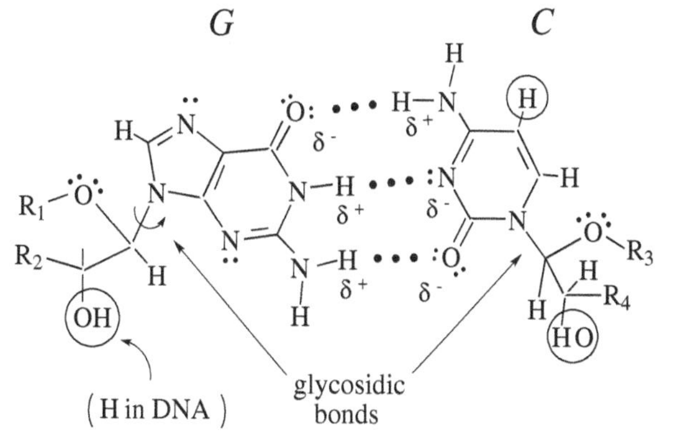

(b) G·C or dG·dC

Review Session for Chapters 16.2—17.6

1. Explain how and why the absorbance at 260 nm (A_{260}) can be used to determine if a double helix forms from 2 single strands of DNA or RNA.

2. (a) Define "base stacking."

(b) Describe the three predominant types of forces that contribute to stabilization of stacked bases in a double helix.

3. (a) What are counterions and why do they bind all nucleic acids?

(b) How do histones serve this function in the case of most chromosomal DNAs?

4. Give two reasons why G•C base pairs are more stable than A•T (or A•U) base pairs. Rank these contributions according to importance in inducing stability.

5. Describe four differences between A and B forms of DNA.

6. (a) Draw the mechanism of alkaline hydrolysis of RNA.

(b) Why is DNA not subject to this mechanism?

7. Why and how does an antisense oligonucleotide functionally inactivate a mRNA for use in translation by a ribosome?

8. Name the four classes of RNA and briefly explain their functional significance. In what reaction(s) do they participate?

9. List four distinctive features of most eukaryotic mRNAs.

10. What is the primary use of a DNA probe and how is this process accomplished?

11. Why are restriction endonucleases required to produce, manipulate and clone specific pieces of DNA?

12. What are the two functional ends of transfer RNA, what are their functions and how do they accomplish them?

Review Session for Chapters 16.2—17.6: Key

1. When two strands of DNA anneal to form a double helix, their absorbance at 260 nm decreases relative to the sum of A_{260} of the individual strands. This hypochromicity is used to determine the extent of duplex formation.

2. (a) Bases "stack" on each other, face to face, like coins in a roll. Base-pair stacking is the primary stabilizing force maintaining the double helical structure.

   (b) Forces listed according to relative contributions to stabilization are:
   (1) the hydrophobic effect – in which water encages the bases at low T.
   (2) London dispersion forces – transient dipole-dipole attractions that involve the π electrons.
   (3) hydrogen bond formation – which helps reinforce the stacked geometry.

3. (a) Nucleic acids have one electronegative phosphodiester phosphate, of which 0.1 unit of negative charge is present per nucleotide. Ninety percent of the negative charge is neutralized by the cation charges, which achieves near electroneutrality, and thereby maintain thermodynamic stability.

   (b) Histones are rich in the electropositive (basic) amino acid residues lysine and arginine. These positive charges neutralize the negative charge of chromosomal DNA, which allows the strands to become condensed and packaged within the chromosome.

4. The bases are ranked according to hydrophobicity as follows: G > A > C > T (or U), so G-C is more hydrophobic than A-T (or A-U). The hydrophobic effect is the predominant stabilizing force in base pairing. In addition, G-C base pairs have 3 hydrogen bonds while A-T (or A-U) only have 2. Due to more extensive electronic overlap, London dispersion forces and stacking are also more stabilizing for G-C versus A-T or A-U.

5. A-DNA versus B-DNA:

A-DNA	B-DNA
(a) C3' endo sugar conformation	(a) C2' endo sugar conformation
(b) base pairs tilted relative to a plane located perpendicular to the helical axis	(b) base pairs are approximately perpendicular to the helix axis
(c) base pairs are splayed away from center of helix, leaving a central axial cavity down the center	(c) base pairs cross the center of the helix since no central axial cavity is present
(d) slightly shorter and wider (squatter) helix than B-form	(d) slightly longer and narrower than A-form

6.

Hydroxide abstracts the 2' hydroxyl hydrogen, producing a 2' hydroxylate nucleophile, which then attacks the adjacent 3' phosphate to produce the 2', 3' cyclic phosphodiester structure. As a result, the attached 5' hydroxyl oxygen of the attached 3'-terminal nucleotide is eliminated, resulting in strand cleavage.

(b) DNA does not have a 2' hydroxyl group, so it cannot form the interal nucleophile that attacks the 3' phosphate group.

7. An antisense oligonucleotide forms a double helix with the mRNA, preventing it from being usable (in single-stranded form) in the decoding process of translation (protein synthesis).

8.

RNA class	*Functional Importance*
(a) ribosomal RNA (rRNA)	rRNA is the predominant structural and catalytic component of ribosomes
(b) transfer RNA (tRNA)	tRNA plays a key role in decoding the mRNA by encoding the placement a specific amino acid in a defined location in a synthesized protein.
(c) messenger RNA (mRNA)	mRNA is produced by transcribing the permanently encoded chromosomal DNA. It carries the trinucleotide-based message to the ribosome, where it is decoded during translation, resulting in insertion of a defined amino acid into a protein.
(d) "small" RNA (snRNA, snoRNA, gRNA, siRNA, miRNA)	These species serve a number of roles, usually involved in catalysis and guiding processes required for RNA maturation.

9. The 5'-terminal m^7G^+ cap is linked by a 5' to 5' bond to the 5'end of the mRNA. The poly (A) "tail" is at the 3' end. These mRNAs often contain introns, which are not usually intended to encode amino acid sequences, and are removed by splicing. If only one protein product results from the mRNA, it is said to be monocistronic (1 gene/message per mRNA).

Eukaryotic mRNA structure:

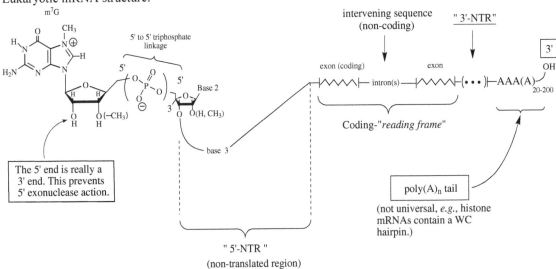

cap	(a) 5'-terminal m^7G cap structure 5' to 5' linked to 2nd nucleoside
1 gene per mRNA	(b) generally monocistronic. Some intervening sequences contain a separate encoded gene.
introns *exons*	(c) genes in parts. Introns are removed from mRNAs by the splicing process, catalyzed by the spliceosome, a snRNAs/protein complex.
poly (A)$_n$	(d) The poly (A) "tail" is added enzymatically and helps determine the lifetime of a given mRNA. This process is analogous to that of punching a train ticket.

10. A labeled probe DNA is used to detect the presence of a specific complementary nucleic acid sequence by annealing/hybridization to form a (radioactively or otherwise) labeled duplex.

11. Restriction enzymes allow one to cut out and classify DNA fragments with specific terminal sequences, then splice them specifically into plasmids/vectors for designed uses.

12. The *anticodon* forms base pairs with the triplet codon of a mRNA during the decoding process of translation.

The *acceptor end* is composed of the 2' or 3' hydroxyl of the 3' terminal ribose of the tRNA. The amino acid is esterified to the carboxyl group of a (cognate) amino acid, which is then transferred into a growing protein chain during translation.

Review Session for Chapters 19—20

1. (a) Name and briefly describe the purpose (point) of the three most central catabolic pathways of intermediary metabolism. (The third has two coupled parts. Name both of them.)

(b) What is (are) the product(s) of each of these pathways? (list according to pathway)

2. Describe the two major ways energy is captured in a chemically usable form by metabolic pathways.

3. The following imaginary reaction from wizard Merlin's notebooks occurred with a standard free energy ($\Delta G^{\circ\prime}$) of -8.5 kJ per mol:

$$\text{carborandum} + \text{gold} \leftrightarrow \text{essence of life}$$

(a) Assume that the poise of the equilibrium does not change (which is often referred to as being "at equilibrium"). A sample contains 44 mM carborandum, 0.1 mM gold and 45 μM essence of life. What is the actual free energy (ΔG) of the reaction under these equilibrium conditions at room temperature (298 K)? Show your work.

(b) A different sample contained 0.9 M carborandum, 0.1 mM gold and 45 μM essence of life. What is ΔG?

(c) How did adding more carborandum affect the equilibrium poise of the reaction? What was the mass action ratio (Q) in each case?

4. (a) Why do only three steps in glycolysis control most of the flux through the pathway under actual cellular conditions?

(b) What are the three steps, the three enzymes that catalyze the reactions? What do the reactions have in common?

5. (a) Define the concept 'metabolically irreversible.'

(b) Define concept 'near equilibrium.'

6. Explain why the kinetics of an enzyme reaction are most easily controlled when K_M is approximately equal to the concentration of the reactant.

7. Consider the following data (where E is the standard state reduction potential):

Reaction				$E^{\circ\prime}$
Acetyl CoA	$+ CO_2 + H^+ + 2\ e^-$	$\rightarrow$ Pyruvate	$+ CoA\text{-}SH$	-0.48
NAD^+	$+ 2\ H^+ + 2\ e^-$	$\rightarrow$ NADH	$+ H^+$	-0.32
FAD	$+ 2\ H^+ + 2\ e^-$	$\rightarrow FADH_2$		-0.22

(a) Under *standard state conditions* (T = 298 K; all component concentrations = 1 M), will it require more energy if the breakdown of pyruvate to acetyl CoA is coupled to FAD formation or to NAD^+ formation? Show work.

(b) Does the answer change if pyruvate is present at 1 mM, while the other species are still present at 1 M?

8. (a) What are the reactant and products of the reaction catalyzed by Triose Phosphate Isomerase?

(b) What do the products get converted to next? Explain the reason this occurs.

9. (a) Why should citrate negatively regulate (discourage) the Phosphofructokinase-1 reaction? What is the general name for this phenomenon?

(b) Why should fructose-1, 6-bisphosphate stimulate the pyruvate kinase reaction? What is the general name for this phenomenon?

Review Session for Chapters 19—20: Key

1. (a) i. Glycolysis: conversion of glucose ("glc") to 2. pyruvate, then to two acetyl CoA
ii. Krebs Cycle: fixing OAA to acetate, makes citrate, which fuels the Krebs Cycle
iii. electron transport/oxidative phosphorylation – conversion of "reducing power" (NADH, $FADH_2$, $CoQH_2$) to water, via the electron transport chain (cytochromes, CoQ) and using O_2 as the terminal electron acceptor.

(b) i. Glycolysis and the Bridge Reaction: 2 ATP per glc, 4 NADH (+2 acetyl CoA) = 12 ATP equivalents
ii. Krebs Cycle: CoA-SH, 3 NADH + $CoQH_2$ + GTP + 2 CO_2 + 2 H^+ = 20 ATP equivalents
iii. oxidative phosphorylation: NAD^+, fumarate, H_2O [½ ($XH_2 \rightarrow O_2$)], proton gradient

2. Nucleoside triphosphates: $X + P_i \leftrightarrow X\text{-}P_i$ (*e.g.*, $X + ATP \leftrightarrow X\text{-}P_i + ADP$)
Energy is used in the course of the bond formation.

Phosphorylation of a substrate energizes it in preparation for subsequent reactions, in which dephosphorylation of that substrate drives product formation.

Reducing power of reduced coenzymes is used to reduce a compound, creating the energized species ½ YH_2. That substrate is subsequently oxidized, which drives that reaction, analogous to the role played by dephosphorylation above.
Also, energy is accrued as the reduced material ½ YH_2, that is, NADH or $CoQH_2$. They are used in electron transport to produce the proton gradient between the mitochondrial inner membrane space and the matrix. This gradient drives the protonmotive force, which powers ATP formation by the F_0F_1 ATPase complex.

$$2\ H^+,\ 2\ e^\ominus$$

$$Y \rightleftharpoons YH_2$$

$$2\ H^+,\ 2\ e^\ominus$$

3. (a) Using the equation from Chapter 20, $\Delta G_{reaction} = \Delta G^{o'} + RT \ln Q$
$$= \Delta G^{o'} + RT \ln \{[\text{essence of life}]/[\text{carborandum}][\text{gold}]\}$$

Using RT = 2477.9 J/mol:

$$\Delta G_{reaction} = \text{-8.5 kJ/mol} + (8.315\ J\ mol^{-1}\ K^{-1})(298\ K)\ \ln\ [(45 \times 10^{-6}\ M)/(44 \times 10^{-3}\ M)(0.1 \times 10^{-3}\ M)]$$
$$= \text{-2.74 kJ/mol}$$
$$\updownarrow$$
The value of Q = 10.22

(b) $\Delta G_{reaction} = \text{-8.5 kJ/mol} + RT \ln [(45 \times 10^{-6}\ M)/(0.9\ M)(0.1 \times 10^{-3}\ M)]$
$$= \text{-10.22 kJ/mol}$$
$$\updownarrow$$
The value of Q = 0.5

(c) Adding carborandum decreases the *partition coefficient* (Q), meaning the reaction has a much higher likelihood of occurring; this is reflected in the much more negative ΔG value, (-10.2 << -2.7) kJ/mole.

4. (a) Steps catalyzed by hexokinase, phosphofructokinase-1 and pyruvate kinase are "metabolically irreversible," with ΔG values of -33, -22 and -17 kJ/mol, while the other 7 reactions are "near equilibrium," with ΔGs of ~0.
(b) The 3 steps are:

Reactions	*Coupled Reactions*
(i) glucose $\rightarrow$ G6P (enzyme: Hexokinase)	ATP $\rightarrow$ ADP, H^+
(ii) fructose-6-$P_i \rightarrow$ fructose-1, 6 $(P_i)_2$ (enzyme: PFK-1)	ATP $\rightarrow$ ADP, H^+
(iii) PEP $\rightarrow$ pyruvate (enzyme: Pyruvate Kinase)	ADP, $H^+ \rightarrow$ ATP

All three enzymes are kinases; note that reaction (iii) is phosphorylation of ADP by PEP, while the other two are sugar phosphorylation reactions in which the phosphate contributor is ATP.

5. (a) Having a large negative ΔG; highly spontaneous (thermodynamically) (and therefore kinetically predisposed to product formation, too).

(b) Having a ΔG approximately equal to zero; flux is approximately the same in both directions (*i.e.,* if the [reactant] is equal to the [product], the K_{eq} is close to or equal to 1).

6. When K_M is approximately equal to substrate concentration, the reaction is occurring at a rate of $\frac{1}{2} V_{max}$. The rate is not maxed out, yet is occurring at a rate $\gg 0$. As a result both negative and positive regulation are possible within a relatively small range of [S] changes.

Consider what happens if the concentration is either 100 times K_M or 1/100 of K_M.

7. (a) The net reactions, constructed by combining the pyruvate (pyr) oxidation reaction with either NAD^+, H^+, 2 electron reduction or FAD, 2 H^+, 2 electron reduction are:

(i) pyr + CoA-SH + NAD^+ + 2 H^+ + 2 e^- $\leftrightarrow$ Acetyl CoA + CO_2 + H_2 + 2 e^- + NADH

net: pyr + CoA-SH + NAD^+ $\leftrightarrow$ Acetyl CoA + CO_2 + NADH
(A_{red})   (B_{ox})   (A_{ox}) (B_{red})

$\Delta G^{o\prime} = $ -n $\mathcal{F} \Delta E^{o\prime} = $ -n $\mathcal{F}$ [(-0.32) – (-0.48)]V = -n $\mathcal{F}$ (0.48–0.32) V
= -(2)(96.48 kJ mol^{-1} V^{-1})(0.16 V)
= -30.87 kJ/mol (Q = 1) ; (n = 2)

(ii) pyr + CoA-SH + FAD + 2 H^+ + 2 e^- $\leftrightarrow$ Acetyl CoA + CO_2 + H_2 + 2 e^- + $FADH_2$

net: pyr + CoA-SH + FAD + 2 H^+ $\leftrightarrow$ Acetyl CoA + CO_2 + $FADH_2$
(A_{red})   (B_{ox})   (A_{ox}) (B_{red})

$\Delta G^{o\prime} = $ -n $\mathcal{F} \Delta E^{o\prime} = $ -n $\mathcal{F}$ [(-0.22) V – (-0.48) V] = -(192.96)(0.26V) = -50.17 kJ/mol (Q = 1); (n = 2)
This situation is more spontaneous than (a) (i).

(b) (i) $\Delta E = \Delta E^{o\prime} – (0.026/n) \ln \{ [A_{ox}][B_{red}]/[A_{red}][B_{ox}] \}$ (Q = 1000); (n = 2)
= 0.07 V = {(0.16 V) – [0.09 V]} = [(1 M)(1 M)/(0.001 M)(1 M)]

$\Delta G^{o\prime} = $ -n $\mathcal{F} \Delta E^{o\prime} = $ -13.50 kJ/mol (Q = 1000)

(ii) $\Delta E = \Delta E^{o\prime} – (0.026/n) \ln \{[A_{ox}][B_{red}]/[A_{red}][B_{ox}]\}$ (Q = 1000); (n = 2)
= 0.17 V = {(0.26 V) – [0.09 V]} = [(1 M)(1 M)/(0.001 M)(1 M)]

$\Delta G^{o\prime} = $ -n $\mathcal{F} \Delta E^{o\prime} = $ -32.80 kJ/mol (Q = 1000) This situation is more spontaneous than (b) (i).

Summary of the results:

Calculation	*Reaction*	*Q*	$\Delta E^{o\prime}$	ΔE	$\Delta G^{o\prime}$	ΔG
(a) (i)	pyr, CoA, NAD^+, 2 H^+, 2 e^- passed	1	0.16 V		-30.87 kJ/mol	
(b) (i)	pyr, CoA, NAD^+, 2 H^+, 2 e^- passed	1000		0.07 V	*Compare* ↓	-13.50 kJ/mol
(a) (ii)	Pyr, CoA, FAD, 2 H^+, 2 e^- passed	1	0.26 V		-50.17 kJ/mol	*Compare* ↓
(b)(ii)	Pyr, CoA, FAD, 2 H^+, 2 e^- passed	1000		0.17 V		-32.80 kJ/mol

The two conclusions are:
(*1*) The FAD reaction remained more spontaneous than the NAD^+ reaction under both [pyruvate] conditions.
(*2*) Lowering the [pyruvate] from 1 M to 1 mM made the reaction less spontaneous with both NAD^+ and FAD.

8. Fructose-1, 6-bisphosphate is dissociated to form dihydroxyacetone phosphate (DHAP) and glyceraldehyde-3-phosphate (G3P) by aldolase (rxn 4). DHAP is converted to G3P by triose phosphate isomerase (rxn 5). Finally, G3P is converted to 1, 3-bisphosphoglycerate by G3P dehydrogenase (rxn 6).

9. (a) The negative regulation of PFK-1 by citrate is an example of (negative) feedback inhibition of glycolysis by a key Krebs Cycle intermediate. Glycolysis is purposefully turned off when plenty of citrate is already present.

(b) Stimulation of pyruvate kinase by fructose-1, 6-bisphosphate is an example of feed-forward stimulation. The presence of too much fructose-1, 6 $(P_i)_2$ indicates that it is accumulating, so more of the later reaction should occur in order to use the accumulated intermediate.

Review Session for Chapters 21—25

1. We discussed three branching catabolic fates of pyruvate. Draw the three reaction (reaction sequences), including cofactors and enzymes.

2. (a) Why does the absence of alcohol dehydrogenase produce even more scurrilous behavior than if the person has the enzyme?

(b) How is the blocked catabolic intermediate related to the typical role of pyridoxal phosphate in enzymatic catalysis?

3. (a) Why does "carbonation" accompany ethanol production?

(b) How does the staff of life of the western world benefit (and us as a consequence)? [Bread, not beer!]

4. Dihydrolipoamide Acetyl Transferase uses a coenzyme not previously described. (a) What is it and how does it function in fueling the Krebs Cycle?

(b) What other coenzyme is also involved in this process? How?

5. (a) What "symport" reaction accompanies import of pyruvate into the mitochondrion and what enzyme catalyzes the reaction?

(b) Does the pH of the cytoplasm increase or decrease as a result?

6. List the reactions, coenzyme(s), cofactor(s) and enzymes involved in the two "oxidative decarboxylation" reactions of the Krebs Cycle.

7. List the reactions, coenzyme(s), cofactor(s) and enzymes involved in the "substrate-level phosphorylation" reaction of the Krebs Cycle.

8. List (only once each) all of the "energy conserving" compounds formed by the Krebs Cycle accompanied by the number of "ATP equivalents" finally accrued after oxidative phosphorylation of 1 molecule of each of these compounds.

9. (a) How do fumarase and malate dehydrogenase "fix" a carbonyl group on succinate in the production of oxaloacetate (OAA).

(b) What crucial 2 carbon compound is then "fixed" to OAA?

(c) How is the product used in food flavoring?

10. What amino acid and what product of pyruvate metabolism are the principle substrates for gluconeogenesis in mammals?

11. What is *protonmotive force* and what enzyme complex uses this phenomenon as the driving energy for ATP synthesis in oxidative phosphorylation? (Contrast it with electromotive force, *i.e.,* emf, voltage.)

12. How does electron transport drive production of the protonmotive force?

Review Session for Chapters 21—25: Key

1. Three fates of pyruvate:

2. (a) The intermediate between pyruvate and ethanol, acetaldehyde, is not broken down effectively by people who are ADH-deficient. As a result, they form acetaldehyde as the ethanol accumulates via the reversible ADH reaction. Accumulated aldehyde-containing acetaldehyde reacts to form Schiff base compound with amine-containing compounds. These compounds have the problematical pharmacological properties that can induce drunken behavior. As a result, when ADH-deficient people consume ethanol they are much more vulnerable to the effect.

(b) Recall the way in which a pyridoxal phosphate in Aspartate Transaminase is neutralized by Schiff base formation with an active site lysine.

3. In ethanol fermentation, CO_2 is evolved by pyruvate decarboxylase as acetaldehyde is made (see Answer 1). The released CO_2 produced by the activated yeast culture is captured by the moistened wheat (*etc.*) fibers of the dough.

4. (a) Dihydrolipoamide is a lipid-based hydrocarbon that is covalently attached to the pyruvate dehydrogenase E_2 subunit (PD-E_2). It accepts ethanol from HETPP to form a thioacetate

then transfers the acetyl group to CoA-SH. In pyruvate dehydrogenase (PD), CO_2 is evolved as HETPP (hydroxyethylthiamine pyrophosphate) is formed from (1) pyruvate and TPP (thiamine PP$_i$). This is one of three cycles that couples to the dihydrolipoamide acetyltransferase (DLAAT) cycle.

The next steps involve transfer of HETPP-bound (2) acetate to (3) lipoic acid then transfer to CoA, making (4) acetyl CoA. The cofactors, co substrates, H^+, and so forth are omitted.

Indigestion, a medically significant problem and money-maker for pharmaceutical companies.

(b) Thiamine pyrophosphate (TPP) (on PD-E$_1$) accepts ethanol and transfers it to the lipoic acid (on PD-E$_2$). Five coenzymes are used: lipoamide (amide), thiamine pyrophosphate, CoA-SH, FAD and NAD$^+$.

5. (a) Pyruvate translocase imports 1 H$^+$ and pyruvate simultaneous (in the same direction). This coordinate translocation mechanisms defines a "symport".

An "antiport" exchanges two species coordinately in opposite directions:

(b) Since 1 H$^+$ leaves the cytoplasm for each translocase reaction, the cytoplasm becomes less acidic and the pH increases;–log [H$^+$] is larger.

6 and 7.

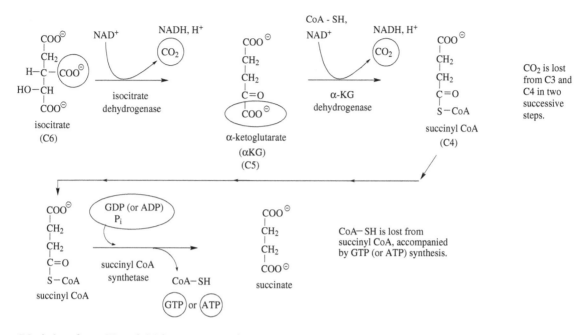

CO$_2$ is lost from C3 and C4 in two successive steps.

8.

Energy source	ATP equivalents/molecule
NADH	2.5
GTP (or ATP)	1
CoQH$_2$	1.5

9.

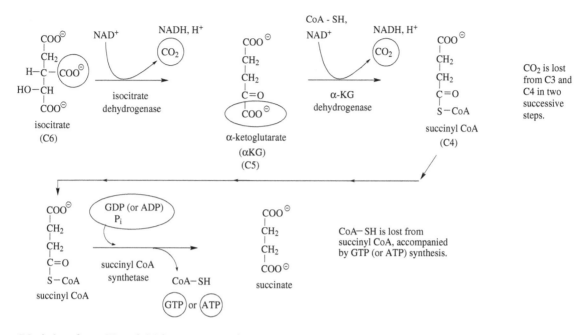

The O is derived from water which is fixed after C=C formation and oxidized from R-OH. For the role of FAD in the CoQ → CoQH$_2$ process, see the Review Session for Chapter 27, question 3.

(b) Acetate is "fixed" to OAA, producing citrate.

(c) Citrate is a common flavoring as a result of its acidic tartness. Its acidity and redox properties lead to food processing events that change the consistency and preservation of foods.

10. Alanine and lactate are the principle gluconeogenic substrates. Compare pyruvate and alanine to determine how one is made from the other. (*Hint.* How are oxaloacetate and glutamate interconverted?)

11. Oxidative loss of reducing equivalents due to a proton gradient across a membrane (of the mitochondrion).

$$XH_2 + \tfrac{1}{2} O_2 + 2\,e^- + 2\,H^+ \leftrightarrow H_2O + 2\,e^- + 2\,H^+ + X$$

The electrons are the connecting (driving) "wire."

Oxidative phosphorylation is driven by a *proton gradient*, which is produced by the active transport of protons from the mitochondrial matrix to the inner membrane space, through the mitochondrial inner membrane.

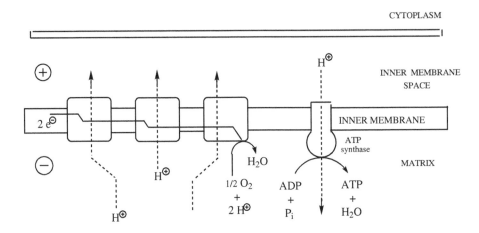

This proton movement creates a gradient, operating as an *aqueous circuit* that connects the electron transport chain to *ATP Synthase*. This is similar to the action of a wire in an electrochemical reaction in that the electrons are passed from the reducing agent, NADH or $CoQH_2$, through the electron transport chain, and finally to the terminal oxidizing agent O_2. This series of *coupled redox reactions* pass the electrons, which drives the proton pumps. The transformed *free energy* is captured by phosphorylation of ADP to produce ATP.

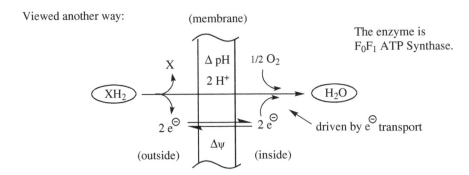

Mathematically, there are two contributions to the protonmotive force (Δp), due to the difference in $[H^+]$ inside versus outside, and the change in membrane potential ($\Delta\psi$). ($\mathcal{F}$ is the Faraday constant)

$$\Delta p = \Delta G/n\,\mathcal{F} = \Delta\psi + (0.059\text{V})\,\Delta pH_{in\text{-}out}$$

12. The electron transport reactions pump H^+ out of the mitochondrion matrix and build up the $[H^+]$ in the inner membrane space, which drives ATP synthesis by F_0F_1 ATPase. Four H^+ are pumped per ATP generated.

Review Session for Chapter 27

1. (a) How many reactions does each round of β-oxidation of a fatty acid require?

(b) What are the products of one round of β-oxidation and what's the tally in terms of ATP equivalents of energy conserving products?

2. (a) Draw the coupled cofactor regeneration cycles that siphon off reducing equivalents then fix them into coenzyme Q in reactions that are coupled to the first oxidative step of fatty acid β-oxidation.

(b) Write down the names and a short definition for the cofactors involved in this siphon.

3. Which three steps of the Krebs Cycle do the first three steps of the fatty acid β-oxidation cycle resemble? Draw (horizontally) the three analogous reactions from (i) β-oxidation, (ii) the Krebs Cycle, and (iii) the three generic chemical transformations that occur beneath the four analogous substrates.

4. Draw the analogous (but backward) [reduction-dehydration-reduction] series of reactions of fatty acid synthesis starting from reactants and going to products beneath your three reaction series from problem 3. Point the arrows in the opposite direction drawn for problem 3.

5. Draw the reactions for the condensation sub-step in Acyl-Carrier Protein (ACP)-mediated fatty acid synthesis.

Note. In the absence of proper exercise and portion control, this pathway can lead to obesity, a modern human calamity.

Review Session for Chapter 27: Key

1. (a) Four reactions occur during each β-oxidation cycle: (1) oxidation #1, (2) hydration, (3) oxidation #2, and (4) thiolysis.

(b) Each cycle yields 1 CoQH$_2$, 1 NADH, H$^+$, 1 acetyl CoA and 1 fatty acyl CoA shortened by 2 carbons. In ATP equivalents: 1 CoQH$_2$ = 1.5 ATP eq.; 1 NADH = 2.5 ATP eq. and 1 acetyl CoA = 10 ATP (*i.e.,* because it feeds one round of the Krebs cycle).

2. (a)

(b) (i). FAD/FADH$_2$ – oxidized/reduced in the Acyl CoA Dehydrogenase reaction.
(ii). Fe-S $^{3+/2+}$ – oxidized/reduced iron-sulfur center in ETF-Ubiquinone Reductase
(iii). CoQ/CoQH$_2$ – oxidized/reduced CoQ; ubiquinone (dione) and ubiquinol (diol).

3.

4.

(iv) Fatty acid
      biosynthesis :

                                                    Note : Not a CoQ / CoQH$_2$ reaction.

5. Initial stage of fatty acid biosynthesis:

Practice Exam 1

1. Draw the predominant structures of the following molecules at the pH indicated. Be sure to include all carbon and hydrogen atoms.

a. (1) urea at pH 12

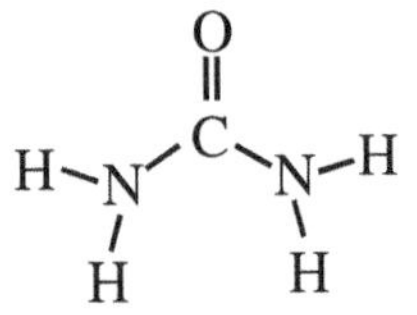

b. (1) L-aspartate at pH 3.9

(no pKa)

pK$_a$ = 3.9

50%

50%

← Also affected →

(no pKa)

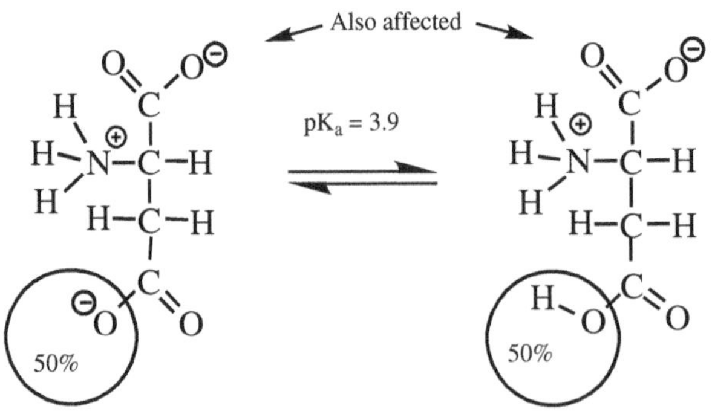

c. (5) histidyl-argininyl-glycyl-tryptophanyl-methioninyl-tyrosine at pH 7.

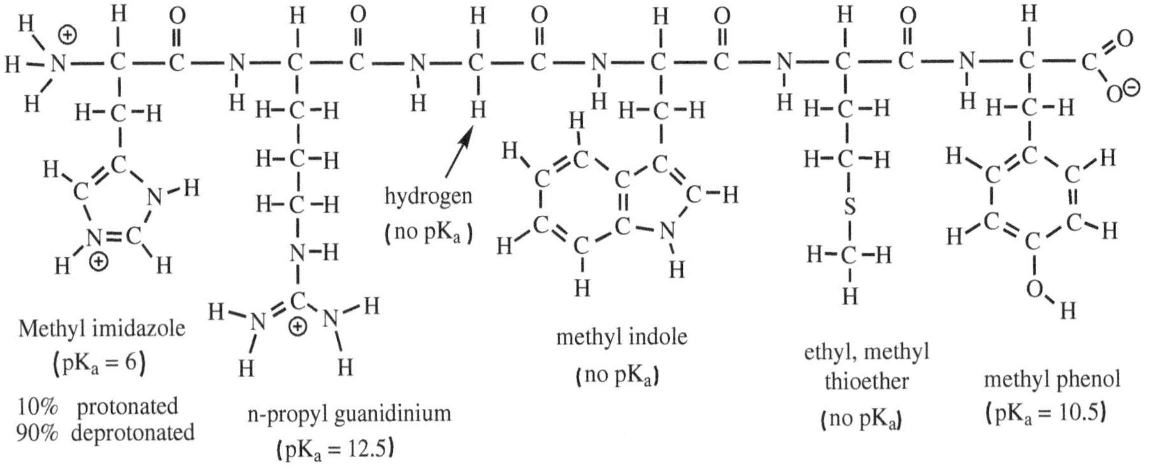

Methyl imidazole
(pK$_a$ = 6)

10% protonated
90% deprotonated

n-propyl guanidinium
(pK$_a$ = 12.5)

hydrogen
(no pK$_a$)

methyl indole
(no pK$_a$)

ethyl, methyl
thioether
(no pK$_a$)

methyl phenol
(pK$_a$ = 10.5)

(Note that the upper carboxylate in 1b will also adopt two forms: ~ 95% deprotonated and ~5% protonated.)

2. (0.25 each) List and name the entire functional group (R) and pK$_a$ (if appropriate) under each amino acid side chain you drew in #1a, 1b and 1c.
(answers shown above)

3. a. (1) What is the charge of the fully protonated molecule in problem #1c?   +3

b. (1) Name each ionizable functional group in the *deprotonated* form.

histidyl:	– imidazole
argininyl:	– guanidine
cysteinyl:	– thiolate
tyrosine:	– phenolate

c. (4) Draw a titration curve for the oligopeptide in problem #1**c**. Assume the pK_a of the C-terminal carboxylate is the same as that on a typical protein C-terminus. Label your axes and include a scale and units.

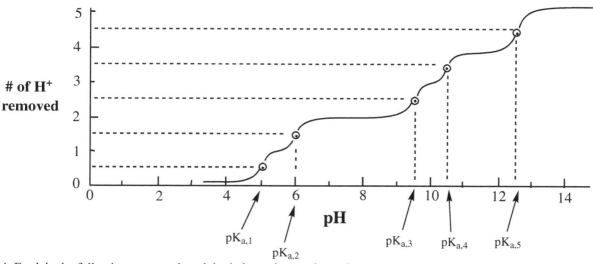

4. Explain the following terms and explain their use in protein work:

a. (0.75) Gel filtration chromatography:

It is a protein purification technique in which molecules flow through a column containing porous beads, which separates the proteins due to size differences. Smaller molecules are included (enter) in the beads and take longer to reemerge (and finally elute from the column) than larger molecules, which don't enter the beads (as often).

b. (0.75) Sodium dodecyl sulfate (SDS):

SDS is an amphipathic molecule with a hydrophilic sulfate and hydrophobic C12 hydrocarbon tail. Used in gel electrophoresis to coat and denature proteins so they electrophorese as rod-shaped particles. Separation is generally predominantly based on differences in protein chain length because the SDS charges overwhelm those of the protein.

c. (0.75) BCA assay:

A method derived from the Biuret reaction in which Cu^{2+} is reduced to Cu^{1+} in proportion to the amount of peptide bonds present (protein concentration). A dye is added to the protein that turns deep purple in proportion to Cu^{+1} produced, thereby amplifying the signal which is indirectly due to [protein].

d. (0.75) A_{280}:

Absorbance at 280 nm, which is the result of trp, tyr (and to a lesser extent, phe) in proteins. If the number of these amino acids in the protein is known, one can calculate the molar concentration (c) of protein using Beer's law ($A = \varepsilon c l$), where ε is the molar extinction coefficient (which can be looked up) and l is the cuvette pathlength.

e. (0.75) Chymotrypsin:

A protease that cleaves the chain, at peptide bonds adjacent to aromatic side chains, producing free carboxyl-termini.

f. (0.75) Ramachandran plot:

A plot of the two peptide torsion angles of φ (phi, pronounced "fy") and Ψ (psi, pronounced "sigh") as the two axes. A point on the plot corresponds to 1 type of conformation since the peptide bond is constrained to planarity. Each of the 2° structures falls in a unique part of the plot. This characterizes the structure of the molecule.

5. (3) Explain why a peptide bond has partial double and partial single bond character. Give the structures of the two contributing forms.

The carboxyl group and peptide bond undergo enol ↔ keto tautomerism. The enol contribution produces the "partial double bond" character, the keto produces the partial single bond character.

$$R_1-\underset{R_2}{\overset{O}{\underset{|}{\overset{||}{CH-C}}}}-\underset{H}{\overset{|}{N}}-\underset{R_3}{\overset{|}{CH}}-R_4 \quad\longleftrightarrow\quad R_1-\underset{R_2}{\overset{|}{CH}}-\overset{\overset{O-H}{|}}{C}=N-\underset{R_3}{\overset{|}{CH}}-R_4$$

keto   enol

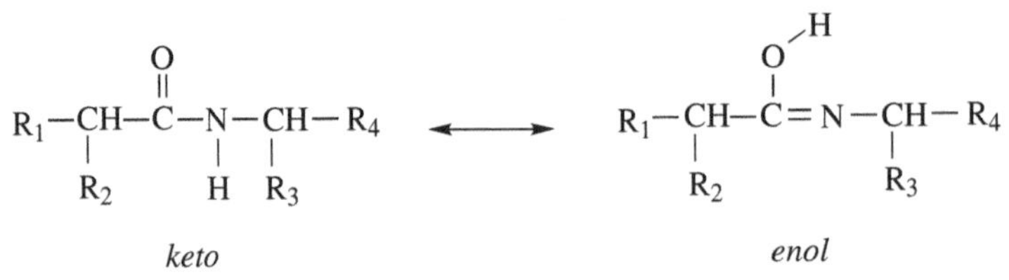

6. (10) Multiple choice: One point maximum each. Circle the single number that corresponds to the best answer. Read every answer carefully and completely. Answers can receive either positive or negative partial credit. (*Note.* Plus signs indicate correct answers.)

A. The following functional groups can form hydrogen bonds:
i. methyl and amide
ii. sec-butyl and thiol
iii. amide and hydroxyl (+)
iv. benzyl and phenol
v. alcohol and phenylisothiocyanate

B. The torsion angle of a bond defines
    i. the rise over the run
ii. the rotational conformation (+)
iii. the distance
iv. the juxtaposition of bound ligands
v. the temperature

C. Thermolysin cuts
i. lys and arg leaving a free C-terminus
ii. ile, leu, val (+)
iii. ile, leu, tyr (partial credit)
iv. phe, tyr, trp
v. lys and arg leaving a free N-terminus

D. Methionines are identified using
i. DTT and β-mercaptoethanol
ii. urea and H_2O
iii. CNBr and carboxypeptidase B
iv. PITC and guanidinium
v. DNFB and CNBr (+)

E. The hydrophobic effect involves
i. hydrophobic bonds
ii. H_2O and peptide bonds (partial credit)
iii. aliphatic residues (partial credit)
iv. conjugate bases
v. nonpolar residues and H_2O (+)

F. The ratio of conjugate bases and acids is calculated using the Henderson-Hasselbalch equation and ___.
i. pO_2 and pK_a
ii. pK_a
iii. pO_2
iv. pH and pK_a (+)
v. pH and log ($[A^-][HA]$)

G. Zwitterions formed by amino acids
i. are present when pH = pI
ii. denature easily
iii. are present at neutral pH
iv. are called dipoles
v. contain both positive and negative charges (+)

H. Chaotropic molecules include
i. ether, H_2O and sulfates
ii. thiols and ammonium
iii. urea and ethanol
iv. guanidinium and urea (+)
v. SDS and ammonium

I. Hemoglobin binds
i. H^+
ii. O_2
iii. thiols
iv. i. and ii.
v. all of the above

J. What controls the hydrophobic effect
i. enthalpy
ii. entropy
iii. free energy
iv. pH
v. all of the above (+)

7. (PUZZLER) (3) Why do proteins typically denature at more acidic and alkaline pHs. Illustrate your answer.

Increasing the pH results in deprotonation of side chains. Decreasing the pH results in hyperprotonation of side chains.

In either case, salt bridges or hydrogen bonds that require both types of charge or protonation state to form cannot form. The result of these disruptions is loss of secondary and tertiary structure, and therefore denaturation.

Practice Exam 2

(*Note.* The letters Ω, Γ, and Ψ are used to indicate different versions of the test.)

1. Draw the predominant structures of the following molecules (corresponding to your Greek letter) at the pH indicated. Be sure to include all carbon and hydrogen atoms.

a. (1)   b. (1)

Ω: 5-phosphoribosyl-1-pyrophosphate at pH 7.4   Ω: carbamoyl aspartate at pH 7.4

Γ: NAD⁺ at pH 7.4   Γ: α-D-glucose-6-phosphate at pH 7.4

Ψ: α-D-glucose at pH 7.4   Ψ: NADP⁺ at pH 7.4

Practice Exam 2   (key 2 of 2)

(*Note.* The letters Ω, Γ, and Ψ are used to indicate different versions of the test.)

1. Draw the predominant structures of the following molecules (corresponding to your Greek letter) at the pH indicated. Be sure to include all carbon and hydrogen atoms.

Ω: 5-phosphoribose-3-pyrophosphate at pH 7.4   Ω: carbamoyl aspartate at pH 7.4
Γ: NAD⁺ at pH 7.4   Γ: α-D-glucose-6-phosphate at pH 7.4
Ψ: α-D-glucose at pH 7.4   Ψ: NADP⁺ at pH 7.4

c. (5) Draw the catalytic triad in a generic serine protease. Order the residues from left to right; label the residues from left to right 1, 2, and 3 for question 2. Leave room for the substrate.

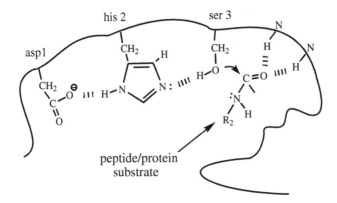

peptide/protein
substrate

2. a. (1) Draw a generic substrate juxtaposed next to the key catalytic atom in your drawing in 1c.
b. (0.5) Explain which atom on which residue in 1c attacks the substrate.

The hydroxyl oxygen atom of serine 3 attacks the carbonyl carbon in nucleophilic reaction.

c. (0.5) Explain which atom on the substrate is attacked and why.

The carbonyl carbon is attacked by the hydroxyl oxygen of serine 3 since it has a partial positive charge.

d. (3) Explain how the residues in the "catalytic triad" in question 1c collaborate to induce the first step in the catalytic mechanism. Use the letter 3 residue nomenclature and numbers assigned in your drawing in your explanation. Name the type of chemical reaction with the appropriate term.

(Useful terms: "charge relay" and "acid-base catalysis")

The electronegatively charged oxygen of aspartate 3 hydrogen bonds with the secondary amino hydrogen of histidine 2. This enhances the electronegitivity of the lone pair electrons on the other ring nitrogen on histidine 2, thereby strengthening the hydrogen bond with the hydroxyl hydrogen of serine 3 and eventually removing it. The resulting negatively charged serine oxygen is then activated to attack the substrate in a nucleophilic attack.

4. Explain the following terms and explain their use in the study of enzyme catalysis. Use one or a few complete sentences.

a. (1) Ω, Γ: Acid-base catalysis (show the generic enzymatic reaction):

catalytic transfer of a proton (*e.g.,* between E-H and B); $E\text{-}H + B \rightarrow E^{-} + B\text{-}H^{+}$

Ψ: Covalent catalysis (shown the generic enzymatic reaction):

Catalytic transfer of a group X between enzyme and substrate; substrate 1, enzyme & substrate 2; etc.

$$e.g., \ E\text{-}X + S \rightarrow E + S\text{-}X$$
$$S_1\text{-}X + E \rightarrow S_1 + E\text{-}X; \ EX + S_2 \rightarrow S_2\text{-}X + E$$

b. (1) Ω, Γ: Michaelis-Menten constant (show the generic enzymatic reaction):

$$E + S \overset{k_1}{\underset{k_{-1}}{\leftrightarrow}} [ES] \overset{k_{cat}}{\rightarrow} E + P; \; K_M = (k_{-1} + k_{cat})/k_1$$

Ψ: Catalytic constant (show the full reaction):

$$E + S \leftrightarrow [ES] \overset{k_{cat}}{\rightarrow} E + P$$

k_{cat} is the number of catalytic events per second.

c. (1) Ω, Γ: Cyclin kinase:

Kinase activity of cyclin heterodimer that phosphorylates substrates and controls cell cycle "checkpoints".

Ψ: Allosterism:

Allosteric effector binds, often to a regulatory subunit, and changes the binding constant or catalytic rate of a linked reaction. Can involve feedback, but may also be induced by an apparently unlinked effector (not synthetically related).

d. (1) Ω, Γ: a zymogen (include a reaction scheme):

Typically described as an inactive enzymes precursor (usually prior to proteolysis) that is covalently modified (typically proteolyzed) to become activated.

Ψ: an enzymatic cascade (a picture would help): A series of linked catalytic reactions, in which the first stage releases a population of activated enzymes, each of which go out and create a population of activated enzymes of a second type. This stage-by-stage amplification leads to a very large set of activated species in later stages.

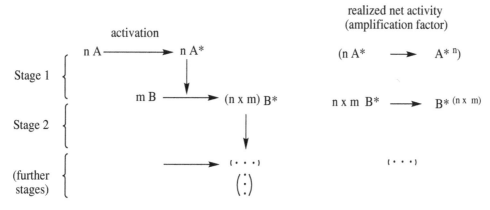

e. (1) Ω, Γ: an uncompetitive enzyme inhibitor (Draw the relevant enzymatic reactions):

$$E + S \leftrightarrow [ES] \rightarrow E + P$$

$$+$$

$$I$$

$$\updownarrow$$

$$[ESI]$$

Uncompetitive – bind to the [ES] complex only and block product formation.

Ψ: a competitive enzyme inhibitor (Draw the relevant enzymatic reactions).

$$E + S \leftrightarrow [ES] \rightarrow E + P$$

$$+$$

$$I$$

$$\updownarrow$$

$$[EI]$$

Competitive inhibitors bind E. No [ES] or P form.

f. (2) Ω, Γ: a Schiff base. Explain its utility. (Draw a generic chemical structure, include as much detail as you can. Include all chemical bonds and atoms.

(Ω, Γ)

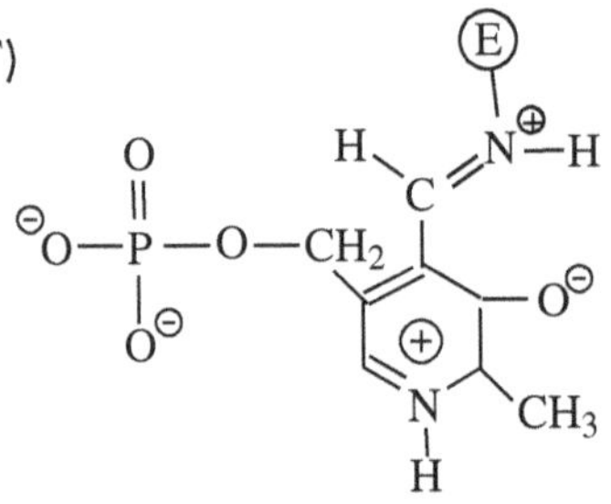

PLP forms Schiff bases with enzymes and substrates. This weakens bonds to the alpha carbon, leading to bond breakage, electron shifts and isomerizations.

Ψ: the redox center of FAD and FADH$_2$. Explain its utility. (Draw the key parts of both chemical structures and include the number of rings. Include all chemical bonds and atoms.)

(Ψ)

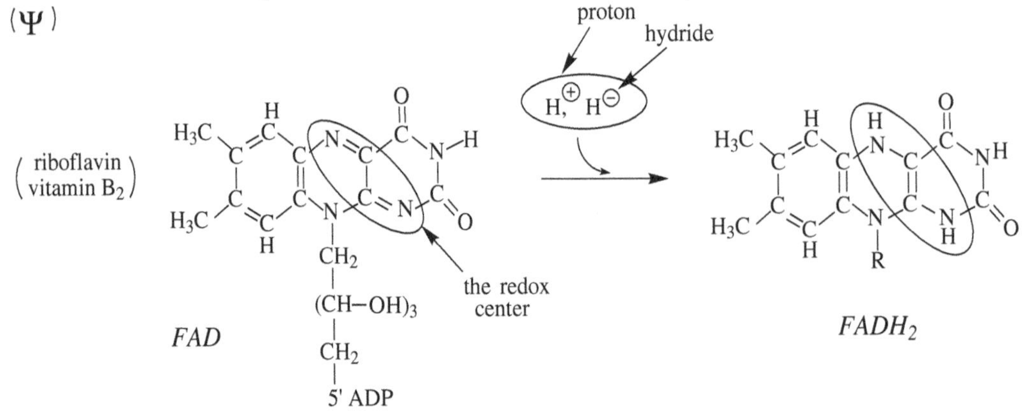

Used in redox reactions in intermediary metabolisms. Transports electrons (reducing equivalents) to the electron transport apparatus in the mitochondria.

5. (3) What chemical does Coenzyme A carry in its typical biochemical function? Draw the abbreviated structure as you've been taught. What biochemical pathway or cycle is this covalent compound used to fuel?

CoA-SH carries an acetyl group, derived by decarboxylation of pyruvate, which is then transferred to oxaloacetate from the Krebs Cycle to form citrate, thereby fueling it. This reaction links liver glycogen stores, through Glycolysis to the Krebs Cycle (and many other pathways).

6. (10) Multiple choice: One point maximum each. Circle the single number that corresponds to the best answer.

A. Lactose synthesis requires:
i. UDP-glucose and fructose
ii. UDP-glucose and galactose
iii. UDP-galactose and glucose (+)
iv. ATP, glucose and galactose
v. ATP, glucose and fructose

B. The following coenzyme and protein are required for our eyes to absorb visual photons:
i. $FADH_2$ and isocitrate dehydrogenase
ii. ubiquinone and rhodopsin
iii. trans-retinal and opsin
iv. cis-retinal and opsin (+)
v. trans-retinal and rhodopsin

C. The following are examples of ketoses:
i. D-ribose and dihydroxyacetone
ii. D-glucose and D-galactose
iii. D-fructose and D-ribose
iv. D-fructose and D-lactose
v. dihydroxyacetone and D-fructose (+)

D. Fructose ribopyranose formation involves:
i. an enamine and acid-base catalysis
ii. an intermolecular Haworth projection
iii. a hemiacetal and acid catalysis (+)
iv. reducing sugar and RNA
v. sucrose and glucose

E. The following are examples of second messengers in signal transduction pathways discussed in class:
i. cyclic AMP and glucose
ii. inositol and vitamin E
iii. inositol diphosphate and steroids
iv. cyclic AMP and phospholipase C
v. inositol triphosphate and diacylglycerol (+)

F. Acetylcholine esterase:
i. contains an active site serine and is inactivated by the nerve gas (+)
ii. contains an active site serine and is activated by the nerve gas antidote CoA
iii. DEP contains an active site threonine and is inactivated by cAMP
iv. contains a "catalytic triad" and is inactivated by the nerve gas DFP
v. is embedded in the finger nails of frogs

G. The proximity effect and transition state stabilization
i. are confusing and hard to separate
ii. enhance the rate of an enzymatic reaction (+)
iii. involve catalytic transfer of H^+
iv. i. and ii. above
v. i. through iii. above

H. The following metals occur naturally in biological cofactors
i. calcium, molybdenum and cesium
ii. sodium, iron, phosphorus and cobalt
iii. potassium, molybdenum and zinc (+)
iv. all of the above
v. ii. and iii. above

I. Coenzyme Q
i. can transform between diol and dione forms
ii. carries reducing equivalents *in glycolysis*
iii. can be reduced by NADH or FADH$_2$
iv. i. and ii
v. i. and iii. (+)

J. D-ribose is not
i. illustrated as a Haworth projection
ii. a pyranose in solution
iii. a furanose in solution
iv. present as the sugar acid in chewing gum (+)
v. in RNA

7. PUZZLER (5 total) Trypsin catalyzes proteolysis, The 'specificity site' has a high affinity for arginine so the protease cuts specifically adjacent to this amino acid.

a. (2) Explain why the compound benzamide inhibits the reaction.

It is an active site-directed mimic of arginine (*e.g.*, see Horton *et al.*)

b. (1.5) Draw out the likely reaction scheme, including the effect of benzamide.

Practice Exam 3

(*Note*. The letters Ω, Γ, and Ψ are used to indicate different versions of the test.)

1. Draw the predominant structure of the molecule corresponding to your Greek letter at the pH indicated. Be sure to include all carbon and hydrogen atoms (or "get it right" if you use abbreviated structural nomenclature).

a. (1)

    Ω: lauric acid at pH 7.4
    Γ: palmitate acid at pH 7.4
    Ψ: arachidonic acid at pH 7.4

b. (1)

    Ω: 2'-O-methyl guanosine-5'-monophosphate at pH 7.4
    Γ: ribothymidine-3'-monophosphate at pH 7.4
    Ψ: cytidine-2', 3'-diphosphate at pH 7.4

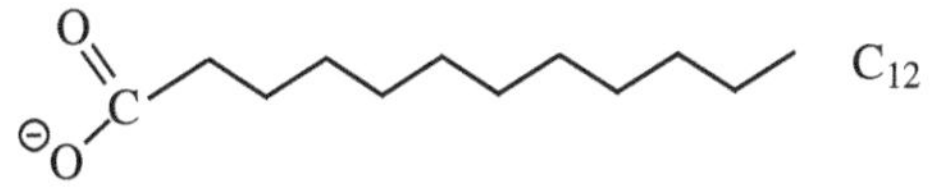

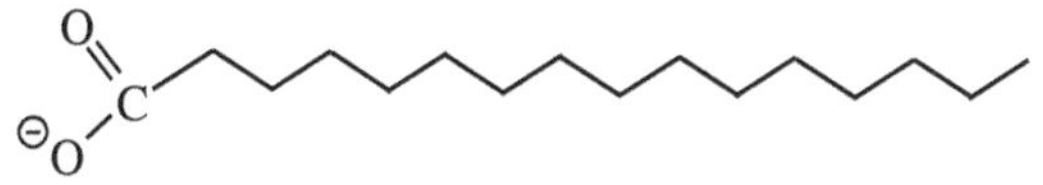

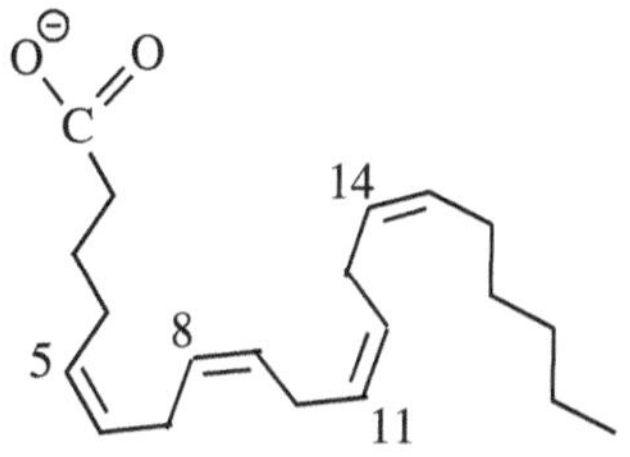

b. (1)

Ω: *syn* 2'-O-methyl guanosine-5'-monophosphate at pH 7.4
Γ: ribothymidine-3'-monophosphate at pH 7.4
Ψ: cytidine-2', 3'-diphosphate at pH 7.4

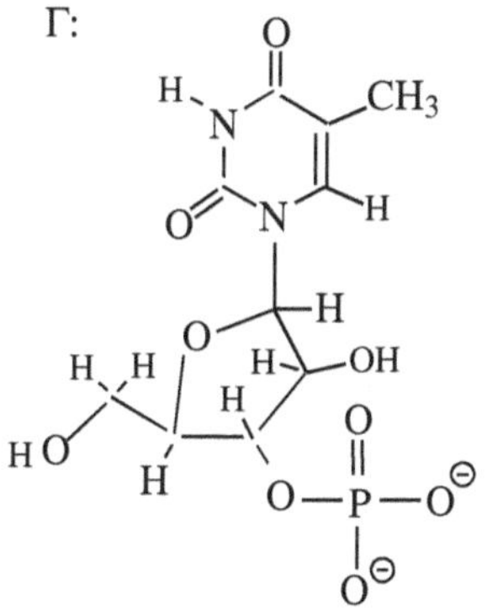

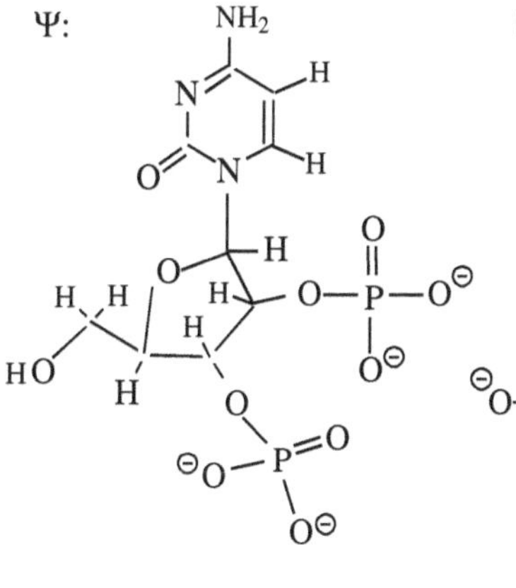

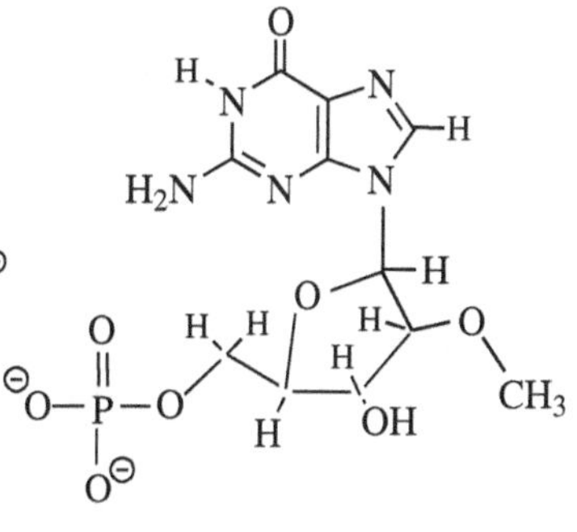

c. (5) Draw the structure of a generic "fluid mosaic" membrane. Label each of the five typical components with appropriate name. Number them 1 to 5.

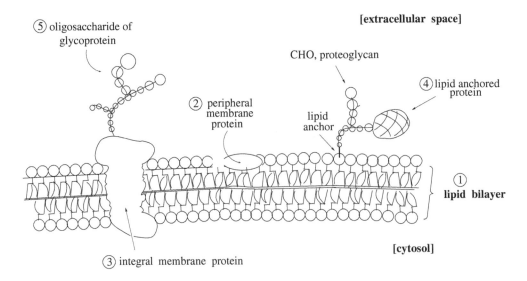

2. (2) Briefly describe 4 of the 5 components of the fluid mosaic pointed out in your drawing in Question 1. Make the numbers match.

(i). lipid bilayer (1)

This refers to a mixed composition liquid-crystal, interrupted by integral membrane proteins, lipid derivatives, anchors, glycosphingolipids, and so forth. The head groups are composed of ethanolamine, inositol, choline, *etc*. The "tails" are composed of even-numbered carbon-containing hydrocarbon chains.

(ii). peripheral membrane proteins (2)

These are non-covalently membrane-bound proteins. They adhere to the membrane at the H_2O interface, predominantly by electrostatic interactions (salt bridges, H-bonds).

(iii). integral membrane proteins (3)

These proteins are embedded in the lipid bilayer with one or both faces exposed to the external solution. They often contain attached tree-like carbohydrate chains.

(iv). lipid anchored proteins (4)

These proteins are connected to lipid tail, for example, the farnesyl group, which is embedded within the lipid bilayer.

(v). oligosaccharide trees (5)

These structures are connected to proteins, proteoglycans, and glycosphingolipids.

3. Explain the following terms with respect to their purpose in lipid or nucleic acid biochemical pursuits.

a. (1) Ω, Γ: Sodium deoxycholate (What it's used for and why does it work?):

It is a detergent that is commonly used to solubilize membrane proteins. It's also produced naturally by the gall bladder to facilitate bile excretion, which is used to facilitate lipid absorption by the gut.

Ψ: a seven-helix transmembrane protein (Explain why it embeds and orients within the membrane bilayer as it does.):

A seven-helix transmembrane protein contains seven alpha helices that span the bilayer region of a membrane. The connecting loops extend out into the extracellular space or into the cytoplasm. The alpha helices are hydrophobic, so they remain embedded with the hydrocarbon-rich interior of the membrane.

b. (1) Ω, Γ: H_2O-octanol partition coefficients (explain the practical significance):

This coefficient characterizes the ratio of a molecule (*e.g.,* an amino acid) that resides in the aqueous phase divided by the amount of the same molecule that enters the octanol phase in an equilibrated two-phase partitioning sample. This is related to the free energy for a partitioning into water versus octanol, a generic mimic for the hydrophobic interior of a protein (or membrane). The results are used to predict whether amino acids in a protein will lodge within a membrane, or inside the protein, versus surface exposure to solvent.

Ψ: farnesyl anchor (explain why it works):

Lipid anchored membrane-bound proteins can be linked to an isoprenoid molecule called farnesene (which contains 3 isoprene units). The farnesyl anchor is hydrophobic and therefore prone to becoming embedded in the lipid bilayer on the extracellular surface of a cell. This anchors the protein to the cell.
(*Note*: Two farnesyl pyrophosphates react to form squalene, which, through a series of reactions, zips up to form cholesterol.)

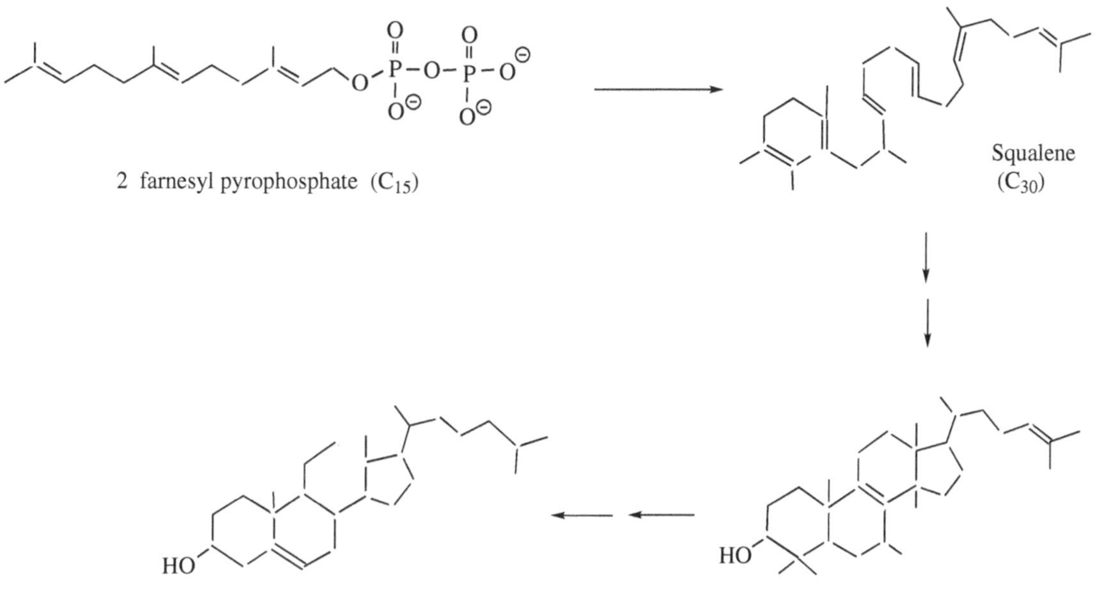

2 farnesyl pyrophosphate (C_{15})

Squalene (C_{30})

Cholesterol

Lanosterol

c. (1) Ω, Γ: hypochroism (Explain the practical significance.)

Hypochroism is a decrease in absorbance (at 260 nm) that occurs when two single strands form a double helix. This effect has been used to determine relative stabilities of different sequences and lengths of double helix constructs. By measuring equilibrium constants, one can calculate ΔG values. By doing this at different temperatures, one can calculate the ΔH and $-T\Delta S$ values at any given temperature. Thus, one can calculate ΔG, ΔH, and $-T\Delta S$ for duplex formation. Doing so with the right set of molecules has produced a database which is used extensively to predict structures from sequences.

(*Note.* One important application is used when one selects optimal PCR primers. Primers predicted to form unwanted secondary structures are avoided.)

Ψ: phosphodiester versus carboxymonoester (also draw the generic structures):

The backbones of DNA and RNA are composed of phosphodiester bonds, in which *two* different R groups are connected to the phosphoryl group. A carboxymonoester occurs in the linkage that connects a fatty acid to glycerol. Only *one* R group is connected to the single carboxyl oxygen.

$$\begin{array}{cc} \overset{\displaystyle O}{\underset{\displaystyle O^{\ominus}}{R_1-O-\overset{\|}{\underset{|}{P}}-O-R_2}} & \overset{\displaystyle O}{R_1-\overset{\|}{C}-O-R_2} \\[2em] \text{phosphodiester} & \text{carbon monoester} \end{array}$$

d. (1) Ω, Γ: counterions (draw a picture):

The phosphodiester backbone carries about 1 negative charge per nucleotide so electropositive counterions such as Na^+, K^+, Mg^{2+}, polyamine^{n+} must bind to the external surface of DNA or RNA in order to achieve (almost) electroneutrality.

Ψ: major and minor grooves (Explain how these structures are *decoded* by proteins.):

The double helical A- and B- form structures have 2 "sides," of which one is larger than the other, so the two are referred to as major and minor grooves. Different functional groups are presented to the two grooves, and as such they act as specific protein recognition elements. This is the nature of the structural code that allows regulated use of nucleic acids in their biochemical functions.

e. (2) Ω, Γ: Z-DNA repeating unit (also explain why the unit is as your state; illustrate your point(s) with a simplified structure):

Z-DNA has a zig-zagged phosphodiester backbone because the duplex has a dinucleotide repeating unit. The typical sequence, alternating CG has: (*1*) two different phosphodiester backbone conformations, (*2*) *anti* C, glycosidic torsion angle and a *syn* G, and (*3*) both C2' *endo* and C3' *endo* deoxyribose conformations.

Ψ: Duplex melt analysis (Draw the relevant reaction. Draw a simple graph that illustrates the results.):
One can determine the ΔG, ΔH, and TΔS for duplex dissociation to form 2 single strands by raising the sample temperature and following the increase in A_{260} (and varying the nucleic acid concentration). This is used to understand how different sequences affect the stabilities of the corresponding duplex structures.

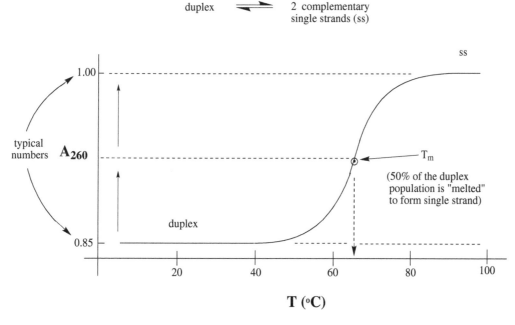

f. (2) Ω, Γ: antisense mRNA. (Explain the functional utility and draw the functioning structure.):

Antisense mRNA is the Watson-Crick complementary sequence corresponding to a "sense" mRNA. The former binds the latter to form a duplex. Since mRNA must be single stranded to be used in translation, antisense binding prevents translation of the bound mRNA.

$$mRNA + antisense\ mRNA \leftrightarrow mRNA \cdot antisense\ mRNA$$

       active   *inactive*

Ψ: Hoogsteen base pairing. [Illustrate the relevant functional groups and explain how the bases interact with each other.]

Hoogsteen base pair formation involves the top/back side of one of the purines, specifically the N7 atom on the imidazole ring. This sort of base-base interaction also occurs in triple- and quadruple-strand complexes.

4. (2) Explain and illustrate the difference between *anti* and *syn* glycosidic torsion angles using guanosine. *Anti* and *syn* conformations interconvert by rotation of the base relative to the (deoxy) ribose. The *anti* conformation involves a base swung away from the "top" of the sugar; *syn* nucleosides have a base above the sugar.

syn G   *anti* G

5. (10) Multiple choice: One point maximum each. Circle the single number that corresponds to the best answer. Read every answer carefully and completely.

A. Eukaryotic mRNAs contain the following:
i. 5'-O-methyl
ii. m'G cap and exons
iii. introns and codons (+)
iv. introns and histones
v. 5'-terminal poly(A)

B. Restriction endonucleases do not
i. cleave specific DNA duplex sites
ii. create "sticky ends"
iii. produce cloned plasmids (+)
iv. define DNA map locations
v. protect bacteria

C. Transfer RNA
i. is typically in the cruciform secondary structure
ii. of pivotal importance in decoding nucleic acid information into protein (+)
iii. contains a 5'-terminal amino acid
iv. has a dinucleotide anticodon
v. is a cloverleaf

D. The hydrophobic effect is not the primary stabilizing force of
i. protein primary structure (+)
ii. lipid micelles
iii. the double helix
iv. stacking
v. transmembrane cation gradients

E. The following molecules are not listed correctly from largest to smallest (from highest to lowest M_r):
i. tryptophan, lysine, leucine, glycine, CO_2
ii. cAMP, cytidine, N^1-methyl uracil, methylphosphate, ammonium
iii. arachidonate, stearate, oleate, palmitate, myristate, laurate
iv. 1-stearoyl-2-oleoyl-phosphatidyl inositol, glucuronate, glycerol
v. [^{3}H-2] erythrose, [(^{2}H-6)$_2$] lactose, [^{14}C-1] glycerol

F. The ratio of absorbances of samples measured at lower and higher temperatures ($A_{\lambda, low} / A_{\lambda, high}$)
i. is proportional to the *ratio of liquid crystal* to *dispersed fatty acids* in re-equilibrated samples.
ii. is proportional to the *ratio of duplexes to single strands* in re-equilibrated samples. (+)
iii. will produce *faster yields* at higher temperatures if the reaction proceeds with Boltzmann kinetics.
iv. i. and ii.
v. i. and iii.

G. Phospholipase C
i. is involved in the tyrosine kinase signal transduction pathway.
ii. catalyzes hydrolysis of phosphatidyl inositide diphosphate to form diacylglycerol and inositol triphosphate in the cAMP-dependent signal transduction pathway.
iii. is a hydrolase and esterase. (+)
iv. is missing in the Eskimo genotype.
v. co-starred with Paul Newman and Grace Kelly in a Elia Kazan-directed off-Broadway psychodrama.

H. The following molecules are commonly incorporated into phosphatidylglycerides
i. serine and cholinesterase.
ii. choline and stearate. (+)
iii. oleate and myrtylistate.
v. octanolamine and inositol.

I. DNA probes
i. cannot be labeled with biotin.
ii. have been used as convincing evidence in many celebrated murder trials. (+)
iii. hybridize with RNA to form a reverse transcript
iv. will all plasmose in the thermonuclear heat of a distant white dwarf Sun.
v. react with arachidonate to form nucleolipids.

J. The following molecules fall within the realm of lipids and derivatives thereof.
i. cerebrosides, steroids and endorphins
ii. isoprenoids, glyceraldehydes and sphingomyelins
iii. phosphatidyl inositides, insect juvenile hormone and lipid-anchored proteoglycans (+)
iv. waxes, pyridoxal phosphate, vitamin E
v. *sn*-(1-oleoyl, 2-myristoyl, 3-palmitoyl) triglyceride, zest of lemon, mannose

6. (PUZZLER #1) (2) A duplex DNA fragment formed by two annealed 13 nucleotide oligonucleotides was obtained from the chloroplast genome upon digestion with the enzyme *Eco*RI. This duplex fragment was treated with the enzymes *Hpa*II and *Bgl*II. Two fragments were obtained.

Specificities of Restrictions Endonuclease Recognition Sites.

Enzyme	Recognition sequence
*Bgl*II	5' A $^\vee$ GATCT $^{3'}$
*Eco*RI	5' G $^\vee$ AATTC $^{3'}$
*Hpa*II	5' C $^\vee$ CGG $^{3'}$

(a) What is the sequence and structure of the original chloroplast *Eco*RI duplex DNA fragment? Explain the logic that led you to your result.

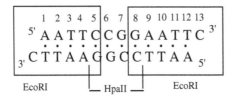

The two ends were created by the *Eco*RI reaction. The center of the molecule can only be composed of a single *Hpa*II site which overlaps the two *Eco*RI site fragments *Bgl*II "treatment" did not have any effect on the fragment since it has not *Bgl*II site.

(b) What are the sequence and structures of the two fragments produced by HpaII and BglII treatments? Explain the logic that led you to your result.

$$5' \ AATTC \ 3' \qquad\qquad + \qquad\qquad 5' \ CGGAATTC \ 3'$$
$$_{3'}CTTAAGGC_{5'} \qquad\qquad\qquad\qquad _{3'}CTTAA \ _{5'}$$

"sticky ends"

*Hpa*II cut the initially isolated *Eco*RI fragment, which already had a 3'C extension, to produce two fragments with both 3'C and 5'CG overhangs. The fragments are the same, so they'd appear to be 1 fragment on a gel.

7. (PUZZLER #2) (2) Draw structures of the following and name the functional groups:

(a) a nucleotide second messenger (Cyclic adenosine monophosphate, cAMP)

(b) a covalent lipid-carbohydrate complex that forms two different second messengers

8. (PUZZLER #3) (4) Similar functional roles are accomplished by cyclic AMP in the Adenylate Cyclase Signaling Pathway and diacylglycerol in the Inositol-Phospholipid Signaling Pathway". Briefly explain their respective functions (Illustrate your answer). Explain how they are analogous.

Cyclic AMP (cAMP) activates *Protein Kinase A*, which produces cellular responses by phosphorylating regulatory, sensory, and growth-controlling proteins.

Phosphatidylinositol triphosphate is lysed by the enzyme phospholipase C to produce *diacylglycerol* (DAG) and *inositol triphosphate* (IP$_3$), which are both *second messengers*. DAG and Ca^{2+} activate *Protein Kinase C* (PKC); IP$_3$ activates a Ca^{2+} channel which imports Ca^{2+} from the extracellular lumen. PKC is shifted from inactive to active (when DAG is present) by an increase in [Ca^{2+}] from 0.5 to 1 mM.

PKC produces a number of cellular responses by phosphorylating regulatory, sensory, and growth-controlling proteins.

Both mechanisms regulate *kinases*, which phosphorylate analogous cellular functions in target cells.

Practice Exam 4

(*Note*. The letters Ω, Γ, and Ψ are used to indicate different versions of the test.)

1. Draw the predominant structures of the following molecules (corresponding to your Greek letter) at the pH indicated. Be sure to include all carbon and hydrogen atoms (or "get it right" if you use abbreviated structural nomenclature).

a. (1)
Ω: glucose-6-phosphate at pH 7.4
Γ: fructose-6-phosphate at pH 7.4
Ψ: fructose-1, 6-bisphosphate at pH 7.4

b. (1)
Ω: phosphoenolpyruvate at pH 7.4
Γ: dihydroxyacetone phosphate at pH 7.4
Ψ: pyruvate at pH 7.4

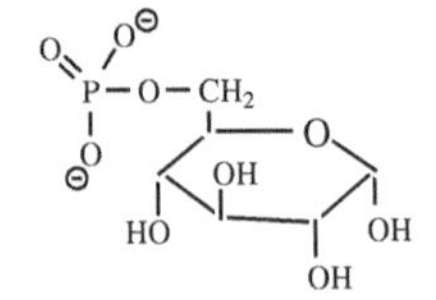

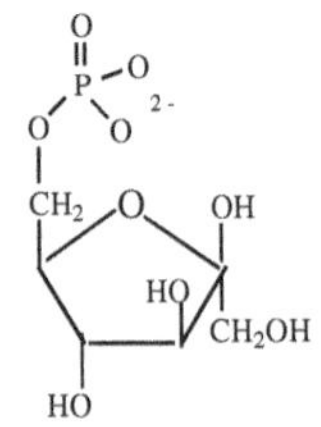

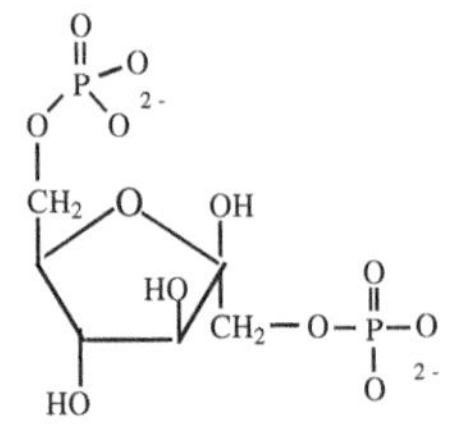

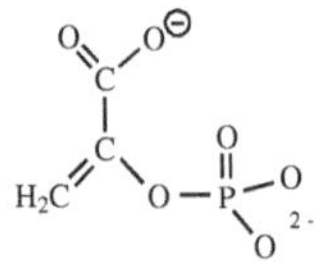

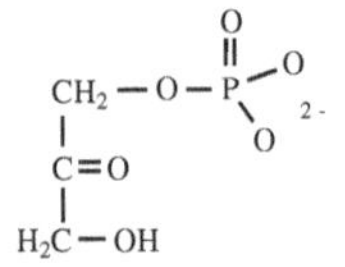

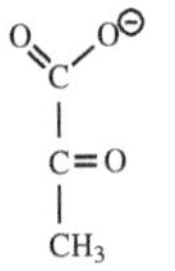

c. (1)
Ω: oxaloacetate at pH 7.4
Γ: citrate at pH 7.4
Ψ: isocitrate at pH 7.4

d. (1)
Ω: succinate at pH 7.4
Γ: fumarate at pH 7.4
Ψ: malate at pH 7.4

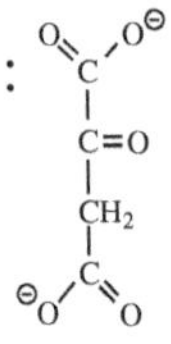

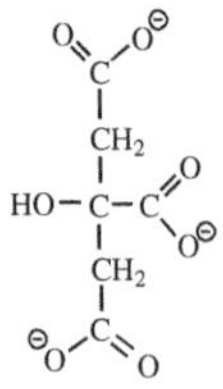

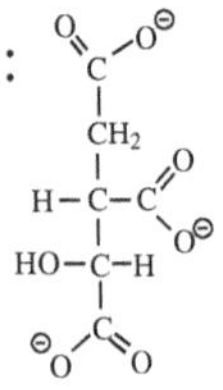

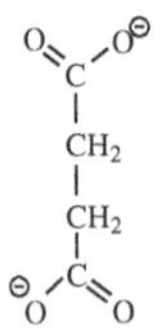

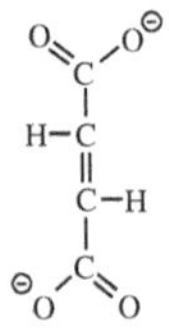

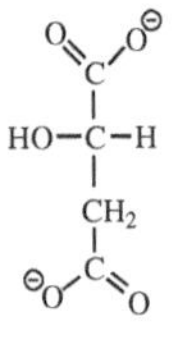

2. a. (4) Name and briefly describe the purpose (point) of the three most central catabolic pathways of intermediary metabolism. (The third has two coupled parts. Name both of them.) Leave room for the answer to part b following each of your definitions.

b. (2) Also list the product(s) of each of these pathways. (See Review Session 11, problem 1.)

(i) Glycolysis + the Bridge Reaction

This involves conversion of glucose to pyruvate (two per glucose) then to 2 acetyl CoA. Products are 2 ATP per glucose, 2 acetyl CoA, 4 NADH, yielding 10 ATP in oxidative phosphorylation.

(ii) Krebs Cycle

Fixing oxaloacetate to acetyl CoA (produced by glycolysis), producing the C6 substrate citrate, which is broken down to oxidatively to oxaloacetate in 1 turn of the cycle. Products include (quantitatively) CoA-SH, 3 NADH (x2 per glucose, yielding 15 ATP), 1 CoQH$_2$ (x2 per glucose, yielding 3 ATP), 1 GTP (x2 per glucose), 2 CO$_2$, and 2 H$^+$.

(iii) oxidative phosphorylation/electron transport

conversion of *reducing equivalents* to H$_2$O (NADH, FADH$_2$, and CoQH$_2$) via the electron transport chain and used O$_2$ as the ultimate electron acceptor. The F$_0$F$_1$ ATPase converts the *protonmotive force* generated by *electron transport* into phosphorylation of ADP to produce ATP. Products are NAD$^+$, FAD, coenzyme Q (all oxidized) and ATP; H$_2$O is produced by the reduction of ½ O$_2$.

3. (2) Describe how the following concept applies to the regulation of glycolysis. Be as detailed as possible.

Ω, Γ: Define "metabolically irreversible."

Having a large negative ΔG; highly spontaneous thermodynamically and predisposed almost completely to product formation.

Ψ: Define "near equilibrium."

Having a ΔG close to or near zero; flux is approximately equal in both directions.

Three reactions in glycolysis are metabolically irreversible; those catalyzed by phosphofructokinase 1, hexokinase and pyruvate kinase. The others are near equilibrium. As a result, they constitute the key control points for metabolic regulation, specifically by allosteric effectors.

(See Review Session 11, problem 5.)

4. (3) Explain why the kinetics of an enzyme reaction are most easily controlled when K$_M$ is approximately equal to the actual concentration of the reactant. Include the appropriate chemical reaction and defining equation in your discussion.

(See Review Session 11, problem 6.)

When K$_M$ is approximately equal to [S], the reaction is occurring at a rate of ½ V$_{max}$; at lower [S], the reaction is not occurring to much of an extent (rate); at higher [S], the reaction is occurring with a rate approaching V$_{max}$.

(Continued)

V_o only changes a lot when [S] is equal or close to K_M, so only then can the rate be regulated upward or downward with small changes in [S]. The relevant equation and reaction are:

$$E + S \leftrightarrow [ES] \xrightarrow{k_{cat}} E + P \quad \text{('}\leftrightarrow\text{' indicates an equilibrium)}$$

with rate constants k_1 and k_{-1}

$$V_o = V_{max} [S]/(K_M + [S])$$

5. (everyone) Explain the following terms and explain their use to study or purpose in metabolism or regulation.

a. (1.5) Why should citrate negatively regulate the phosphofructokinase-1 reaction? What is the general term for this phenomenon? (Review Session 11, problem 9a)

The negative regulation of PFK-1 by citrate is an example of negative feedback inhibition of glycolysis by a key Krebs Cycle intermediate. If plenty of citrate is present, sufficient acetyl CoA is being produced by glycolysis, and the pathway should be inhibited.

b. (1.5) Why should fructose-1, 6-bisphosphate stimulate the pyruvate kinase reaction? What is the general term for this phenomenon? (Review Session 11, problem 9b)

Stimulation of pyruvate kinase by fructose-1, 6-bisphosphate is an example of feed-forward stimulation. The idea is that accumulation of fructose-1, 6-bisphosphate stimulates the later reaction, in order to use the accumulated intermediate.

c. (2) triose phosphate isomerase, including the definition, reactant (s), reaction and product(s):
    (Review Session 11, problem 8.)
Fructose-1, 6-bisphosphate dissociates to form two C3 compounds, dihydroxyacetone phosphate (DHAP) and glyceraldehyde-3-phosphate (G3P) in the Aldolase reaction. Since only G3P can be used in subsequent Glycolytic reactions, DHAP must be converted to G3P. This reaction is catalyzed by Triose Phosphate Isomerase.

DHAP   G3P

d. (2) Aldolase, including the definition, reactant(s), reaction and product(s):
Enzyme that catalyzes aldol condensation in the Gluconeogenic direction, and cleaves fructose-1, 6-bisphosphate to form dihydroxyacetone phosphate and glyceraldehyde-3-phosphate in the Glycolytic direction:

(See Review Session, problem 8.)

6. (2) (everyone): Name the coenzyme used by dihydrolipoamide acetyl transferase and explain how it functions fueling the Krebs Cycle. (A picture would help.)

(See Review Session 12, problem 4.)
Lipoic acid is a lipid-based hydrocarbon that acts as a covalently attached tether between the pyruvate dehydrogenase E_2 subunit and the acetyl group. It functions by accepting the ethanol group from the thiamine pyrophosphate group (on the E_1 subunit), by forming a thioacetate, and finally by transferring the acetate to CoA-SH.

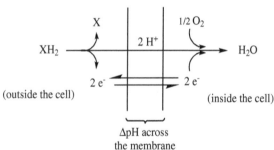

7. (everyone) a. (3) Define "protonmotive force". Explain what enzyme complex uses this phenomenon as the driving energy for ATP synthesis in oxidative phosphorylation. Give a diagram and the relevant equation.

(See Review Session 12, problem 11.)
Oxidative loss of reducing equivalents due to a proton gradient across a membrane (or the mitochondrion). Diagrammatically:

Mathematically, there are two contributions to the protonmotive force (Δp) due to (1) the difference between inside and outside [H^+], and (2) the change in membrane potential ($\Delta \Psi$).

$$\Delta p = \Delta G / n \,\mathscr{F} = \Delta\psi + \Delta pH_{\text{in-out}} \ (0.059V)$$

ATP Synthase catalyzes $H^+ \rightarrow$ back in.

b. (1) How does electron transport drive production of the protonmotive force?

(Review Session 12, problem 11):
The electron transport reactions "pump" H^+ out of the mitochondrial matrix and build up the [H^+] in the inner membrane space, which drives ATP synthesis by the F_oF_1 ATPase.

8. (4) Multiple choice: One point maximum each. Circle the single number that corresponds to the best answer.

A. Coenzyme/substrates in the pyruvate dehydrogenase complex shuttles C2 fragments in the following order:
i. pyridoxal phosphate-E_1, lipoyl-E_2, CoA-SH, oxaloacetate
ii. thiamine pyrophosphate-E_1, lipoyl-E_2, CoA-SH, citrate (+)
iii. thiamine pyrophosphate-E_1, lipoyl-E_2, CoA-SH, oxaloacetate
iv. thiamine pyrophosphate-E_1, lipoyl-E_2, ACP-SH, oxaloacetate
v. thiamine pyrophosphate-E_1, lipoyl-E_2, ACP-SH, citrate

B. Pyruvate translocase
i. is a pyruvate: malate antiport
ii. is a pyruvate: OH^- symport
iii. is a pyruvate: H^+ antiport
iv. exits the mitochondrion
v. fuels the Krebs Cycle (+)

C. The correct order of substrate production in the Krebs Cycle is:
i. succinate, fumarate, malate, oxaloacetate, coenzyme A
ii. glucose-6-phosphate, fructose-6-phosphate, fructose-1, 6-bisphosphate
iii. fumarate, succinate, dihydroxyacetone phosphate, oxaloacetate
iv. succinate, fumarate, malate, oxaloacetate, citrate (+)
v. succinate, succinyl CoA, fumarate, malate, oxaloacetate

D. β-oxidation yields the following:
i. 1 $FADH_2$, 1 $CoQH_2$ (indirectly), 1 NADH, H^+, 1 acetyl CoA and a 2 C unit shorter fatty acid (+)
ii. 1 GTP, 1 $CoQH_2$ (indirectly), 1 NADH, H^+, 1 acetyl CoA and a 2 C unit shorter fatty aldehyde
iii. 1 $FADH_2$, 1 $CoQH_2$ (indirectly), 2 NADH, H^+, 2 acetyl CoA and a 2 C unit shorter fatty acid
iv. 1 $FADH_2$, 1 $CoQH_2$ (indirectly), 2 NADH, 2 H^+, 1 acetyl CoA and a 2 C unit shorter fatty acid
v. fire and water, sufficient supplies for 1 cell for 1 second

9. (PUZZLER) a. (2) Draw the mechanism for acetoacetyl-Acyl Carrier Protein (ACP) formation from acetyl-ACP. Include both key proteins with substrates bound and unbound, reactants and products.

b. (2) Which product do fatty acid synthesis and ethanol fermentation of pyruvate have in common? Explain.

Both processes produce 1 carbon dioxide molecule per single C2 unit added (fatty acid synthesis) or fermented (pyruvate decarboxylation in ethanol production).

Final Test Review

1. Draw the structure of the following amino acids at pH 7. Show all bonds and atoms.
2. Name (below) *all* of the functional groups in the two molecules drawn in #1, including those on the R group. Be precise with your nomenclature.

(a) L-methionine   (b) L-serine

(c) L-histidine   (d) L-glutamate

(e) L-leucine   (f) L-asparagine

(g) L-proline   (h) L-aspartate

(i) L-lysine   (j) L-valine

(k) L-isoleucine   (l) L-cysteine

3. Define a van der Waals interaction. Include a description of the relation between interatomic distances and free energies.

4. Define the following terms:
(a) Protein primary structure

(b) Protein secondary structure.

(c) Protein tertiary structure.

(d) Protein quaternary structure

5. Define a hydrogen bond interaction. Include a description of the relation between interatomic distances and free energies.

6. Define a protein β-sheet; include a drawing of the key hydrogen bonds.

7. Describe Edman degradation. Draw the structure of the key reagent.

8. Define a chaotropic agent. Give two specific examples used with proteins.

9. Define a protein α-helix; include a drawing of the key hydrogen bonds.

10. Define a disulfide bond. Draw one between two fully drawn "residues."

11. Define the purpose of a Ramachandran plot. What are ϕ and ψ?

12. Define a zwitterion. Draw a generic amino acid zwitterion.

13. Define and describe the energetic origins of the hydrophobic effect with respect to protein stability.

14. Define an "initial" rate" in the context of enzyme kinetics. Give the relevant equation and name the terms in it.

15. Does an act of catalysis change an enzyme. Explain and give the enzymatic reaction.

16. Define the term "maximum velocity" in the context of enzymatic catalysis. Give the relevant equation and name the terms.

17. List two requirements for enzymatic catalysis.

18. Write down the Michaelis-Menten equation and name all of the terms in it.

19. (a) What is the definition of K_M?

(b) What is a Michaelis complex?

20. What is the definition of k_{cat}?

21. Write out the linked chemical equilibrium reactions involving E, S and other relevant parameters that describe simple Michaelis-Menten catalysis.

22. Show how the reactions for E, S and I differ for competitive and uncompetitive inhibition of enzyme catalysis.

23. List the three (3) requirements for enzyme catalysis described in class.

24. How do the *initial rate* and *steady state* postulates differ in how they describe enzyme kinetics?

25. If an enzyme catalyzes a reaction extremely quickly, what external factor still limits the velocity of the reaction? Explain.

26. (a) Draw acetylcholine. See part (b) before you start.

(b) What atom on what amino acid in the esterase site of acetylcholine esterase reacts with the substrate? Include a cartoon depiction of the enzyme surface (drawn under your acetylcholine drawing in part a) showing how it interacts with the substrate. Include an arrow to show how the electrons of the key enzyme atom attack the substrate.

27. (a) Explain in general how the nerve gas antidote pyridine aldoximine methiodide (PAM) reactivates acetylcholine esterase.

(b) What is the key functional group on the antidote?

(c) What kind of reaction produces the reactivated enzyme?

28. Draw the generic reaction for the bisubstrate-enzyme ping-pong reaction. Define the letters used.

29. Explain how an enzyme cascade works. Include a drawing and emphasize how amplification of the initial signal is achieved.

30. Blood clotting is initiated by very few enzymes. Explain how a clot composed of many, many, many proteins is produced. Include a reaction or reactions and emphasize how such a large number of products is made. Name this phenomenon.

31. Define a zymogen and give one specific example that is important for digesting food.

32. Describe how feedback inhibition works. Why is this a logical way to regulate a metabolic pathway.

33. Define allosterism. Describe how it is initiated using the appropriate name.

34. Describe one specific example of how kinase and phosphatase reactions regulate either glycolysis or the Krebs Cycle. Name the regulated enzyme and explain which form is active and which is inactive.

35. What does cyclin kinase regulate? What two amino acids are phosphorylated?

36. Give two examples of *reversible* factors that control the catalytic capability of an enzyme. Explain each briefly.

37. Give two examples of *irreversible* factors that control the catalytic capability of an enzyme. Explain each briefly.

38. (a) The Arrhenius equation describes the relation between which physical parameters and constants?

(b) Is it a thermodynamic equation or a thermodynamic equation? Explain.

39. List the two "chemical modes of catalysis." Define each briefly.

40. List the two "binding modes of catalysis." Define each briefly.

41. Explain, providing the appropriate diagrams, how uncatalyzed and catalyzed reaction coordinate versus energy trends differ. Explain each change you show.

42. Define a nucleophilic substitution (S_N2) reaction. Provide a mechanism diagram. Define a nucleophile in your explanation.

43. (a) Define an electrophile.

(b) Explain how to create one via acid-base catalysis.

44. (a) Why is histidine used so often by enzymes to carry out acid-base catalysis?

(b) Explain why enzymes typically have an optimal pH.

45. (a) Draw the catalytic triad of amino acids typically present in serine proteases. Label them.

(b) Explain how they collaborate to accomplish acid-base catalysis.

46. Draw the structure of ATP. Include all atoms and bonds.

47. Why is Mg^{2+} typically required to achieve optimal activity with ATP cosubstrate enzyme reactions?

48. Define a coenzyme. Give one example and describe how the coenzyme is used in a typical biochemical application.

49. Briefly describe one example of a scenario in which a transition metal is used to facilitate a biochemical reaction.

50. What is ATP used to do in most biochemical applications? Give two different examples that involve different parts of the molecule.

51. (a) Draw NAD^+ (NADH) in both the oxidized and reduced form. Use the contraction R to signify the common parts of the structure in the latter drawing (*i.e.*, only show the part that changes, connected to R).

(b) Describe how the coenzyme is used in a typical biochemical application.

52. (a) Draw the business end of FAD ($FADH_2$) in both the oxidized and reduced form.

(b) Describe how the coenzyme is used in a typical biochemical application.

53. (a) Draw pyridoxal phosphate prior to and after forming a Schiff base with the ε-amino group of a lysine residue from an enzyme E.

(b) Describe how the coenzyme is used in a typical biochemical application.

54. (a) What chemical group does coenzyme A typically carry in biochemistry? Draw the covalent complex as you've been instructed in class (*i.e.*, not the entire thing).

(b) Why is this function such a crucial link in intermediary metabolism?

55. Explain how the biotin-avidin noncovalent binding interaction is used to capture ligand-binding entities. Provide a diagram that illustrates your explanation.

56. (a) What is the crucial function of N^5, N^{10} tetrahydrofolate in the production of DNA?

(b) Explain why our understanding of this function can be used in a strategy for anti-cancer chemotherapy.

57. (a) What is UDP-galactose and how is it used to make lactose?

(b) Draw the structure of lactose as given in your notes.

58. How does cis-retinal function in transducing the signal of a photon of light into a chemically recognizable form?

59. (a) Explain how a ketose differs from an aldose and give one example of each.

(b) Explain how a pyranose differs from a furanose and give one example of each.

60. (a) Draw the chair and boat conformations of the β-D-glucopyranose structure.

(b) Explain why a polysaccharide chain would turn directions much more in one case than in the other and which conformational isomer would be expected to produce the most radical change in chain direction.

61. (a) Why is the hexa-atomic ring of inositol more stable than that of galactose?

(b) Given what you know about the biosynthesis of lactose from UDP-galactose and glucose, why would you expect inositol to be less likely to polymerize than glucose or galactose? (*Hint.* What are their relative capabilities in serving as nucleophilic centers?)

62. (a) What does NAG-α(1→6)-NAM-α(1→4)-glc-β(1→4)-gal mean?

(b) Draw a simplified diagram with specific emphasis on the meaning of α and β.

63. (a) How does glycogen differ from the amylopectin in starch?

(b) How would this facilitate the function of glycogen?

64. How does penicillin work selectively on bacteria but not harm us significantly? (What does it do?)

65. (a) How do extra-cellular surface carbohydrates regulate osmotic pressure around cells?

(b) What advantage does this impart to them?

66. Explain how the terms export and clearance apply to the functional status of carbohydrate-bearing proteins.

67. (a) How does Phospholipase C produce two different second messengers in a signal transduction pathway we discussed in class?

(b) Draw the appropriate reactant.

68. What function does chondroitin sulfate serve in cartilage and skeletal joints?

69. (a) Do saturated or unsaturated fatty acids of the same length have a lower melting temperatures (T_m)?

(b) What reactions are followed? What do differences in the T_m values monitor?

70. How do lipid bilayers and micelles differ? Include a simple drawing.

71. (a) Draw the structure of a general phosphatidyl choline molecule.

(b) Name the functional groups on your drawing.

72. Explain five details that describe the "fluid mosaic" membrane model.

73. (a) Draw the structures of the four nucleic acid bases in DNA.

(b) Label the bases drawn in part (a).

74. (a) Draw a RNA trinucleotide using three different bases.

(b) Label the bases and other structural subunits drawn in part (a).

75. (a) Draw the structures of the two Watson-Crick base pairs. Use " ||||| " to designate hydrogen bonds.

76. Define a glycosidic bond in a nucleoside.

77. Explain how and why the absorbance at 260 nm (A_{260}) can be used to determine if a double helix forms from 2 single strands of DNA or RNA.

78. (a) Define "base stacking."

(b) Describe the three predominant types of "forces" that contribute to stabilization of "stacked" bases in a double helix.

79. (a) What are counterions and why do they bind all nucleic acids?

(b) How do histones serve this function in the case of most chromosomal DNAs?

80. Give two reasons why G•C base pairs are more stable than A•T (or A•U) base pairs. Rank these contributions according to importance in inducing stability.

81. Describe four differences between A and B forms of DNA.

82. (a) Draw the mechanism of alkaline hydrolysis of RNA.

(b) Why is DNA not subject to this mechanism?

83. Why and how does an antisense oligonucleotide functionally inactivate a mRNA for use in translation by a ribosome?

84. Name the four classes of RNA and briefly explain their functional significance. In what reaction(s) do they participate?

85. List four distinctive features of most eukaryotic mRNAs.

86. What is the primary use of a DNA probe and how is this process accomplished?

87. Why are restriction endonucleases required to produce, manipulate and clone specific pieces of DNA?

88. What are the two functional ends of transfer RNA and how do they work to accomplish these functions?

89. (a) Name and briefly describe the purpose (point) of the three most central catabolic pathways of intermediary metabolism. (The third has two coupled parts. Name both of them.)

(b) What is (are) the product(s) of each of these pathways? (list according to pathway)

90. Describe the two major ways that energy is captured in a chemically usable form by metabolic reaction pathways.

91. The following reaction from Merlin's notebooks occurred with a standard free energy ($\Delta G^{\circ\prime}$) of -8.5 kcal per mol:

$$\text{carborandum} + \text{gold} \leftrightarrow \text{essence of life}$$

(a) At equilibrium, a real sample of these three solutes contained 44 mM carborandum, 0.1 mM gold and 45 μM essence of life. What is the actual free energy (ΔG) of the reaction under these equilibrium conditions at room temperature (298 K)? Show your setup and work.

(b) A different sample contained 0.9 M carborandum, 0.1 mM gold and 45 μM essence of life. What is ΔG?

(c) How did adding more carborandum affect the equilibrium poise of the reaction? What was the mass action ratio (Q) in each case?

92. (a) Why do only three steps in glycolysis control most of the flux through the pathway under actual cellular conditions?

(b) What are the three steps, the three enzymes that catalyze the reactions and what do the reactions have in common?

93. (a) Define metabolically irreversible

(b) Define near equilibrium.

94. Explain why the kinetics of an enzyme reaction are most easily controlled when K_M is approximately equal to the actual concentration of the reactant.

95. Consider the following data:

Reaction				$E^{\circ\prime}$
Acetyl CoA	$+ CO_2 + H^+ + 2\,e^-$	--> Pyruvate	$+ \text{CoA-SH}$	-0.48 V
NAD^+	$+ 2\,H^+ + 2\,e^-$	--> NADH	$+ H^+$	-0.32 V
$FAD + 2\,H^+$	$+ 2\,e^-$	--> $FADH_2$		-0.22 V

(a) Under standard state conditions (T = 298 K), will it require more energy if the breakdown of pyruvate to acetyl CoA is coupled to FAD formation or to NAD^+ formation? Show your setup and work.

(b) Does the answer change if pyruvate is present at 1 mM while the other species are still present at 1 M.

96. Where do the reactant/product of triose phosphate isomerase derive from (directly) and what do they get converted to catabolically.

97. (a) Why should citrate negatively regulate (discourage) the phosphofructokinase-1 reaction? What is the general name for this phenomenon?

(b) Why should fructose-1,6-bisphosphate stimulate the pyruvate kinase reaction? What is the general name for this phenomenon?

98. We discussed three branching catabolic fates of pyruvate. Draw the three reaction (reaction sequences), including cofactors and enzymes.

99. (a) Why does the absence of alcohol dehydrogenase produce even more scurrilous behavior than if the person has the enzyme?

(b) How is the blocked catabolic intermediate related to the typical role of pyridoxal phosphate in enzymatic catalysis?

100. (a) Why does "carbonation" accompany ethanol production?

(b) How does the staff of life of the western world benefit (and us as a consequence)? [Bread, not beer!]

101. Dihydrolipoamide acetyl transferase uses a coenzyme we did not discuss earlier. (a) What is it and how does it function in "fueling" the Krebs Cycle?

(b) What other coenzyme is also involved in this process? How?

102. (a) What "symport" reaction accompanies import of pyruvate into the mitochondrion and what enzyme catalyzes the reaction?

(b) Does the pH of the cytoplasm increase or decrease as a result?

103. List the reactions, coenzyme(s), cofactor(s) and enzymes involved in the two "oxidative decarboxylation" reactions of the Krebs Cycle.

104. List the reactions, coenzyme(s), cofactor(s) and enzymes involved in the "substrate-level phosphorylation" reaction of the Krebs Cycle.

105. List (only once each) all of the "energy conserving" compounds formed by the Krebs Cycle accompanied by the number of "ATP equivalents" finally accrued after oxidative phosphorylation of 1 molecule of each of these compounds.

106. (a) How do fumarase and malate dehydrogenase "fix" a carbonyl group on succinate in the production of oxaloacetate (OAA).
(b) What crucial 2 carbon compound is then "fixed" to OAA?

(c) How is the product used in food flavoring?

107. What amino acid and what product of pyruvate metabolism are the principle substrates for gluconeogenesis in mammals?

108. What is "protonmotive force" and what enzyme complex uses this phenomenon as the driving energy for ATP synthesis in oxidative phosphorylation? (Contrast it with electromotive force, emf, voltage.)

109. How does electron transport drive production of the protonmotive force?

110. (a) How many reactions does each round of β-oxidation of a fatty acid require?

(b) What are the products of one round of β-oxidation and what is the tally in terms of ATP equivalents of energy conserving products?

111. (a) Draw the coupled cofactor regeneration cycles that siphon off reducing equivalents then fix them into coenzyme Q in reactions that are coupled to the first oxidative step of fatty acid β-oxidation.

(b) Write down the names and a short definition for the cofactors involved in this "siphon".

112. Which three steps of the Krebs Cycle do the first three steps of the fatty acid β-oxidation cycle resemble? Draw (horizontally) the three analogous reactions from (i) β-oxidation, (ii) the Krebs Cycle, and (iii) the three generic chemical transformations that occur beneath the four analogous substrates.

113. Draw the analogous (but backward) [reduction-dehydration-reduction] series of reactions of fatty acid synthesis starting from reactants and going to products beneath your three reaction series from problem 4. Point the arrows in the opposite direction drawn for problem 4.

114. Draw the reactions for the condensation substep in acyl-carrier protein-mediated fatty acid synthesis.

Final Test Review (with answers)

1. Name the molecules drawn below. Be as exact with your nomenclature as you can.

(a) L-Methionine

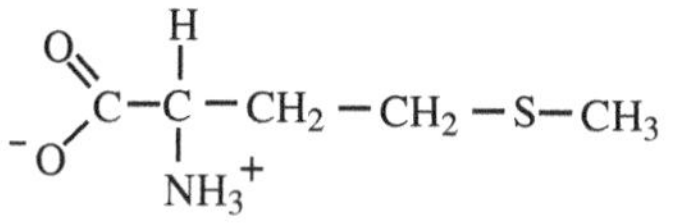

$^-O_2C-CH(NH_3^+)-CH_2-CH_2-S-CH_3$

(b) L-Serine

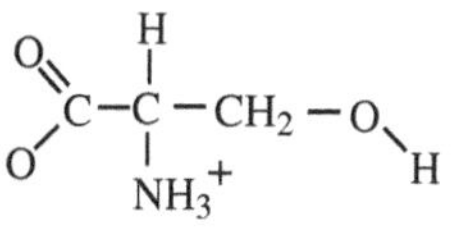

$^-O_2C-CH(NH_3^+)-CH_2-O-H$

(c) L-Histidine

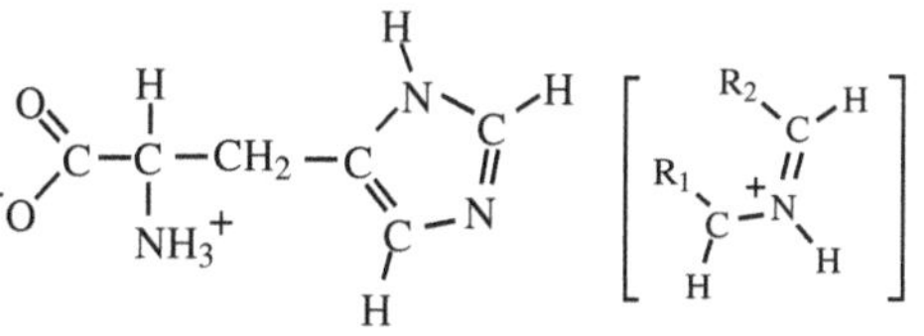

$^-O_2C-CH(NH_3^+)-CH_2-$ (imidazole ring)

(d) L-Glutamate

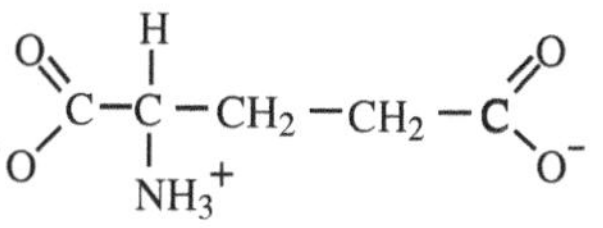

$^-O_2C-CH(NH_3^+)-CH_2-CH_2-CO_2^-$

(e) L-Leucine

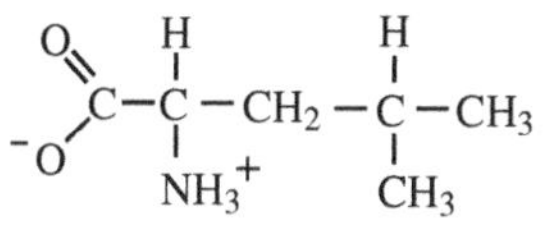

$^-O_2C-CH(NH_3^+)-CH_2-CH(CH_3)-CH_3$

(f) L-Asparagine

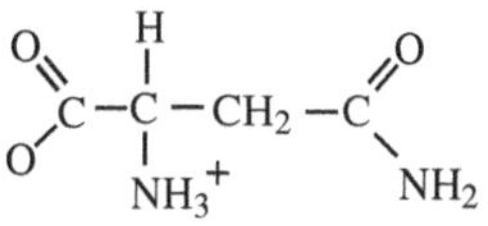

$^-O_2C-CH(NH_3^+)-CH_2-C(=O)NH_2$

(g) L-Proline

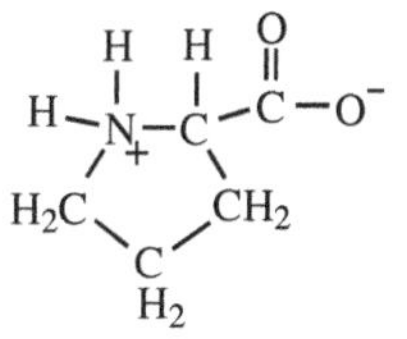

$H-N^+(H)-CH(-C(=O)-O^-)$ (pyrrolidine ring with H_2C, CH_2, CH_2)

(h) L-Aspartate

$^-O_2C-CH(NH_3^+)-CH_2-CO_2^-$

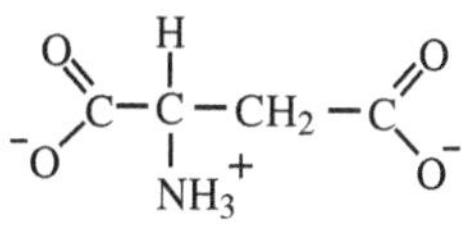

(i) L-Lysine

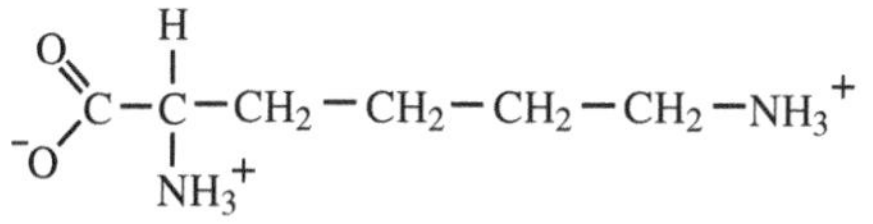

$^-O_2C-CH(NH_3^+)-CH_2-CH_2-CH_2-CH_2-NH_3^+$

(j) L-Valine

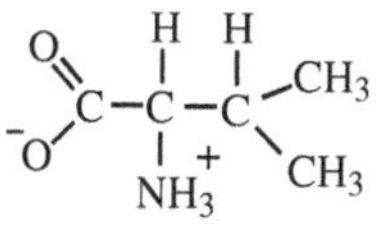

$^-O_2C-CH(NH_3^+)-CH(CH_3)(CH_3)$

(k) L-Isoleucine

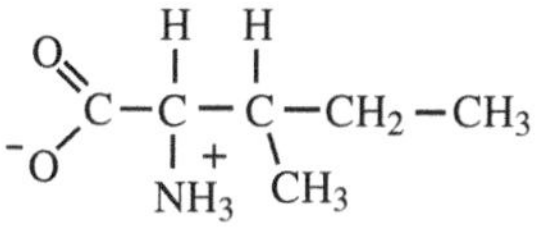

$^-O_2C-CH(NH_3^+)-CH(CH_3)-CH_2-CH_3$

(l) L-Cysteine

$^-O_2C-CH(NH_3^+)-CH_2-SH$

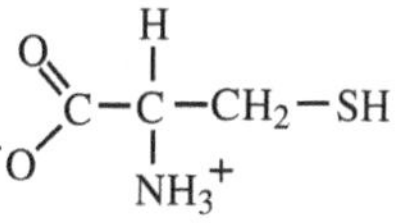

2. The <u>van der Waals interaction</u> describes the relation between interatomic distances, electronic charge, solution dielectric and free energies.

3. (a) Protein <u>quaternary structure</u> defines the relation among subunits in a multisubunit lattice.
(b) Protein <u>primary structure</u> defines the amino acid sequence.
(c) Protein <u>tertiary structure</u> defines the packing of helices, sheets, turns, etc.
(d) Protein <u>secondary structure</u> defines the motifs formed by short-range interactions between amino acids.

4. A <u>hydrogen bond</u> interaction involves polar O, N or both and the atom for which it is named, and constitutes one of the important protein stabilization elements.

5. Name the following protein structure: <u>β-sheet</u>.

6. <u>Edman degradation</u> is used to determine the sequence of a protein based on sequential chemical reactivity.

7. A <u>chaotropic agent</u> induces denaturation of proteins by disturbing the hydrophobic effect.

8. Name the following protein structure: <u>α-helix</u>.

9. Name the following protein structure: <u>disulfide bond</u>.

$$R_1 - CH_2 - S - S - CH_2 - R_2$$

10. A <u>Ramachandran plot</u> is a graph of the conformational torsion angles ϕ and ψ for the residues in a protein or peptide, a map of the structure of the polypeptide backbone.

11. A <u>zwitterion</u> has two charges which neutralize each other.

12. The <u>hydrophobic effect</u> is the primary "force" of protein structural stabilization.

13. The <u>initial rate</u> is the characteristic speed of an enzyme's kinetics extrapolated to the time when a defined amount of substrate is added to the enzyme solution.

14. An act of <u>catalysis</u> does not change an enzyme and lowers the transition state free energy of the associated reaction.

15. The <u>maximum velocity</u> of an enzymatic catalysis reaction is the rate achieved when it is saturated with substrate.

16. The <u>Lineweaver-Burk</u> (or double reciprocal) equation defines parameters that are used to characterize the kinetics of an enzyme.

17. K_m is the substrate concentration when $V_0 = V_{max}/2$, or <u>Michaelis-Menten constant</u>.

18. A <u>Michaelis complex</u> is the enzyme-substrate combination formed during an enzyme catalysis event.

19. The catalytic rate constant of an enzyme is abbreviated as k_{cat}.

20. <u>Competitive inhibition</u> of enzyme catalysis occurs when an inhibitor binds to the active site of the enzyme.

21. <u>Uncompetitive inhibition</u> of enzyme catalysis occurs when the inhibitor only binds to the enzyme-substrate complex.

22. The <u>steady state approximation</u> postulates that a constant input feed of substrate is supplied whose rate equals that of product formation.

23. Two internal factors that limit the velocity of an enzymatic reaction are _____ and _____.

[hydrophobic effect, H-bonding, disulfide bonds, van der Waals forces, ionic bonds (salt bridges) or dipole-dipole interactions (actually underlying all of the others)]

24. Two external factors that limit the velocity of an enzymatic reaction are _____ and _____.

[pH, solvent polarity, temperature, salt concentration(s) and types, presence of chaotropes, osmolytes, others]

25. What amino acid and functional group in the esterase site of acetylcholine esterase reacts with the substrate?   serine, <u>hydroxylate</u>

26. Pyridine aldoximine methiodide (PAM) reactivates acetylcholine esterase, functioning as a

     <u>nerve gas antidote</u>.

27. What kind of reaction produces the reactivated enzyme? <u>nucleophilic substitution</u>

28. The bisubstrate-enzyme <u>ping-pong</u> reaction is used by transaminases in the exchange of an amino group for a carbonyl group between two progressively binding substrates.

29. An <u>enzyme cascade</u> works by amplifying an initial signal via several linked protease cleavage reaction stages. (*e.g.*, blood clotting)

30. A <u>zymogen</u> is a protein that is converted from inactive to active forms by a covalent modification, typically protease cleavage.

31. A decrease in the activity of an enzyme as a result of binding of a product from the reaction in question or subsequent reactions is referred to as <u>feedback inhibition</u>.

32. <u>Allosterism</u> involves binding of a regulatory molecule at a site other than the active site.

33. <u>Kinase</u> and <u>phosphatase</u> reactions, involving phosphate addition and removal respectively, regulate both glycolysis and the Krebs Cycle.

34. <u>Cyclin kinase</u> regulates entry and exit from mitosis by catalyzing a covalent modification reaction.

35. Which two amino acids are modified in the reactions catalyzed by the enzyme in question 35?
                                        <u>tyrosine</u>, <u>threonine</u>.

36. Two examples of *reversible* factors that control the catalytic capability of an enzyme are:

_____, _____

[noncovalent modifications, pH and pK_a changes, [salt] changes, possibly others]

37. Two two examples of *irreversible* factors that control the catalytic capability of an enzyme are:

_____, _____

[covalent modification, proteolysis, irreversible inhibitors, possibly others]

38. The Arrhenius equation accounts for the temperature dependence of the rate of a reaction.

39. List the two "chemical modes of catalysis". acid-base, covalent

40. List the two "binding modes of catalysis". proximity effect, transition-state stabilization

41. A nucleophile attacks an electropositive site in its role in a chemical (enzymatic) reaction.

42. A common process used to produce the species in problem 41 is: acid-base catalysis.

43. The most common amino acid used by enzymes to carry out acid-base catalysis is histidine.

44. A catalytic triad of amino acids is typically present in (enzyme class name) serine proteases.

45. The amino acids collaborate to accomplish acid-base catalysis.

46. The most typically cited currency of energy in metabolism is (abbreviation) ATP.

47. Mg^{2+} is typically required to achieve optimal activity with (answer 46)-cosubstrate enzyme reactions?

48. A coenzyme is either a loosely bound cosubstrate or strongly bound prosthetic group.

49. The heavy metal molybdenum is used to facilitate the biochemical reaction in xanthine oxidase, a key enzyme in purine catabolism.

50. When ATP used in some biochemical applications it yields AMP and pyrophosphate.

51. The (vitamin) nicotinamide is required to synthesize coenzyme NAD^+ for use in metabolic redox reactions.

52. The other key redox coenzyme is abbreviated FAD.

53. The coenzyme pyridoxal phosphate often forms a Schiff base with the ε-amino group of a lysine residue in the enzyme.

54. What chemical group does coenzyme A typically carry in the course of its biochemical function? acetate

55. The biotin-avidin noncovalent binding interaction is used to capture ligand-binding entities in the "affinity capture" technique.

56. The coenzyme N^5, N^{10} methylenetetrahydrofolate is required to incorporate the methyl group into thymidine, a necessary prerequisite for the production of DNA.

57. Our understanding of this function can be used in a strategy for (treatment technique) anticancer chemotherapy.

58. The coenzyme bound carbohydrates UDP-galactose and glucose are required to synthesize lactose?

59. Cis-retinal functions in <u>transducing</u> the signal of a photon of light into a chemically recognizable form?

60. The two important straight-chain forms of carbohydrate structure are the <u>ketose</u> and <u>aldose</u>.

61. The two important ring forms of carbohydrates are the <u>pyranose</u> and <u>furanose</u>.

62. The two important ring conformations of β-D-glucopyranose are the <u>chair</u> and <u>boat</u>.

63. The cyclohexane ring containing the compound <u>inositol triphosphate</u> is released by phospholipase C in the phospholipid signal transduction mechanism.

64. The acronym NAG is used to abbreviate the name of the compound <u>N-acetyl glucosamine</u>.

65. The key polysaccharide in starch is <u>amylopectin</u>.

66. The key polysaccharide in the liver is <u>glycogen</u>.

67. The antibiotic <u>penicillin</u> selectively inhibits cell wall peptidylglycan synthesis in bacteria.

68. Extra-cellular surface <u>carbohydrates</u> regulate the osmotic pressure around cells.

69. Phospholipase C produces two different second messengers in the phospholipid signal transduction pathway. The lipid-containing second messenger is <u>diacylglycerol</u>.

70. The compound chondroitin sulfate <u>lubricates</u> cartilage and skeletal joints.

71. Saturated / <u>unsaturated</u> (circle one) fatty acids of the same length have a lower melting temperature (Tm).

72. Lipid T_m values monitor the transformation from <u>liquid crystal</u> to dispersed forms.

73. Lipid <u>bilayers</u> are composed of two face-to-face monolayers while lipid <u>micelles</u> form a biphasic sphere.

74. The most popular model for a biological membrane is called the <u>fluid mosaic</u> model.

75. The four nucleic acid bases in RNA are <u>adenine</u>, <u>guanine</u>, <u>cytosine</u> and <u>uracil</u>.

76. The two normal base pairs in DNA and RNA are called <u>Watson-Crick</u> base pairs.

77. The <u>glycosidic</u> bond in a nucleoside connects the base to the sugar.

78. The <u>absorbance at 260 nm</u> can be used to determine if 2 single strands of DNA or RNA form a double helix.

79. The face-to-face interaction between nucleic acid bases is called <u>base stacking</u>.

80. Counterions bind all nucleic acids and are required to neutralize the <u>phosphodiester phosphates</u>.

81. Protein complexes called <u>histones</u> serve this counterion function in the case of most chromosomal DNAs.

82. G•C / <u>A•T (or A•U)</u> (circle one) base pairs are less stable than <u>G•C</u> / A•T (or A•U) (circle one) base pairs.

83. Two differences between A and B forms of DNA are

	A-form		B-form
1.	_____	versus	_____ .
2.	_____	versus	_____ .

3' endo sugar conformation	2' endo sugar
base pairs tilted 20° from helix axis	bp's perpendicular to helix axis
central axial cavity in the helix	base pairs cross center of helix
shorter, squatter helix	longer, narrower helix

84. The 2'-hydroxyl group catalyzes <u>alkaline hydrolysis</u> of RNA, a good example of anchiomeric assistance in a non-protein biomolecular mechanism.

85. An antisense oligonucleotide functionally inactivate a mRNA for use in translation by a ribosome by forming a double helix with it and precluding <u>tRNA anticodon</u> binding.

86. Name the two most prevalent of the four classes of RNA. <u>ribosomal RNA</u> and <u>transfer RNA</u>

87. Two distinctive features of most eukaryotic mRNAs are

_____ and _____

[m^7G^+ (5'-5') cap, monocistronic, contains introns and exons, poly(A) tail]

88. A <u>DNA probe</u> is used to detect the presence of a specific complementary nucleic acid sequence.

89. <u>Restriction endonucleases</u> are required to produce, manipulate and clone specific pieces of DNA?

90. The two functional ends of transfer RNA are the anticodon and <u>amino acid acceptor</u>.

91. The three most central catabolic pathways of intermediary metabolism are
<u>glycolysis</u>, <u>Krebs Cycle</u>, and <u>electron transport/oxidative phosphorylation</u>

92. The four major compounds in which energy is captured in a chemically usable form by metabolic reaction pathways are <u>ATP</u>, <u>NADH</u>, <u>$FADH_2$</u>, and <u>Coenzyme QH_2</u>.

93. The <u>mass action ratio</u> (Q) corrects for deviations from standard state concentrations (1 M).

94. <u>Three</u> (number) steps in glycolysis control most of the flux through the pathway under actual cellular conditions?

95. What do the reactions in problem 94 have in common? They are <u>metabolically irreversible</u>.

96. In contrast, the rest of the reactions are <u>near equilibrium</u>.

97. The kinetics of an enzyme reaction are most easily controlled when K_M is approximately equal to <u>the actual concentration of the reactant</u>.

98. The enzyme triose phosphate isomerase converts <u>dihydroxyacetone phosphate</u> into glyceraldehyde-3-phosphate.

99. When citrate negatively regulates (discourage) the phosphofructokinase-1 reaction, the general name for this phenomenon is <u>feedback inhibition</u>.

100. When fructose-1, 6-bisphosphate stimulates the pyruvate kinase reaction, the general name for this phenomenon is <u>feed-forward activation</u>.

101. The three possible catabolic fates of pyruvate are <u>acetyl CoA</u>, <u>ethanol</u> and <u>lactate</u>.

102. The enzyme alcohol dehydrogenase converts <u>acetaldehyde</u> to ethanol.

103. <u>Dihydrolipoamide acetyl transferase</u> uses the coenzyme lipoic acid in fueling the Krebs Cycle?

104. What symport reaction accompanies import of pyruvate into the mitochondrion and what enzyme catalyzes the reaction? (enzyme name) <u>pyruvate translocase</u>

105. The two oxidative decarboxylation reactions of the Krebs Cycle are catalyzed by <u>isocitrate dehydrogenase</u> and
<u>α-ketoglutarate dehydrogenase</u>.

106. List the reactions, coenzyme(s), cofactor(s) and enzymes involved in the "substrate-level phosphorylation" reaction of the Krebs Cycle. <u>succinyl CoA synthetase</u>

107. The enzymes <u>fumarase</u> and malate dehydrogenase "fix" a carbonyl group on succinate in the production of oxaloacetate.

108. What crucial 2 carbon compound is then "fixed" to OAA? <u>acetate</u>

109. What amino acid and what product of pyruvate metabolism are the principle substrates for gluconeogenesis in mammals? <u>alanine</u> and <u>lactate</u>

110. What energy sources are used to produce the protonmotive force? <u>NADH, CoQH$_2$, FADH$_2$</u> (indirectly)

111. What enzyme complex uses this phenomenon as the driving energy for ATP synthesis in oxidative phosphorylation? <u>f$_o$ f$_1$ ATP synthase</u> (ATPase accepted, but not synthetase)

112. How does electron transport drive production of the protomotive force? <u>exports H$^+$ from mitochondrion</u> (which creates a gradient, making them predisposed to flowing back in).

113. How many reactions does each round of β-oxidation of a fatty acid require? <u>four</u> (oxidation #1, hydration, oxidation #2, thiolysis)

114. What are the products of one round of β-oxidation and what's the tally in terms of ATP equivalents of energy conserving products? <u>1 CoQH$_2$, 1 NADH, H$^+$, 1 acetyl CoA, 1 fatty acid (minus 2 Cs)</u>

115. A set of coupled cofactor regeneration cycles siphon off reducing equivalents then fix them into coenzyme Q in reactions that are coupled to the first oxidative step of fatty acid β-oxidation. Write down the names of the four cofactors involved in this siphon. <u>CoA, FAD/FADH$_2$, Fe-S$^{2+/3+}$, CoQ/CoQH$_2$</u>

116. Which three steps of the Krebs Cycle do the first three steps of the fatty acid β-oxidation cycle resemble? <u>succinate dehydrogenase, fumarase, malate dehydrogenase</u>

Index

Printed in the USA/Agawam, MA
August 24, 2022

797532.019